GASEOUS ION CHEMISTRY AND MASS SPECTROMETRY

GASEOUS ION CHEMISTRY AND MASS SPECTROMETRY

Edited by

Jean H. Futrell

University of Utah

A WILEY-INTERSCIENCE PUBLICATION

JOHN WILEY & SONS

New York · Chichester · Brisbane · Toronto · Singapore

Library of Congress Cataloging in Publication Data:

Main entry under title:

Gaseous ion chemistry and mass spectrometry.

   ''A Wiley-Interscience publication.''
   Includes bibliographies and index.
   1. Chemical reaction, Conditions and laws of
Congresses. 2. Gases, Ionized—Congresses. 3. Mass
spectrometry—Congresses. I. Futrell, J. H.

QD501.G3244 1986   541.3′9   85-29589
ISBN 0-471-82803-3

Printed in the United States of America

10 9 8 7 6 5 4 3 2 1

CONTRIBUTORS

A. Welford Castleman, Jr.
Department of Chemistry
The Pennsylvania State University
University Park, Pennsylvania

Jean H. Futrell
Department of Chemistry
University of Utah
Salt Lake City, Utah

Werner Lindinger
Institut für Experimentalphysik
Leopold-Franzens-Universität
Innsbruck, Austria

Tilmann D. Märk
Institut für Experimentalphysik
Leopold-Franzens-Universität
Innsbruck, Austria

James D. Morrison
Physical Chemistry Department
La Trobe University
Melbourne, Australia

Randall B. Shirts
Department of Chemistry
University of Utah
Salt Lake City, Utah

David L. Smith
Department of Medicinal Chemistry
Purdue University
West Lafayette, Indiana

Austin L. Wahrhaftig
Department of Chemistry
University of Utah
Salt Lake City, Utah

v

PREFACE

This book has its roots in the organization of the annual workshop conducted by the Department of Chemistry at the University of Utah as the inaugurating event of the 1983/84 academic year. This workshop annually brings to our campus a number of international experts in the field chosen for discussion, who join the local community of experts in that topic to constitute the teaching faculty for this concentrated special-topics short course. The workshop is run in a "Gordon Conference" format at a local ski resort and involves the active participation of 50 to 100 students and faculty from our university and neighboring campuses. Each workshop begins with a series of introductory lectures and concludes with a discussion of the latest research topics and applications; these more formal instructional settings are complemented by round-table discussions and informal sessions designed to clarify any topics brought forward that were unclear or not widely understood by the participants. The workshop theme chosen by the Departmental Advisory Committee for the September 1983 conference was gas-phase ion chemistry and mass spectrometry. I was the designated organizer for that workshop.

Mass spectrometry is among the most interdisciplinary of science specialties, with applications in fields ranging from archeology and astrophysics to zoology. The 1983 workshop emphasized a small but important subset of this discipline concerned with fundamental aspects of ionization and ion reactions plus selected applications and extensions. The faculty for the workshop included Professors Austin Wahrhaftig, Randall Shirts, William Breckenridge, Joseph Michl, David Smith, H. L. C. Meuzelaar, and me from our campus, plus Professors James Morrison (La Trobe University, Melbourne, Australia), Robert McIver (University of California, Irvine), Barney Ellison (University of Colorado, Boulder), Will Castleman (The Pennsylvania State University), and Tilmann Märk and Werner Lindinger (Leopold-Franzens-Universität, Innsbruck, Austria). The selection of these specialists as faculty emphasized the fundamentals of ion chemistry and physics, instrumentation for the investigation of ion–neutral reactions, and the extension of mass spectrometry techniques, methods, and theory to research questions of special interest to our department.

A feature of the annual Utah Chemistry Workshop that is, in part, responsible for the high quality of these annual retreats is the designation of a 2-hour time slot at the conclusion of the workshop for a critique of what was done and recommendations for improvements. Participants in the critical review sessions are divided into two groups: students and senior investigators (Postdoctoral Fellows, Visiting Scholars, and faculty). These main groups are further subdivided to isolate the comments of those working within and outside of the discipline reviewed at the workshop. A surprising result of the critique of the workshop in question was the nearly unanimous recommendation that the lecture notes be reorganized and published as a reference work on the topic. The Chemistry Advisory Board of John Wiley & Sons concurred, and this book is the direct result.

Space limitations, demands of other duties, and other factors have led to a reduction in the number of contributors compared with the number at the workshop, and a corresponding reduction in the number of topics covered by the present work. From the viewpoint of presenting a comprehensive treatment of gas-phase ion chemistry we especially regret the loss of Bob McIver and Barney Ellison as contributors to this volume. These two authors felt that the topics that they discussed are adequately reviewed in the recent literature* and that most of their contributions would not differ substantially from those reviews. Because ion cyclotron resonance has played such a central role in work on gas-phase ion chemistry in the past decade, a brief discussion of this technique by the editor is included in this volume as a replacement for the more detailed treatment by McIver at our workshop.

The organization of this volume parallels that of the 1983 Utah Workshop. The fundamental theories that underlie gas-phase ion chemistry are presented first, then instrumentation, ionization methods, techniques, and ion–molecule reaction kinetics and dynamics. The formation of cluster ions and nucleation phenomena are then discussed as a frontier research topic. A chapter on biomedical applications as a representative example of the applications of mass spectrometry concludes this volume. Some 20% of the workshop was devoted to applications of mass spectrometry to the physical and life sciences; David Smith's chapter summarizes selected topics illustrating the impact of mass spectrometry on biology and medicine.

I appreciate very much the active cooperation of all the contributing authors, who suffered through one to three stages of revision of their manuscripts. None of them seriously underestimated the time required and the volume was completed without destroying our collective enthusiasm for the project. We thank the editorial staff of John Wiley & Sons, and especially Chemistry Editor Theodore P. Hoffman, for their patience and active cooperation. My Utah colleague Professor Chuck Wight kindly read and critiqued my chapters. A special note of thanks is extended to my secretary, Cheryl Gabbott, for much typing, cheerfully done, plus her able assistance in organizing this book and helping with the many details that required careful

*Robert T. McIver, Jr., *American Laboratory,* 18 (November 1980); Veronica M. Bierbaum, G. Barney Ellison, and Stephen R. Leone, in *Gas Phase Ion Chemistry,* Vol. 3, T. Bowers, Ed., Academic Press, New York (1984), p. 1; Robert T. McIver, Jr., *Scientific American* **243,** 186 (1980).

follow-up. Finally, I thank my wife, Nancy, for being too busy as a resident physician in neurology to notice the nights and weekends devoted to this effort.

JEAN H. FUTRELL

Salt Lake City, Utah
March 1986

CONTENTS

GASEOUS ION
CHEMISTRY
AND MASS
SPECTROMETRY

Introduction

JEAN H. FUTRELL

Department of Chemistry
University of Utah
Salt Lake City, Utah

The history of mass spectrometry begins in 1910 with J. J. Thomson's parabola spectrograph, which he described in his 1913 monograph (1). Until the end of World War II, the field was limited to a few physicists scattered throughout the world. The mass spectrometers of that period were esoteric instruments severely limited by the vacuum technology and crude electronics of that time. Nevertheless, the isotopic composition of the elements was mapped out, and the measurement of accurate isotopic masses led to the discovery of packing fractions, and in turn to an understanding of the binding forces within the atomic nucleus. As early as 1921, Thomson made the following remarkable prediction (2): "Another subject on which Positive Rays may, I think, be expected to throw light is that of the structure of the molecule. For, as we have seen, when a compound gas is in the discharge tube there are among the positive rays not only the individual atoms which went to make up the molecule, but also unsaturated combinations of these atoms. The proportions in which these combinations are present yield information about the configuration of the molecule."

The first widespread realization of this prediction came during World War II, which sparked a dramatic development in the field of electronics, and during which the needs of the Manhattan Project for isotope separations and analyzers led to production of the first commercial mass spectrometers. Production of high-octane aviation gasoline depended on the mass spectrometer for the rapid analysis of complex mixtures of hydrocarbons.

The development of both instruments and applications to chemistry has continued from 1945 and shows no sign of diminishing. The main emphasis in mass spectrometry has always been on chemical analysis. Proof of molecular structure is the

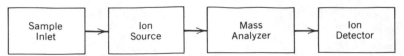

Figure 1.   Block diagram showing components of a mass spectrometer.

second most important application (judging from number of publications), with biomedical applications of structural studies receiving major emphasis. Fundamental studies of ion chemistry and physics rank third.

The last topic is the one emphasized in this book. As soon as chemists had access to these instruments it was demonstrated that not only was the mass spectrometer an extremely powerful analytical tool, but also one could obtain data on the fundamental processes of ion formation and on the structures and energy states of the ions themselves. Further, the reactions of gaseous ions with neutral species constituted a completely new realm of chemistry. Studies of ion formation led to a variety of new ion sources, while study of the unimolecular processes of ion fragmentation provided insight into the interpretation of mass spectra. Thus all aspects of the field—both fundamental and applied—are synergistically coupled and centered about the instrumentation used for these studies.

A functional block diagram representative of all mass spectrometers is shown in Fig. 1. Central to all mass spectrometric investigations is the ion source. White (3) has listed the requirements of an ionization source for analytical applications as follows:

1.   The supply of ions needs to be compatible with the analyzer geometry and detector sensitivity.
2.   The energy spread in the ion beam should be small compared with the ion-accelerating voltage. A large spread makes a double-focusing mass analyzer mandatory.
3.   In magnetic mass spectrometers the width of the ion-source exit slit must be small compared with the radius of the mass analyzer.
4.   The half-angle of divergence of ions from the source should be small.
5.   Small samples require a high-efficiency source.
6.   The ionization source should be selective against unwanted background ions.
7.   The source should not suffer from memory effects or interaction between samples.
8.   Ion emission should be stable, and not noisy.
9.   Minimal chemical preparation of samples should be necessary.
10.   The source should not suffer from nonlinearities resulting from space-charge effects.
11.   The source should be free from mass discrimination.
12.   Samples should not decompose in the source.

TABLE 1 Ion Sources in Current Use for Analysis

Type	Ion Energy (V)	Sample Type	Sensitivity	Current Stability	Uses
Electron impact	0.1–0.5	Gases, vapors of liquids and solids	High	Good	General purpose
Photoionization	0.05–0.2	Gases, vapors	Fair	Good	Molecular spectroscopy
Chemical ionization	0.5–3.0	Organic vapors	Varies	Fair	Organic analysis
Thermal ionization	0.2	Inorganic solids	Very high to very low	Fair	Isotope abundance
Vacuum spark	1000	Metals, semiconductors, insulators	Very high	Poor	Trace impurities
Gas discharge	100–1000	Gases, vapors	High	Poor	Packing fractions
Penning discharge	100–1000	Gases, vapors	High	Fair	Leak detection
Arc source	5–20	Vapors	Very high	Fair	Isotope separators
Plasma ionization	0.5–4	Aqueous solutions	High	Fair	Trace metal analysis
Field ionization	0.5–100	Organic vapors	Varies	Fair	Organic analysis
Field desorption	0.5–5.0	Organic solids	Varies	Fair	High-MW organic analysis
Fast atom bombardment	1–100	Organic solids	High	Fair	High-MW organic analysis
Thermo spray ionization	—	Organics in solution	Fair	Fair	HPLC effluent analysis
Ion bombardment	5–100	Inorganic solids	High	Poor	Surface studies
Laser ionization	1–50	Solids	Fair–good	Poor	Surface studies
Fission fragment ionization	—	Solids, organics	Fair	Fair	High-MW organics

HPLC, high-pressure liquid chromatography; MW, molecular weight.

3

Table 1 lists most of the different kinds of ionization sources currently used and comments on their relevant characteristics and applications.

The requirements of an ionization source for ion chemistry and physics studies are essentially the same as for analytical applications. Of those listed in Table 1, only electron impact, photoionization, chemical ionization, and fast atom bombardment sources are extensively used. Electron impact ionization (Chapter 3) is the most widely applied ionization method, while photoionization (Chapter 4) permits the most precise control of the ionization process. High-pressure or chemical ionization sources vary enormously in their properties and operating characteristics and defy simple, general descriptions. The phenomena involved usually may be interpreted for each case by considering the primary ions generated to be those characteristic of electron impact (Chapter 3) followed by the plasma relaxation processes characteristic of swarm experiments (Chapter 7). The relatively recently introduced and not fully characterized fast atom bombardment (FAB) method is described briefly in Chapter 13.

Once ions are created it is necessary to manipulate them intelligently using the functional blocks of Fig. 1 labeled "mass analyzer" and "detector," to obtain the desired information. These topics are discussed generally in Chapter 5 on ion optics and mass analyzers, while Chapter 6 describes the special technique of ion cyclotron resonance. The variety of techniques for characterizing ion abundances and the precise and detailed information that may be obtained are largely responsible for the extensive fundamental and applied applications of mass spectrometry, a sampling of which is presented in the remainder of this book.

The organization follows approximately the pedagogical order of the 1983 Utah Workshop (see Preface). First the underlying theory is developed in Chapters 1 and 2, which treat unimolecular dissociation and collision theory, respectively. The fundamentals of ion formation processes are then discussed for electron and photon impact sources in Chapters 3 and 4. Fundamental considerations for mass analysis and detection are discussed in Chapters 5 and 6, while swarm techniques and beam methods for the investigation of collision processes (which emphasize kinetics and dynamics, respectively) are described in Chapters 7 and 8. Applications are described in the remaining chapters: ion photodissociation spectroscopy (Chapter 9), state-to-state dynamics studies (Chapter 10), ion–molecule reaction kinetics (Chapter 11), chemistry and physics of cluster ions (Chapter 12), and biomedical mass spectrometry (Chapter 13).

REFERENCES

1. J. J. Thomson, *Rays of Positive Electricity and Their Application to Chemical Analysis,* Longmans Green and Company, London (1913).

2. Ref. 1, 2nd ed. (1921).

3. F. A. White, *Mass Spectrometry in Science and Technology,* John Wiley & Sons, New York (1968), Chapter 3.

PART 1

THEORETICAL FOUNDATIONS

CHAPTER 1

Unimolecular Dissociation
of Gaseous Ions

AUSTIN L. WAHRHAFTIG
Department of Chemistry
University of Utah
Salt Lake City, Utah

As stated in the Introduction, this chapter is derived from an introductory paper presented at a workshop on gaseous ion chemistry. As such, it has three aims—to present a very brief introduction to the major ideas of classical theories of reaction kinetics; to present the basic theory of the formation of a mass spectrum of a polyatomic molecule; and to comment on a number of points that the author believes have been well known for many years but that are often ignored, and sometimes "rediscovered," by newcomers to studies involving ions. There is no attempt at completeness. A few references are given that might aid the reader who wishes to study the more recent work.

The chemistry discussed in this volume generally is of the form

$$A + B \rightarrow (AB) \rightarrow C + D \tag{1}$$

There will be a missing (or extra) electron somewhere, but the presence of a charge may be ignored in this discussion. If the species (AB) exists for a significant time, then $(AB) \rightarrow C + D$ and $(AB) \rightarrow A + B$ are unimolecular reactions. The principle of microscopic reversibility implies that a study of $(AB) \rightarrow A + B$ can give useful information regarding $A + B \rightarrow (AB)$. To study ion chemistry, one first needs ions, so we consider the quite general mechanism

$$S \xrightarrow[\text{process}]{\text{Ion formation}} X^+ \xrightarrow{\text{Ion chemistry}} P.$$

Here, S denotes the starting material, X^+ the ion to be studied, and P the products. Each arrow may denote a multiplicity of steps possibly involving other species. Often, X^+ is formed by a unimolecular dissociation of S^+; unless X^+ is monatomic, the unimolecular dissociations of X^+ are possible reactions that must be considered in any discussion of its chemistry. This chapter considers singly charged positive ions, but much of the discussion will be more generally applicable. The phrase "polyatomic ion" should generally be taken to exclude small polyatomic ions, which are likely to behave in part like diatomic ions, and to refer to ions with eight or more atoms.

Unimolecular reactions are the reactions of a species A

$$A \rightarrow A' \quad \text{(rearrangement, isomerization),}$$

$$A \rightarrow B + C \quad \text{(dissociation),}$$

where the reaction involves only A, a *single* polyatomic species. The rate of reaction may be given by

$$\frac{d[A]}{dt} = -k^{(1)}[A] \quad \text{(a first-order reaction),}$$

$$\frac{d[A]}{dt} = -k^{(2)}[A]^2 \quad \text{(a second-order reaction),}$$

$$\frac{d[A]}{dt} = -k([A],t)[A],$$

where the (assumed) first-order "rate constant," $k([A],t)$, is not a constant.

Early study of such reactions in the gas phase was limited to systems at thermal equilibrium. The techniques and the instrumentation required for the detailed study of gaseous ion chemistry are all new, mostly developed in the last 10 or 20 years, although some types of radiochemical and nonthermal activation techniques are far older. Many theories of importance in the interpretation of modern experiments, such as the RRK, TST, and RRKM theories (discussed below), were based primarily on studies at thermal equilibrium and were developed 30 and more years ago. Hence, we start with a brief discussion of thermal reaction kinetics.

Early theorists were baffled by the existence of unimolecular gas-phase reactions, which as first studied were all first order. The dependence of the rate constants for such reactions on temperature was of the Arrhenius form,

$$k = Ae^{-E_0/RT},$$

indicating that molecules with high energy were reacting. However, if the energy for reaction were acquired by collision, then the reaction should *not* be first order, or so it was thought. One of the earliest proposals, discussed over the years 1913–

1919 by F. Perrin, was that the absorption of radiation of frequency E_0/h by individual molecules led to reaction, but this theory did not prove satisfactory. The basic idea that remains the basis for theories today is that of Lindemann in 1922. His explanation, as given in almost every physical chemistry and kinetics text, is as follows. An activated species A* is postulated, with reactions

$$A + A \underset{k_d}{\overset{k_a}{\rightleftarrows}} A* + A, \qquad A* \overset{k_r}{\to} P(+ Q + \cdots).$$

Then,

$$\frac{d[A]}{dt} = -k_a[A]^2 + k_d[A*][A], \qquad \frac{d[A*]}{dt} = k_a[A]^2 - k_d[A*][A] - k_r[A*].$$

Assume that A* is very reactive: When formed, it quickly undergoes either deactivation or unimolecular reaction to products (or product). Then $[A*] \ll [A]$, $d[A*]/dt \approx 0$ (steady-state approximation), and $k_a[A]^2 \approx k_d[A*][A] + k_r[A*]$. Hence,

$$[A*] = \frac{k_a[A]^2}{k_d[A] + k_r}$$

and

$$\frac{d[A]}{dt} = -\frac{k_a k_r [A]^2}{k_d[A] + k_r} \begin{cases} -\dfrac{k_a k_r}{k_d}[A] & \text{when } k_d[A] \gg k_r, \\[2em] -k_a[A]^2 & \text{when } k_d[A] \ll k_r. \end{cases}$$

The trend of the first-order rate "constant," $k^{(1)} = k_a k_r/k_d$, to decrease with decreasing [A] (pressure of A for a gas-phase reaction) was observed shortly after its prediction. For many years, a primary goal of theorists in chemical kinetics was to develop a theory that, when applied to known reactions, would yield calculated results in agreement with experiments for both the temperature and the pressure dependence of reaction rates.

Two general schemes were developed for this purpose. The first, associated with the names Hinshelwood, Rice and Ramsperger, and Kassel, treats a molecule as a collection of harmonic oscillators. The several variants of this scheme differed in the assumptions made about such factors as:

1. Nature of the oscillators (all of one frequency, ν, or multiples of one frequency).

2.   Whether classical or quantum model oscillators were used.
3.   Requirement for reaction:
    a.   Critical energy in an oscillator.
    b.   Critical energy in a bond.
    c.   Critical internal configuration.

The major problems with the theories were:

1.   The harmonic oscillator model is not a realistic approach to a complete potential surface.
2.   The classical approximation for a system of oscillators is very poor for the values of E_0, v_i of the real polyatomic molecules studied.
3.   Quantum mechanical versions led to equations that were too complex to evaluate (in precomputer days) except by replacing the real molecular frequencies and other real values by those of an oversimplified model.

The developers of the second general scheme for the treatment of reaction rates, transition-state theory (TST), included a number of major scientists—F. London, H. Pelzer, M. Polanyi, and E. Wigner—but the principal proponent by far was Henry Eyring. The important ideas, in oversimplified form, are as follows:

1.   Most molecules will have configurations near the equilibrium configuration, so a harmonic oscillator model (perhaps with some internal rotations) is reasonable.
2.   Reaction results from the very small fraction of the molecules that have configurations close to a critical state, called the activated complex configuration. This critical state is the state at a saddle point on the potential-energy surface, about which point the potential surface is approximated by

$$V = E_0 - \frac{1}{2}b_1q_1^2 + \frac{1}{2}\sum_{j=2}^{n} b_j q_j^2 \qquad b_1, b_j > 0.$$

Here, q_1 is the reaction coordinate. In general the normal coordinates q_j and the values of b_j are *not* the same as the coefficients and normal coordinates for the molecules with near-equilibrium configurations.

3.   The reaction rate is slow enough that it does not significantly perturb the thermodynamic equilibrium that would otherwise exist between molecules in the activated complex configuration and molecules in more common configurations.

Then, one obtains the well-known equation

$$k' = \frac{kT}{h}\frac{Q^{\ddagger}}{Q}\exp\left(\frac{-E^{\ddagger}}{RT}\right), \tag{3}$$

where $Q^{\ddagger}$ and Q are the partition functions for the molecule in its activated complex and normal configurations, respectively, and $E^{\ddagger}$ is the potential energy of the saddle point on the potential surface relative to the minimum for the reactant equilibrium configuration. Remember, the k' to the left of the equal sign denotes the rate constant; the k in kT/h refers to Boltzmann's constant.

Transition-state theory as formulated above has the advantage of simplicity in application. It was applied by Eyring and his students and associates to an unbelievable range of problems, far too many to discuss here. However, the assumption of high-pressure conditions is implicit in the assumption of an equilibrium population of activated complexes. Discussion of the low-pressure fall-off of reaction rate is thus very difficult. It is not applicable to systems of molecules with energy distributions far from thermal equilibrium—that is, to most systems discussed elsewhere in this volume. A fairly complete history of transition-state theory by Laidler and King (1) and an excellent review article (2) on the "Current Status of Transition State Theory" may be found in the 1983 issue of the *Journal of Physical Chemistry* dedicated to Henry Eyring (3).

Let us now consider some aspects of mass spectrometry, particularly the early work, as it affected theories of reaction kinetics. The earliest work on the mass spectra of gases gave very complex results—sample peaks mixed with background peaks from sealing wax, water vapor, and whatever else was present in the "high vacuum" systems of the 1920s (4). Complicating the spectra were many peaks resulting from ion–molecule reactions. The aim of mass spectroscopists, for many years, was to build a mass spectrometer that would give a "true" mass spectrum— the spectrum of peaks, m/z, due solely to ionization of single molecules by electrons, with minimal background from impurities and from subsequent ion–molecule reactions. Instruments that met this requirement were constructed in the period 1930–1940, and in 1942 the first mass spectrometers designed and constructed for general use in gas analysis came on the market (5). By 1948, the 75-V electron ionization (EI) mass spectra of a large number of hydrocarbons, including all 18 of the octanes, had been published. Very few mass spectra had been run on organic compounds other than hydrocarbons; alcohols and amines, if volatile enough to be introduced, contaminated the instruments of that time. Other ionization techniques, mentioned elsewhere in this volume, were either unknown or not used in the mass spectrometry of gases.

In an EI source, the pressure is low, so the mean free paths of ions and molecules are large compared with the dimensions of the ionization region. The electron–molecule interaction time is of the order of 10^{-15} sec, short relative to the time for any molecular vibration. Hence, ionization is "vertical," occurring with essentially no change in internuclear distances. Pressures in the mass spectrometer are sufficiently low that each ion is an isolated system. Then,

$$e + M \xrightarrow{\text{fast}} M^+(E) \xrightarrow{\text{slower}} \text{Products.}$$

Energy is conserved: The total (internal + relative translational) energy is the same in each set of products as in the ion, $M^+(E)$, from which they were formed.

The early mass spectroscopists were mostly physicists; a few were physical chemists. They considered simple systems, monatomic gases and H_2, N_2, CO, O_2. The major aspects of the mass spectra of the diatomic molecules were understood in 1941, when Hagstrum and Tate showed that the positive ions formed and their kinetic-energy distributions were consistent with the vertical transitions from the ground state of the molecule to the various potential surfaces determined for the diatomic ions by molecular spectroscopists.

No discussion or explanation of the mode of formation of the positively charged fragment ions observed in the mass spectrum of an organic molecule was presented until about 10 years later, in 1950, when a qualitative discussion was given by Eyring at a symposium on radiobiology. He noted that there were both experimental data and a theoretical basis for using a statistical model for the decomposition of polyatomic ions (except small ions) rather than the model applicable to diatomic molecules and small polyatomic ions, which required a detailed knowledge of the potential surfaces of the ions. That discussion, with revisions and additions, led to the theory now referred to as the quasiequilibrium theory (QET) of mass spectra, first published in 1952 (6). The essential ideas, many of which seem obvious now but were new then, are as follows:

1.   Unlike diatomic molecular ions, which generally are either stable or else dissociate in the time needed for one vibration, polyatomic molecular ions generally dissociate on a time scale long compared with the time for most vibrations.

2.   The removal of one electron from a large molecule will in general leave it in a stable state, but with excess vibrational and electronic energy relative to the potential minimum (or minima) for dissociation(s) of the ion.

3.   Randomization of this excess energy over all degrees of freedom is rapid relative to the rate of ion dissociation.

4.   As a result, it is appropriate to describe the rearrangement and dissociation reactions of a polyatomic ion by a statistical theory.

5.   It is appropriate to describe the mass spectrum of a molecule as the result of a series of consecutive sets of competing reactions, starting with the parent molecular ion.

6.   Application of the basic ideas of TST to systems of a specified energy, rather than systems at a specified temperature, leads to the equation

$$k(E) = \frac{1}{h} \frac{W^{\ddagger}(E - E_0)}{dW(E)/dE} = \frac{1}{h} \frac{W^{\ddagger}(E - E_0)}{\rho(E)}, \tag{4}$$

where $W(E)$ is the number of energy states of the ion having energy less than or equal to E. Then, $\rho = dW/dE$ is the density of energy levels. The definition of $W^{\ddagger}$ is the same as that for W, except that the configuration of the transition state is assumed for the ion; an activation energy, E_0 or ϵ_{act}, is required for the ion to reach the transition state, so the internal energy in the ion is only $E - E_0$. This is the

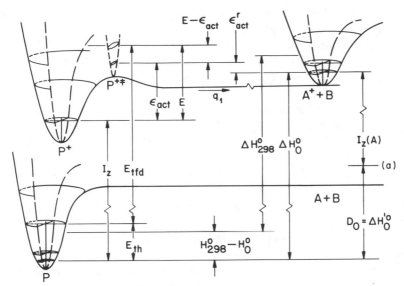

Figure 1. Schematic representation of the energy terms relating to the ionization and dissociation of a polyatomic molecule: I_z, Ionization energy of the species P; E_{th}, thermal excitation of a molecule P prior to ionization; E_{tfd}, energy transferred to P by the incident electron or photon; E, resulting internal energy in the ion; q_1, reaction coordinate for $P^+ \rightarrow A^+ + B$; ϵ_{act}, activation energy for $P^+ \rightarrow A^+ + B$; ϵ_{act}', activation energy for the reverse reaction, $A^+ + B \rightarrow P^+$; ΔH_0^0, ΔH_{298}^0, standard enthalpy changes for $P \rightarrow A^+ + B + e$ at 0 K and 298 K, respectively; $D_0 = \Delta H_0^{'0}$, standard enthalpy change for the dissociation $P \rightarrow A + B$; $I_z(A)$, ionization energy of the fragment A. Only two of the many vibrational degrees of freedom of P, P^+, $P^{+\ddagger}$, and A^+ are represented; the potential surfaces of A and B are omitted. The spacing of vibrational levels is greatly exaggerated relative to D_0 and ϵ_{act}: I_z is typically 2–4 times D_0. The level (a) represents the energy of the dissociated neutral system, A + B, with the species in their ground states. Most of these energy terms refer to the differences in energy between specific levels; the exceptions are ΔH_{298}^0 and ΔH_{298}^0, which are the usual thermodynamic quantities.

general statement of the original QET rate expression. See Fig. 1 for a schematic representation of the energy terms involved.

The calculation of a mass spectrum requires knowledge of the energy distribution of the parent ions initially formed in the ionization chamber, the values of E_{0i} for all reactions involved in the formation of product ions, and appropriate expressions for $\rho(E)$ and for $W_i^{\ddagger} (E - E_{0i})$ for all activated complex configurations. Then, over the energy range of interest, the rate constants for all reactions may be calculated as functions of energy. From these rate constants and the mass spectrometer time scale, the relative abundances of the different ions collected may be calculated, also as a function of energy. Finally, integration over energy with the appropriate probability distribution curve yields the calculated mass spectrum for comparison with experiment. Data obtained by other methods, such as charge exchange and photoelectron–ion coincidence measurements, give directly data on fragmentation at a given energy for comparison with theory (6,7).

Obviously, no exact or close to exact calculations of mass spectra have been made. Many approximate calculations have been reported, based on models of widely varying sophistication. It now appears clear that the QET description of polyatomic ion dissociations is in agreement with most of the fragmentations of most polyatomic ions. The theory does not pretend to consider all possible products resulting from the ionization of a polyatomic molecule; rather, it proposes that the *bulk* of the ionization generally yields products interpretable in terms of the statistical model.

In 1952, shortly before publication of the paper containing Eq. (4), R. A. Marcus published the paper that led to the "M" being attached to the RRK (Rice, Ramsperger, and Kassel) model to give the more exact RRKM (8) model for unimolecular reactions. Marcus adapted the Eyring transition-state approach to ordinary thermal reaction kinetics and combined it with the basic ideas of Lindemann and RRK theory concerning collisional activation–deactivation. The resulting equation is the product of two expressions, one giving the probability of the system having a specified energy, the other giving the rate of reaction of a system having a specified energy. The latter expression is essentially identical to Eq. (4) above. Thus, there is no difference in the basic formulation used when RRKM theory or QET is applied to any collection of species having a nonthermal energy distribution. However, there are far more scientists working in kinetics than the restricted number specializing in the kinetics of ion decompositions. Thus, expressions for the evaluation of $W(E)$ and $\rho(E)$; suitable models for molecules; elaborations of the equations to consider problems of conservation of angular momentum, centrifugal barriers to dissociation, and so on have been developed primarily by scientists who consider themselves to be working with RRKM theory.

There were at least two problems with the initial QET paper: (1) The need for a very large number of low-lying electronic states of the ion and the importance of transitions between these states in energy randomization were overemphasized. (2) The classical approximation was used for $W^{\ddagger}(E - E_0)$ and $\rho(E)$; this approximation is poor for $\rho(E)$ and horrible for $W^{\ddagger}(E - E_0)$ for the values of vibrational frequencies and activation energy that are appropriate for most molecules (recall that we are describing the precomputer era of physics and chemistry). Computer programs are now available for the accurate counting of states for low values of $E - E_0$ and E and for the reasonably precise evaluation of $W^{\ddagger}$ and ρ at higher values of E. Two major limitations appear when one wishes to make a calculation on a real system: (1) the lack of precise knowledge of the many parameters, frequencies, and activation energies of the ion and, especially, the structure and frequencies of the activated complexes; and (2) the assumptions inherent in QET (and the RRKM theory), only some of which have been eliminated in more elaborate versions.

Energy randomization that is rapid relative to reaction plays a part in any statistical theory of reaction rate, be it RRKM theory, QET, or any other theory. In reactions in solution or in a gas at not too low a pressure, collisions not only transfer energy between molecules but also rearrange, and hence randomize, the internal energy distribution over the degrees of freedom within molecules. In thermal reactions, there is no clear-cut way to distinguish between inter- and intramolecular energy transfer; by definition, only the latter is possible in isolated systems.

The QET furnished a theoretical basis for discussing ion dissociations as conventional reactions. The early calculations showed that it appeared to be consistent with experimental results. The major developments in the study of unimolecular dissociation of ions, however, came from the increasingly large number of organic chemists who entered the field starting around 1955, as commercial instruments became more dependable, available, and capable of running samples other than high-vapor-pressure hydrocarbons. The organic chemists quickly saw that many of the reactions of mass spectral dissociations were interpretable in terms of the concepts of organic chemistry. In general, they did not find it necessary to worry over the arguments of several physical chemists in the field for ion dissociation mechanisms that correlated reactions with excitation to specific electronic states, which should not give reactions that can be discussed similarly to thermal dissociations.

The tremendous volume of literature on the relationship of mass spectra to structures of organic molecules is *not* the subject of our discussions here (9). However, the fact that very similar language can be used to describe isolated ion dissociations and thermal molecular dissociations is of interest. The TST equation of Eyring for reaction rate can be expressed in the thermodynamic (more correctly, pseudothermodynamic) form

$$k' = \frac{kT}{h}\frac{Q^{\ddagger}}{Q}\exp\left(-\frac{E^{\ddagger}}{RT}\right) = \frac{kT}{h}\exp\left(-\frac{\Delta G^{\ddagger}}{RT}\right)$$

$$= \frac{kT}{h}\exp\left(-\frac{\Delta H^{\ddagger}}{RT}\right)\exp\left(\frac{\Delta S^{\ddagger}}{R}\right). \tag{5}$$

Here, $\Delta G^{\ddagger}$, $\Delta H^{\ddagger}$, and $\Delta S^{\ddagger}$ represent the free energy, enthalpy, and entropy of activation, respectively. Every undergraduate chemistry student now learns how changes in $\Delta H^{\ddagger}$ and $\Delta S^{\ddagger}$ affect k, the rate constant.

Can the same terms, $\Delta H^{\ddagger}$ and $\Delta S^{\ddagger}$, be applied to the very low-pressure dissociations of ions, or are there other quantities that can be used in similar fashion? The use of a temperature, T, implies thermal equilibrium, and T should not be used to describe a system unless that system is in, or essentially in, thermal equilibrium. *If* an energy distribution could reasonably be approximated by a Boltzmann distribution, one could *define* an *effective* temperature for the reacting system. The energy distribution obtained on ionization of polyatomic molecules by electrons (or photons) usually is far from thermal, however, and there is no meaningful definition of effective temperature. The energy distribution curves implied by photoelectron spectra often have several maxima, sometimes separated by minima at which the probability of finding the corresponding energy is near zero. McLafferty has discussed some experimental mass spectral results explainable only in terms of a quite sharp minimum in the energy distribution curve (10). Still, the well-known arguments on relative yields of products in thermal reactions based on loose vs. tight activated complexes (which controls the sign and magnitude of $\Delta S^{\ddagger}$) and activation energy (which is directly related to $\Delta H^{\ddagger}$) generally apply equally well to mass spectral reactions.

In Eq (4), $W^{\ddagger} (E - E_0)$ is a very rapidly increasing function of $E - E_0$,

$$W^{\ddagger} (E - E_0) \propto (E - E_0)^n,$$

where n typically is less than, but approaches, the number of internal degrees of freedom. If two similar competing reactions with approximately the same functions for the number of activated complex states, $W_1^{\ddagger}$ and $W_2^{\ddagger}$, have different activation energies, E_{01} and E_{02}, with $E_{01} < E_{02}$, then $W_1^{\ddagger} (E - E_{01}) \gg W_2^{\ddagger} (E - E_{02})$ for all $E > E_{01}$ (all energies at which either reaction is possible). From Eq. (4), the rate of the reaction with the smaller E_0 will be larger than that of the competing reaction at any internal energy for the dissociating ion. Consequently, the relative abundance of the product having the smaller activation energy will be greater than that of the product of the competing reaction for any energy distribution of the dissociating ions. In the thermal reaction equivalent, $\Delta H_1^{\ddagger} < \Delta H_2^{\ddagger}$ and $\Delta S_1^{\ddagger} \approx \Delta S_2^{\ddagger}$; from Eq. (5), the product from the reaction with the smaller $\Delta H^{\ddagger}$ will be formed in the larger amount.

Now consider the case where $E_{01} \approx E_{02}$, but the two functions $W_1^{\ddagger} (E - E_{01})$ and $W_2^{\ddagger} (E - E_{02})$ are different due to differences in the activated complexes for the two reactions. A tight activated complex, that is, one in which the vibrational frequencies are as high as or higher than in the equilibrium configuration, will have relatively widely spaced energy levels compared with those for a loose complex, that is, one for which several degrees of freedom have appreciably lower frequencies than in the equilibrium configuration. Both $W^{\ddagger}$ and $\Delta S^{\ddagger}$ depend directly on energy-level spacing, so if reaction 1 proceeds via a tight complex and reaction 2 via a loose complex,

$$W_1^{\ddagger} (x) \ll W_2^{\ddagger} (x) \qquad \text{and} \qquad \Delta S_1^{\ddagger} < \Delta S_2^{\ddagger}.$$

Also, if $E_{01} \approx E_{02}$, then $\Delta H_1^{\ddagger} \approx \Delta H_2^{\ddagger}$. It then follows from Eq (4) that

$$\frac{k_1(E)}{k_2(E)} = \frac{W_1^{\ddagger} (E - E_{01})}{W_2^{\ddagger} (E - E_{02})} \ll 1$$

at all energies, so that for *any* energy distribution the ratio of the rates integrated over E will also be much less than unity. The equivalent expression for the ratio of the thermal rates, from Eq. (5), is

$$\frac{k_1 \text{ (thermal)}}{k_2 \text{ (thermal)}} = \exp \left(\frac{\Delta S_1^{\ddagger} - \Delta S_2^{\ddagger}}{R} \right) \ll 1.$$

The calculations in the two cases are different but, qualitatively, the final results are the same.

If one reaction is favored by having both a loose complex and a low activation energy, corresponding to a low $\Delta H^{\ddagger}$ and a positive $\Delta S^{\ddagger}$, it will predominate even more strongly.

The most interesting cases are those where reaction 2 of a pair of competing reactions proceeds via a loose activated complex and reaction 1 proceeds via a tight complex, but $E_{02} > E_{01}$. The thermal analog is well known: One has

$$\Delta H_2^{\ddagger} < \Delta H_1^{\ddagger}, \qquad \Delta S_2^{\ddagger} < \Delta S_1^{\ddagger},$$

and, from Eq. (5),

$$\frac{k_1 \text{ (thermal)}}{k_2 \text{ (thermal)}} = \exp\left[-\frac{(\Delta H_1^{\ddagger} - \Delta H_2^{\ddagger})}{RT}\right] \exp\left[(\Delta S_1^{\ddagger} - \Delta S_2^{\ddagger})R\right].$$

Then $k_1 = k_2$ at the temperature given by $\Delta H_1^{\ddagger} - \Delta H_2^{\ddagger} = T(\Delta S_1^{\ddagger} - \Delta S_2^{\ddagger})$ and reaction 1 (lower activation energy) is favored at lower temperatures; reaction 2 (loose complex, more positive activation entropy) is favored at higher temperatures.

A very similar effect is observed in many mass spectra obtained as a function of ionizing electron voltage. Simple bond breaking reactions generally proceed by way of a loose activated complex. The lowest-energy dissociation is often one that yields a stable molecule plus the ion of a stable molecule with low ionization potential. Consider, for example, the pair of reactions

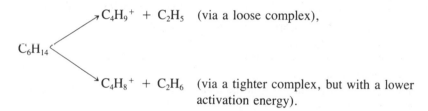

$$C_4H_9^+ + C_2H_5 \quad \text{(via a loose complex)},$$

$$C_6H_{14}$$

$$C_4H_8^+ + C_2H_6 \quad \text{(via a tighter complex, but with a lower activation energy)}.$$

At low electron energies such that parent ions can be formed with an excess energy only slightly above the minimum for dissociation, the predominant ion is $C_4H_8^+$. As the electron voltage is increased, the fraction of parent ions with higher energies increases, so the relative abundance of butyl ion increases and soon dominates.

When mass spectra are obtained as a function of electron energy with a conventional mass spectrometer, it is generally observed that only the parent molecular ion is obtained at the lowest electron energies. As the electron energy is slowly increased, the various product ions make their appearance and it is a simple matter, so it seems, to note the lowest potential at which a particular ion is observed and assign that "appearance potential" to the ion. Experimental parameters that must be considered in such an experiment are discussed elsewhere in this volume (see pages 95 and 98). Only after consideration of the following can the absolute value of the number so obtained be correlated with an ionization potential or activation energy:

1. The ion source of a conventional mass spectrometer uses a hot filament as the electron source. Thus, the electrons have a spread in energy corresponding to

an emitter temperature of perhaps 1500°C, producing a "high-energy tail" to the electron distribution that is often observable over a 1-V or larger range. The distribution is broadened by the ion draw-out potential and electron collimating magnetic field usually used.

2. Contact potentials and surface potentials can—and usually do—introduce errors.

3. The probability of ionization may approach zero at the ionization potential, or at some higher potential.

4. The kinetics of the formation of fragment ions is usually such that the probability of formation of a fragment ion approaches zero as the energy in the ion approaches the activation energy from higher energies.

Problem 1 can be largely eliminated by the incorporation of a source of monoenergetic electrons and other design features into the ion source, but with a loss by a factor of the order of 10^3 in electron current, and hence in sensitivity of measurement. Much better, but experimentally more complex, is the use of photon ionization. A number of other techniques involving modified sources or mathematical treatment (deconvolution) of data also can be used, sometimes in combination, to obtain the result expected for monoenergetic electrons or photons. Problem 2 can be addressed by mixing a calibration gas, typically argon, with the substance under study; determining the ionization efficiency curves (ion abundance vs. electron or photon energy) simultaneously for sample and reference; and then correcting the voltage scale using the known ionization potential of the reference (usually a highly accurate spectroscopic value).

Problems 3 and 4 are inherent in the measurement. Let us consider some of the several ways isolated positive ions may be formed (11,12):

$$A + h\nu \rightarrow A^* \rightarrow A^+ + e, \tag{6}$$

$$A + h\nu \rightarrow A^+ + e, \tag{7}$$

$$A + e \rightarrow A^* + e \rightarrow A^+ + 2e, \tag{8}$$

$$A + e \rightarrow A^+ + 2e. \tag{9}$$

In reactions (6) and (8), the excited state, A^*, is assumed to have a lifetime long enough ($> \sim 10^{-12}$ sec) for the initial excitation and subsequent autoionization to be treated as a two-step process.

Consider the transition from one specific level of A to *one specific level* of A^* or A^+. To a first approximation, the probability of ionization depends on the energy of the photon ($h\nu$) or the electron as indicated in Fig. 2 for the processes described by Eqs. (6)–(9).

For reaction (6), the amount of A^+ formed from one level of A^* is described essentially by a delta function. For reactions (7) and (8), the curve is a step function equal to zero for $E < E_0$ and having a constant value for $E > E_0$. For reaction (9), the ion abundance is a linear function of excess energy in the electron:

$$I_{A^+} = K(E - E_0).$$

These statements are not exact but appear to be reasonable approximations up to at least $E = 2E_0$; more detailed discussions appear in other chapters in this volume. At higher electron or photon energies, ion production by transition to any one specific state of A* or A^+ by reaction (7), (8), or (9) becomes a decreasing function of E.

Except for a hydrogen atom, the number of states for the ion is greater than one; a large polyatomic molecule has an enormous number of levels both for A and for A^+ (or A*). Then, for reaction (9),

$$I_{A^+} = \sum_i \sum_j K_{ij}(E - E_{ij}),$$

where K_{ij}, E_{ij} refer to the probability of ionization from the ith level of A to the jth level of A^+ (assumed to be additive) and to the energy difference between those two states, respectively. If one further assumes all molecules A are in their ground state,

$$I_{A^+} = \sum_j K_j(E - E_j) \approx \int_0^{E - I_z} P(E_{int}) (E - I_z - E_{int}) \, dE_{int}.$$

The total energy added to the molecule, E_j, is equal to the ionization potential, I_z, plus the internal energy in the ion; the remainder of the initial energy, E, in the incident electron, given by $E - I_z - E_{int}$, is carried away by the two departing electrons. The function $P(E_{int})$ is the continuous approximation to the sum of the K_j over all levels having an energy within an energy range E to $E + \delta E$ and so depends both on the density of levels at each internal energy and on the transition probabilities to those levels. The density of states of a polyatomic ion increases very rapidly with increasing vibrational energy, while the transition probabilities generally depend on Franck–Condon factors in the same way as optical transition probabilities. The

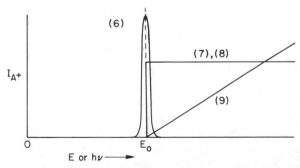

Figure 2. Typical first-approximation ionization efficiency curves for ionization from one level of a species A to one level of A^+. See text for description of process corresponding to each curve.

transitions to the few states at and very close to the minimum of the ion potential surface will be a very small fraction of the total. Also, if the equilibrium configuration of the lowest state of the ion is at all different from that of the molecules, the Franck–Condon factors for transitions to the lowest ion levels will be relatively small.

The result is that, experimentally, the ion intensity falls off asymptotically to zero as electron energy is decreased, even with monoenergetic electrons. Depending on the molecule and the experimental conditions, the lowest potential at which ions are observed could be only millivolts above the true ionization potential or it could be half a volt or more too high (11). The step-function behavior for ionization by photons eliminates one of the causes of the nearly asymptotic approach to zero ions observed with electrons. Hence, where Franck–Condon factors are large at the threshold, photon ionization will yield good ionization potentials. Low (or zero) Franck–Condon factors obviously imply that the adiabatic threshold E_0 value may not be observable even for photoionization.

Now, let us consider the energetics of fragment ion formation, for simplicity using an ion formed by a simple bond break:

$$R_1R_2 + e \rightarrow R_1^+ + R_2 + 2e.$$

Then, by the law of conservation of energy,

$$KE(R_1R_2) + IE(R_1R_2) + KE(e) = D_0(R_1-R_2) + I_z(R_1)$$

$$+ KE(R_1^+) + KE(R_2) + KE(e_1) + KE(e_2) + IE(R_1^+) + IE(R_2),$$

where KE(X) is the kinetic energy (translational) of X,
   IE(X) is the internal energy (electronic, vibrational, rotational) of X, and
   $D_0(R_1-R_2)$ is the bond energy.

The effect of $IE(R_1R_2)$ can be fairly well estimated and its effect on the ionization efficiency curve determined; $KE(R_1R_2)$ can usually be neglected. Then, if all the KE and IE terms on the right-hand side of the equation go to zero as KE(e) decreases, one has

$$AP(R_1^+) = KE(e)_{min} = D_0 + I_z(R_1).$$

where $AP(R_1^+)$ is the appearance potential of R_1^+. However, as $KE(e) \rightarrow KE(e)_{min}$, the excess energy in the ion, $(E - E_0)$, over that required for dissociation gocs to zero and so the rate of dissociation goes to zero. Only if $(E - E_0)$ is large enough to give a measurable rate of dissociation on the time scale of the residence time of the ion in the mass spectrometer ion source, typically 10^{-6}–10^{-7} sec, will the fragment ion be observed. If the reverse, bimolecular reaction has an activation energy, then internal or translational energy will necessarily be imparted to the fragments on dissociation of the activated complex. If the fragmentation is not the lowest-energy dissociation path, it also must have a rate large enough for it to be competitive with lower-energy dissociations. The result is that the lowest electron energy at which a fragment ion is observed is typically between 0 and 1.0 eV higher than the theoretical limiting value $D_0 + I_z(R_1)$. This difference is called the "kinetic shift."

In the same fashion as for neutral species, quantities such as D_0 and I_z can be expressed as differences of heats of formation. A tabulation of heats of formation is far more convenient than a tabulation of appearance potentials and ionization potentials for use in discussing properties of ions; hence, data on appearance potentials are usually converted to the equivalent heats of formation. A tabulation of heats of formation of ions by Rosenstock, Draxl, Steiner, and Herron also discusses the various techniques used to determine the ΔH_f values and their probable errors (12).

The question of when appearance potentials obtained with a conventional ion source could reasonably be used in a meaningful way was considered quite carefully by D. P. Stevenson in a series of papers starting in 1942 (13). He wished to use appearance potentials to calculate bond energies from sequences of reactions such as

$$C_3H_8 + e \rightarrow C_2H_5^+ + CH_3 + 2e, \qquad AP_1, \tag{10}$$

$$C_2H_6 + e \rightarrow C_2H_5^+ + H + 2e, \qquad AP_2, \tag{11}$$

$$H_2 \rightarrow 2H, \qquad D_0(H_2), \tag{12}$$

$$CH_4 + C_2H_6 \rightarrow C_3H_8 + H_2, \qquad \Delta H_r, \tag{13}$$

$$CH_4 \rightarrow CH_3 + H, \qquad D_0(CH_3-H). \tag{14}$$

Inspection quickly shows that reaction (14) is given by (10) $-$ (11) $+$ (12) $+$ (13). Hence, the first C–H bond dissociation energy in methane is given by

$$D_0(CH_3-H) = AP_1 - AP_2 + D_0(H_2) + \Delta H_r.$$

Stevenson stated that it was *not* reasonable to expect that the KE and IE terms in reactions (10) and (11) and all the other errors implicit in his use of a standard ion source would be negligible. However, it *was* reasonable to expect that all those errors would be nearly the same for reactions (10) and (11) to the extent that these were similar reactions run on the same mass spectrometer under exactly the same conditions. Then, since the calculation involved the difference $AP_1 - AP_2$, the unknown terms should largely cancel. In applying this principle to, for example, a specific type of dissociation of a basic molecule modified by various substituents, it is essential that none of the substituents cause a significant change in the dissociation scheme of the molecule, for example by introducing a new, lower-energy fragmentation path.

Many workers have published ionization efficiency curves that they approximated, at least at the low-voltage end, as consisting in part of straight line segments with abrupt changes in slope at their intersections. The changes in slope often were small; in many cases later work showed they were nonexistent, with the data properly fitted by a smooth curve. When one is assisted by the random errors in experimental data, it is often very easy to see straight line segments in a smooth curve with the breaks (changes in slope) coming at exactly the correct places, if one believes that there should be breaks. Of course, sometimes such small abrupt changes in slope are real; those reported in the more recent mass spectrometry literature *should* represent the opening of additional channels for the reaction in question.

The energy distribution of ions formed by electrons or photons must be determined if one is to perform a QET calculation and is of interest in many experiments. To the extent that the threshold laws previously stated apply well above threshold for electron or photon ionization, it can be shown that

$$\frac{d^2 I_t(E)}{dE^2} = P_i(E) \quad \text{(for electron ionization)},$$

$$\frac{dI_t(h\nu)}{d(h\nu)} = P_i(E) \quad \text{(for photon ionization)}.$$

Here $I_t(E)$ is the total (parent plus all fragment ions) ionization produced by electrons or photons with energy E, and $P_i(E)$ is the function described previously for the probability of ionization to give parent ions with internal energy $E - I_z$. Also, the relative values of $d^2 I_t(E)/dE^2$ or $dI_i(h\nu)/d(h\nu)$ give the relative amounts of the various fragment ions i produced by fragmentation of parent ions formed with internal energy $E - I_z$. It is difficult to obtain photon ionization data that are sufficiently noise-free to permit very precise derivative curves to be obtained. It is far more difficult to obtain data with monoenergetic electrons that is satisfactory for double differentiation. Consequently, other methods are desired. The photoelectron spectrum has been used as an approximation to $P_i(E)$, with satisfactory results. Other methods, such as photoelectron–photoion coincidence methods, have been developed in recent years. Knowledge of the internal energy distribution of the ions obviously is also very important in any discussion of ion–molecule reactions and so will be discussed in more detail with reference to specific topics in other chapters.

Another topic of interest in mass spectrometry that has a direct relation to kinetics in general is the amount of energy imparted to a fragment ion and its neutral partner on dissociation of the reactant ion. If the reactant ion is a parent ion, its translational kinetic (thermal) energy is small; hence, by conservation of momentum, a measurement of ion kinetic energy also determines the kinetic energy of the neutral fragment. Techniques have been developed for determining the kinetic-energy distribution of fragment ions with high precision. This distribution is related directly to both the magnitude of the reverse activation energy for the reaction, if any, and the shape of the potential surface in the neighborhood of the reaction coordinate on the product side of the transition state. The relative amounts of translational, rotational, and vibrational energy in the two fragments depend on this shape—in particular, on the amount of curvature in the reaction coordinate.

The basic ideas of QET are frequently applicable to the more general topics of this volume. If the collision complex formed in a bimolecular reaction has a long enough half-life for energy randomization among its internal degrees of freedom to occur, QET should be applicable to its dissociations. This appears to be the case in some ion–molecule collision complexes. Also, as previously mentioned, a unimolecular reaction is the reverse of a bimolecular collision. A discussion of one has implications for the other. Any trajectory describing in complete detail the changes of all coordinates and momenta in a dissociation is reversible and may equally well be used to describe an ion–molecule collision.

Finally, it should once again be emphasized that this chapter is a survey of arbitrarily selected topics, primarily from the older literature, and primarily relating to the properties of polyatomic ions that are large enough for statistical considerations to apply to their dissociation.

REFERENCES AND NOTES

1. K. J. Laidler and M. C. King, *J. Phys. Chem.* **87**, 2657–2664 (1983).

2. D. G. Truhlar, W. L. Hase, and J. T. Hynes, *J. Phys. Chem.* **87**, 2664–2682 (1983). This is an excellent review article (323 references) covering the many areas of active investigation in transition-state theory today.

3. Another general reference on the earlier work on unimolecular reactions is N. B. Slater, *Theory of Unimolecular Reactions,* Cornell University Press, Ithaca, New York (1959). This book takes a rather mathematical approach and includes critical discussions of the various reaction rate theories. The work of the author is emphasized, but other theories are not neglected. Nonthermal equilibrium energy distributions are discussed only in relation to the fall-off in reaction rate at low pressure.

4. F. W. Aston, *Mass Spectra and Isotopes,* 2nd ed., Edward Arnold, London (1942). This is the authoritative summary of the early work, from about 1906 to 1941, by the leading scientist in the field over most of those years.

5. H. F. Wiley, "Observations on Some of the Events Leading to the Formation of ASTM Committee E-14 Twenty Years Ago," Speech at the American Society for Mass Spectrometry Annual Banquet, Dallas, Texas, June 7, 1972.

6. H. M. Rosenstock, M. B. Wallenstein, A. L. Wahrhaftig, and H. Eyring, *Proc. Nat. Acad. Sci. USA* **38**, 667–678 (1952).

7. M. Vestal and J. H. Futrell, *J. Chem. Phys.* **52**, 978–988 (1970).

8. R. A. Marcus and O. K. Rice, *J. Phys. Colloid Chem.* **55**, 894 (1951); R. A. Marcus, *J. Chem. Phys.* **20**, 355–359, 359–364, 364–368 (1952).

9. An excellent introduction to the literature up to 1960 and to the field of mass spectrometry in general is J. H. Beynon, *Mass Spectrometry and Its Application to Organic Chemistry,* Elsevier, Amsterdam (1960). For current work, see the journal *Organic Mass Spectrometry.* The status of research in the entire field of mass spectrometry may be seen in Volumes 45 (1982) and 46–48 (1983) of the *International Journal of Mass Spectrometry and Ion Physics,* which contain the papers presented at the 9th International Mass Spectrometry Conference, August 30–September 3, 1982, edited by E. R. Schmid, K. Varmuza, and I. Fogg. In particular, Volume 45 contains the plenary lectures, including review papers giving the current status of many areas of mass spectrometry.

10. F. W. McLafferty, T. Wachs, C. Lifshitz, G. Innorta, and P. Irving, *J. Amer. Chem. Soc.* **92,** 6867–6880 (1970).

11. J. D. Morrison, "Ionisation and Appearance Potentials," in *MTP International Review of Science, Physical Chemistry Series One,* Vol. 5, A. Maccoll, Ed., Butterworths, London (1972), pp. 25–54.

12. H. M. Rosenstock, K. Draxl, B. W. Steiner, and J. T. Herron, *J. Phys. Chem. Ref. Data* **6,** Suppl. 1 (1977).

13. D. P. Stevenson, *Trans. Faraday Society* **49,** 867–878 (1953); *J. Chem. Physics* **10,** 291–294 (1942).

CHAPTER 2

Collision Theory
and Reaction Dynamics

RANDALL B. SHIRTS
Department of Chemistry
University of Utah
Salt Lake City, Utah

1 INTERACTION OF IONS WITH NEUTRALS

Reaction dynamics is the study of how atoms, molecules, and their ions interact on the most fundamental level. We distinguish dynamics from spectroscopy. The latter deals with how molecules behave alone: They move in uniform translational motion, vibrate, rotate, and absorb or emit photons if photons are around. This chapter will discuss collisions, that is, the behavior of atoms, molecules, and their ions when they come near each other. Collisions cause energy or momentum to be exchanged between the interacting species, and they can result in reaction. In condensed phases, molecules interact continuously with many neighbors. The dynamics of reactions in this case is extremely difficult to study on a fundamental level. In the gas phase, however, only a very small part of a molecule's time is spent in collisions; gas-phase collisions are rare events. Nevertheless, these random encounters contain all the chemistry of the gas phase: interactions that break bonds, form new ones, or cause excitation of the colliders. We will deal only with gas-phase collisions.

The exact treatment of chemical reaction dynamics is quantum mechanical. Nevertheless, since atoms and molecules are quite heavy (at least compared with electrons), classical mechanics is sufficient to explain most of the collision and reaction dynamics. Near the end of the chapter (Section 4) we enlist quantum mechanical ideas, but let us begin by using classical mechanics.

Newton's equations describe classical motion in its simplest form:

$$\mathbf{F} = m\mathbf{a} = m\ddot{\mathbf{x}}. \tag{1}$$

This equation may be taken as our definition of a force: A force is something that changes the motion of a particle. If one applies a force through a distance, one does work. Work is a form of energy, so by applying a force one increases either the kinetic energy or the potential energy of the particle. Consider the latter. If a force is applied to a particle, then we must apply an equal and opposite force to keep its kinetic energy from changing. For a differential displacement, the work we must do in resisting the force is

$$dW = dU = -\mathbf{F} \cdot d\mathbf{s} = -F_x\, dx - F_y\, dy - F_z\, dz, \tag{2}$$

where F with a subscript denotes a partial derivative in that direction. We have, on identifying components of Eq. (2), another description of a force:

$$\mathbf{F} = -\nabla U. \tag{3}$$

Thus, we can describe any force (technically, any conservative force) by a potential-energy function. The force as a function of position can be evaluated by taking the derivatives implied in Eq. (3). The most common potential-energy function is the Coulomb potential energy between two charges. It is given in SI units by

$$U = \frac{q_1 q_2}{4\pi\epsilon_0 r}, \tag{4}$$

where q_1 and q_2 are the charges, r is their separation, and ϵ_0 is the permittivity of a vacuum. Equation (4) is derived from the fact that the force falls off as r^{-2}. The factor of $4\pi\epsilon_0$ is required by the choice of units.

How does an ion interact with a neutral atom or molecule? There is no Coulomb force (or potential) between a charged ion and a neutral point particle. To answer this question let us see what force (or potential) exists between a charge and a dipole. A dipole is described by two charges $+q$ and $-q$ separated by a distance R. The dipole moment, μ, is the product Rq. Let us place the charges on the z-axis with $+q$ at $z = R/2$ and $-q$ at $z = -R/2$ and consider the potential of a charge Q at a point $\mathbf{r} = (r, \theta, \phi)$. To get the potential energy, just add up the potential energies due to the interaction of each charge with the test charge, Q:

$$U = \frac{(+q)Q}{4\pi\epsilon_0 r_1} + \frac{(-q)Q}{4\pi\epsilon_0 r_2}. \tag{5a}$$

Now, using a little trigonometry (see Fig. 1), we obtain

$$U = \frac{qQ}{4\pi\epsilon_0}\left(r^2 + \frac{R^2}{4} - R r \cos\theta\right)^{-1/2}$$

$$- \frac{qQ}{4\pi\epsilon_0}\left(r^2 + \frac{R^2}{4} + R r \cos\theta\right)^{-1/2}. \tag{5b}$$

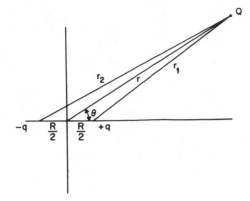

Figure 1. Geometry for calculating the potential of a dipole in terms of r, the distance from Q to the center of the dipole, and θ, the polar angle.

Close enough to real molecules (small values of r), electron–electron repulsions will make this formula inaccurate, so let us assume that $r \gg R$. Factoring out r from the radicals, we obtain

$$U = \frac{qQ}{4\pi\epsilon_0 r}\left[\left(1 + \frac{R^2}{4r^2} - \frac{R}{r}\cos\theta\right)^{-1/2} - \left(1 + \frac{R^2}{4r^2} + \frac{R}{r}\cos\theta\right)^{-1/2}\right]. \quad (5c)$$

Then, since $(1 + x)^{-1/2} = 1 - x/2 + 3x^2/8 + \cdots$, we expand each of the radicals:

$$
\begin{aligned}
U &= \frac{qQ}{4\pi\epsilon_0 r}\left[\left(1 + \frac{R}{2r}\cos\theta - \frac{R^2}{8r^2} + \frac{3R}{8r^2}\cos^2\theta + \cdots\right)\right.\\
&\qquad\qquad \left. - \left(1 - \frac{R}{2r}\cos\theta - \frac{R^2}{8r^2} + \frac{3R}{8r^2}\cos^2\theta + \cdots\right)\right]\\
&= \frac{qQ}{4\pi\epsilon_0 r}\left(\frac{R}{r}\cos\theta + \cdots\right)\\
&= \frac{qRQ}{4\pi\epsilon_0 r^2}\cos\theta\,(1 + \cdots) \approx \frac{\mu Q}{r^2}\cos\theta. \quad (5d)
\end{aligned}
$$

Here we have used $\mu = qR$, and we have neglected terms that have negative powers of r greater than 2. Taking derivatives of Eq. (5d), we conclude that there is a force between a dipolar molecule and an ion, but it falls off as r^{-3}, much faster than does the force between ions. The force on the charge depends on the orientation of the dipole. If Q is positive, it is attracted to the negative end ($\theta \approx \pi$) and repelled from the positive end ($\theta \approx 0$). As the dipole spins, as it usually will be doing, the ion feels both positive and negative force, so the net interaction will be an average of the force over the possible orientations. This average dipole orientation (ADO) force (1) is much weaker than its maximum value, $\mu Q/r^2$. The precise value will depend on the temperature, due to averaging over a Boltzmann distribution of the possible orientations (1).

Some molecules (e.g., atoms) do not even have a dipole moment. How do these interact with ions? The answer to this question is contained in the polarizability of the molecule. A molecule (or atom or ion) is composed of positive and negative particles (i.e., nuclei and electrons). When a charged particle approaches, for example, a positive charge, the nuclei are repelled and the electrons are attracted to it. Because of the strong attractive forces that hold the nuclei and electrons together, they will not come apart easily. Nevertheless, the molecule is temporarily polarized by the presence of the charged particle: A small dipole is formed by the slight "sloshing" of electrons toward the positive charge and the slight retreat of the nuclei. The following equation expresses the result:

$$\boldsymbol{\mu}(t) = \boldsymbol{\alpha}\ \boldsymbol{\epsilon}(t), \tag{6}$$

where the dipole moment, $\boldsymbol{\mu}$, is due to the electric field intensity, $\boldsymbol{\epsilon}(t)$, caused by an ion flying by. These two quantities are related by the polarizability tensor, $\boldsymbol{\alpha}$. The polarizability is simply a number for an atom, since $\boldsymbol{\mu}$ must be parallel to $\boldsymbol{\epsilon}$, but for a molecule, the electrons can slosh more easily in one direction than another. A tensor is then necessary.

We now derive the form of the interaction potential for an ion and an atom. In this case, we have $\mu = \alpha\epsilon$, and the potential energy of a dipole in a field is given by $U = -\boldsymbol{\mu}\cdot\boldsymbol{\epsilon}$, so as we increase the field from zero,

$$dU = -\boldsymbol{\mu}\cdot d\boldsymbol{\epsilon} = -\alpha\epsilon\ d\epsilon. \tag{7a}$$

On integrating from zero field,

$$U = \int_0^U dU' = -\alpha \int_0^\epsilon \epsilon'\ d\epsilon' = -\frac{\alpha\epsilon^2}{2}, \tag{7b}$$

but $\epsilon = Q/4\pi\epsilon_0 r^2$, so

$$U(r) = -\frac{\alpha Q^2}{2(4\pi\epsilon_0)^2 r^4} \qquad \text{(for large } r\text{)}. \tag{7c}$$

Notice that the charge-induced dipole potential falls off faster than the charge–dipole potential (r^{-4} compared with r^{-2}). On the other hand, the charge-induced dipole potential is always attractive, since α, r, and Q^2 are always positive. Equation (7c) gives the correct form of the average interaction between an ion and a neutral nonpolar molecule for sufficiently large distances. If a molecule has a permanent dipole, the charge–dipole interaction provides an additional angular force that may change the speed of rotation in a collision, but as we have seen, the charge–dipole force is ineffective in bringing the species together to react, because it tends to average out. Additional terms such as charge–quadrupole terms and dispersion terms can also be added to the attractive force (2). These additional terms all fall off faster than r^{-4} but become increasingly important at smaller values of r. For sufficiently small separations, all these ideas are inaccurate, because the individual electrons in the molecule can respond independently to individual electrons in the ion. The interaction for very small values of r is repulsive but falls off very quickly. Assuming the fall-off is exponential and neglecting all the attractive terms except Eq. (7c), we get the following form for the potential:

$$U(r) = Ae^{-\gamma r} - \frac{\alpha Q^2}{2(4\pi\epsilon_0)^2 r^4} , \tag{8}$$

where A and γ are parameters that can be determined from theory or experiment. If the neutral species is a molecule with a polarizability tensor, it is still useful to use the spherically averaged polarizability $\alpha = (\alpha_1 + \alpha_2 + \alpha_3)/3$, where the α_i are polarizabilities along the three principal axes.

The potential of E. (8) is plotted in Fig. 2. Note that at some distance that depends on the parameters, A, γ, and α, the repulsive exponential and the attractive r^{-4} tail balance to form a well. If the molecule gets trapped in this well with insufficient energy to escape, a stable, larger ion or complex is formed. This possibility is discussed in Section 5.

2 COLLISION THEORY SIMPLIFIED—CLASSICAL THEORY

2.1 Separation in Center-of-Mass Coordinates

Keeping track of the coordinates and velocities of two particles is a complicated problem. Luckily, this problem can be tremendously simplified by examining it in a special frame of reference: center-of-mass coordinates (note that the following separation is general and does not depend on the form of the potential):

$$\mathbf{r} = \mathbf{r}_2 - \mathbf{r}_1, \qquad \mathbf{v} = \mathbf{v}_2 - \mathbf{v}_1 = \dot{\mathbf{r}}_2 - \dot{\mathbf{r}}_1,$$

$$\mathbf{R}_{cm} = \frac{m_1\mathbf{r}_1 + m_2\mathbf{r}_2}{m_1 + m_2}, \qquad \mathbf{V}_{cm} = \frac{m_1\mathbf{v}_1 + m_2\mathbf{v}_2}{m_1 + m_2} = \dot{\mathbf{R}}_{cm}. \tag{9}$$

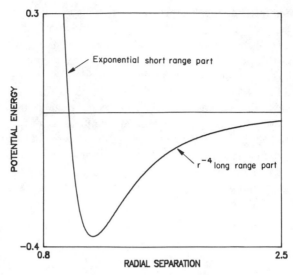

Figure 2. Schematic interaction potential between an ion and an atom. The attractive part is due to the ion-induced dipole interaction. The repulsive part is due to electron–electron and nucleus–nucleus repulsion. At distances where these balance out, a well is formed.

In the new coordinate system defined by the relative coordinate **r**, the combined linear momentum of the particles is zero, because the center of mass $\mathbf{R}_{cm}$, moves along with the particles.

Let us examine the angular momentum and energy of the system. The angular momentum is the sum of the individual angular momentum vectors:

$$\mathbf{L} = \mathbf{r}_1 \times \mathbf{p}_1 + \mathbf{r}_1 \times \mathbf{p}_2 = m_1 \mathbf{r}_1 \times \mathbf{v}_1 + m_2 \mathbf{r}_2 \times \mathbf{v}_2 \tag{10a}$$

Now, converting to center-of-mass coordinates,

$$\mathbf{L} = m_1 \left(\mathbf{R}_{cm} + \frac{m_2}{m_1 + m_2} \mathbf{r} \right) \times \left(\mathbf{V}_{cm} + \frac{m_2}{m_1 + m_2} \mathbf{v} \right)$$

$$+ m_2 \left(\mathbf{R}_{cm} - \frac{m_1}{m_1 + m_2} \mathbf{r} \right) \times \left(\mathbf{V}_{cm} - \frac{m_1}{m_1 + m_2} \mathbf{v} \right)$$

$$= (m_1 + m_2) \mathbf{R}_{cm} \times \mathbf{V}_{cm} + \left[\frac{m_1 m_2^2}{(m_1 + m_2)^2} + \frac{m_2 m_1^2}{(m_1 + m_2)^2} \right] \mathbf{r} \times \mathbf{v}$$

$$= M(\mathbf{R}_{cm} \times \mathbf{V}_{cm}) + \frac{m_1 m_2}{m_1 + m_2} (\mathbf{r} \times \mathbf{v}), \tag{10b}$$

where $M = m_1 + m_2$ is the total mass. Likewise, the total kinetic energy of the two particles is:

$$T = \frac{1}{2} m_1 \mathbf{v}_1^2 + \frac{1}{2} m_2 \mathbf{v}_2^2 = \frac{1}{2} m_1 \left(\mathbf{V}_{cm} + \frac{m_2 \mathbf{v}}{m_1 + m_2} \right)^2$$

$$+ \frac{1}{2} m_2 \left(\mathbf{V}_{cm} - \frac{m_1 \mathbf{v}}{m_1 + m_2} \right)^2$$

$$= \frac{1}{2} (m_1 + m_2) \mathbf{V}_{cm}^2 + \frac{1}{2} \frac{m_1 m_2^2 + m_2 m_1^2}{(m_1 + m_2)^2} \mathbf{v}^2$$

$$= \frac{1}{2} M \mathbf{V}_{cm}^2 + \frac{1}{2} \frac{m_1 m_2}{m_1 + m_2} \mathbf{v}^2. \tag{11}$$

Equations (10) and (11) show that the angular momentum and kinetic energy can be divided into two parts: one due to the motion of the center-of-mass coordinate system and one due to the motion in the relative coordinate system. Motion in the relative coordinate system, however, acts as if it depends on the "reduced" mass

$$\mu = \frac{m_1 m_2}{m_1 + m_2}. \tag{12}$$

By examining the two particles in this new coordinate system, we can ignore the center-of-mass motion and study only the relative motion in which the two particles behave like a single particle of mass μ, with position $\mathbf{r}$, angular momentum $\mu(\mathbf{r} \times \mathbf{v})$, and kinetic energy $\frac{1}{2}\mu \mathbf{v}^2$.

2.2 Separation in Polar Coordinates for a Central Potential—the Centrifugal Force

If the potential of interaction between the particles depends only on the magnitude of the separation, $|\mathbf{r}|$ (a central potential), then the force, $\mathbf{F} = -\nabla U = m\mathbf{a}$, is along the $\mathbf{r}$ direction, so

$$\frac{d}{dt} \mathbf{L} = \frac{d}{dt} \mu(\mathbf{r} \times \mathbf{v}) = \mu(\mathbf{v} \times \mathbf{v} + \mathbf{r} \times \mathbf{a}) = 0. \tag{13}$$

The last equality comes from the fact that the cross product of two parallel vectors is zero. Thus $\mathbf{L}$ is a constant, a conserved quantity. The length $|\mathbf{L}|$ is one constant, but the direction also is constant. The vectors $\mathbf{r}$ and $\mathbf{v}$ define a plane,† and since

---

†If $\mathbf{r} \| \mathbf{v}$, then $\mathbf{L}$ is zero, and no plane is defined, but this exception does not destroy the conclusions, since we are free to use any plane containing the vector $\mathbf{r}$ (and $\mathbf{v}$).

the direction of **L** (which is perpendicular to both of them) is constant, **r** and **v** stay in the same plane unless a third body comes nearby to destroy the two-body description. In the gas phase, molecules have collisions only rarely. It is an *extremely rare* event for three bodies to come near to each other simultaneously, so this reduced description is extremely useful.

Let us choose the direction of **L** to be our z-axis. The relative motion of the particles is then in the x–y plane. If we sit on particle 1, then the relative coordinates, **r**, are just those of particle 2. In its motion, particle 2 behaves as if it has mass μ. We can choose our positive x-axis to be the direction from which particle 2 approaches. Thus, $\dot{r}$ will be in the $-x$ direction initially (before the particles are close enough to interact). Now we can introduce polar coordinates in the x–y plane:

$$x = r \cos \phi, \qquad \dot{x} = \dot{r} \cos \phi - r\dot{\phi} \sin \phi,$$

$$y = r \sin \phi, \qquad \dot{y} = \dot{r} \sin \phi + r\dot{\phi} \cos \phi. \tag{14}$$

Using a little trigonometry, we obtain

$$|\mathbf{L}| = \mu\mathbf{r} \times \mathbf{v} = \mu(x\dot{y} - y\dot{x}) = \mu r^2\dot{\phi}, \tag{15}$$

$$T = \tfrac{1}{2}\mu(\dot{x}^2 + \dot{y}^2) = \tfrac{1}{2}\mu(\dot{r}^2 + r^2\dot{\phi}^2). \tag{16}$$

By eliminating $\dot{\phi}$ from these two equations, and adding $U(r)$ to T to get the total energy, we get

$$E = \frac{1}{2}\mu\dot{r}^2 + \frac{L^2}{2\mu r^2} + U(r). \tag{17}$$

Thus, for a central potential (one that depends only on $|\mathbf{r}|$, and not on angles) the motion can be viewed as a particle with mass μ in an "effective" potential.

$$U_{\text{eff}}(r) = U(r) + \frac{L^2}{2\mu r^2} \tag{18}$$

The repulsive term in $U_{\text{eff}}(r)$ given by $L^2/2\mu r^2$ is called the centrifugal potential. Since the centrifugal potential is proportional to r^{-2}, a particle with nonzero angular momentum can never get too close to the origin. The centrifugal term arises because without it the particle could not maintain constant angular momentum close to the origin, as required by the laws of physics, without going so fast that energy could not be conserved. The centrifugal potential is thus a fictitious force that enforces the requirement that angular momentum be conserved.

The centrifugal potential must be included in the equation for radial motion to correct for the fact that one is ignoring the angular motion. An analogy helps to clarify this idea. Suppose you are standing on the edge of the highway watching cars pass. In the distance, you note that the cars are rapidly approaching you (the relative distance r is decreasing). As they get closer, the rate of approach decreases little by little until, as they pass you, they cease to approach and begin to recede. You would have to say (remember, you were watching only their relative distance) that a force must have acted on them to slow them down and turn them around to make them go away. Actually, there is no such force: The cars never deviate from their original straight line motion, but when one is viewing only the relative distance, the centrifugal force must be invoked in the radial coordinate to account for the behavior that is being ignored in the angular coordinate.

At large separations, before any interaction has taken place, the initial velocity vector is parallel to the x-axis. The angular momentum vector is $\mathbf{L} = m(\mathbf{r} \times \mathbf{v}) = mrv \sin \phi = mvb$, where $b = r \sin \phi$ and is called the impact parameter (see Fig. 3). If there were no interaction between the bodies [$U(r) = 0$], then b would be the distance of closest approach as the particle flies by. When the potential is added, the motion of the incoming particle is governed by Eqs. (15)–(17).

2.3   The Deflection Angle

Unfortunately, the details of an individual trajectory are not yet observable in a molecular experiment. The only observable results of a collision are the energy or velocity of the particle after collision and its direction of motion. (In some cases where the particle has internal structure, it may be possible to examine its internal energy.) For ordinary collisions between spherical particles, the total kinetic energy must be conserved, so the initial and final velocities must be equal at distances large enough that $U(r)$ is negligible. The quantity of interest in these elastic and nonreactive collisions is the deflection angle, θ.

Particle 2 with reduced mass μ approaches parallel (at least initially) to the x-axis with kinetic energy $\frac{1}{2}\mu v^2$ and angular momentum μvb and is affected by the

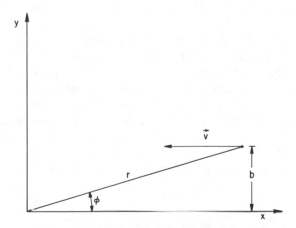

Figure 3. Description of a collision in relative coordinates. The velocity is taken to have magnitude v in the $-x$-direction with initial y-coordinate (impact parameter) b. For polar coordinates r, ϕ, Eqs. (14)–(21) apply.

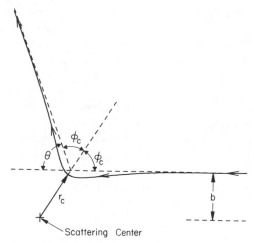

Figure 4.   Definition of the deflection angle θ in terms of the polar angle at closest approach ϕ_c. The scattering center is at the origin.

potential. At some distance, the effective potential energy equals the initial incoming kinetic energy, and the particle can approach no closer. This is due to a repulsive potential, $U(r)$; to the centrifugal term in the effective potential, $L^2/2\mu r^2$; or to a combination of the two. The distance of closest approach, r_c, is called the turning point. Subsequently, the particle begins to recede, and eventually it recedes to far enough away that its direction of motion stops changing. The change in direction of motion is called the deflection angle θ, and is measured from the negative x-axis (see Fig. 4). An undeflected particle has $\theta = 0$. Note that a deflection through $+\theta$ and one through $-\theta$ are experimentally indistinguishable, since we do not know on a molecular scale on which side of the x-axis the velocity vector initially began. Likewise, the deflection $\pm 2\pi N \pm \theta$ (where N is an integer) is indistinguishable from the deflection θ.

Let us begin our measurement of time in a collision at the moment the particle reaches the turning point; in other words, $r(t = 0) = r_c$. We can call the corresponding polar angle ϕ_c; that is, $\phi(t = 0) = \phi_c$. Now, let us define a new angle $\phi' = \phi - \phi_c$. Examining Eqs. (15) and (16), we can see that the equations are unchanged if we change t to $-t$ and ϕ' to $-\phi'$. This invariance means that the solution to these differential equations is symmetric about the turning point. Since the polar angle ϕ_c develops during the time interval $-\infty$ to 0, at time $t = +\infty$ the polar angle is $2\phi_c$. The final scattering angle θ is just $\pi - 2\phi_c$ (see Fig. 4). We can integrate Eqs. (15) and (16) to obtain an expression for ϕ_c and θ. Using Eqs. (15) and (16),

$$d\phi = \frac{d\phi}{dt}\frac{dt}{dr}\,dr = \frac{\dot{\phi}}{\dot{r}}\,dr$$

$$= \frac{L}{\mu r^2}\left(\frac{2}{\mu}\left(E - U(r) - \frac{L^2}{2\mu r^2}\right)\right)^{-1/2}dr. \qquad (19)$$

Integrating and using $b\sqrt{2\mu E} = L$,

$$\int_0^{\phi_c} d\phi = \phi_c = -\int_\infty^{r_c} \frac{L\,dr}{\mu r^2} \left(\frac{2}{\mu}\left(E - U(r) - \frac{L^2}{2\mu r^2}\right)\right)^{1/2}$$

$$= b\int_{r_c}^\infty \frac{dr}{r^2}\left(1 - \frac{U(r)}{E} - \frac{b^2}{r^2}\right)^{-1/2}. \qquad (20)$$

Equation (20) allows us to express the deflection angle as a function of initial energy and impact parameter:

$$\phi(b, E) = \pi - 2b\int_{r_c}^\infty \frac{dr}{r^2}\left(1 - \frac{U(r)}{E} - \frac{b^2}{r^2}\right)^{-1/2}, \qquad (21)$$

where r_c is the outermost solution of $E = L^2/2\mu r_c^2 + U(r_c) = U(r_c) + Eb^2/r_c^2$. Note that if $b = 0$ (head-on collision), then $\theta = \pi$ (unless $U = 0$). In most cases, as b increases from zero, θ decreases from π, and as b becomes large, θ decreases to zero. Let us take an assumed form for the potential and calculate the deflection function for a number of energies and impact parameters. We, in the next section, first take as an example a potential that describes the collisions of hard spheres.

2.4 Hard-Sphere Collisions

So far we have talked as though species will not interact without an attractive force to pull them together. This is false. If two particles of definite size are moving on a head-on path, they will collide even if they have negligible attraction (as many automobile accident victims might attest). Consider the collision of two spheres of radii R_1 and R_2. There is no interaction if $r > R_c$, where $R_c = R_1 + R_2$. If $r = R_c$, the repulsive force is infinite. The hard-sphere potential is then given by (Fig. 5)

$$U(r) = \begin{cases} 0 & (r > R_c) \\ \infty & (r \leqslant R_c) \end{cases}$$

For the hard-sphere potential, if $b < R_c$ then $r_c = R_c$. If $b > R_c$ then $r_c = b$ and no deflection occurs. Figure 6 makes the geometry of hard-sphere collisions clear. It is obvious from this figure that

$$\theta = \pi - 2\phi \qquad \text{and} \qquad b = R_c \sin\phi = R_c \cos\left(\frac{\theta}{2}\right)$$

so

$$\theta = 2\cos^{-1}\left(\frac{b}{R_c}\right) \qquad (b < R_c). \qquad (22a)$$

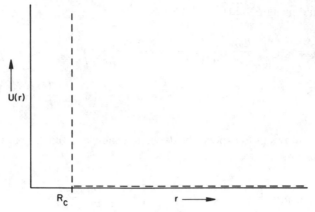

Figure 5. The potential of interaction between hard spheres. The potential is infinite at the small distances and zero at large distances.

We can recover the same formula by substituting into Eq. (21) and making the substitution $r = b \cosh x$, where $r_c = b \cosh x_c$:

$$\theta = \pi - 2b \int_{R_c}^{\infty} \frac{dr}{r^2} \left(1 - \frac{b^2}{r^2} \right)^{-1/2}$$

$$= \pi - 2 \int_{x_c}^{\infty} \frac{dx}{\cosh x} = 2 \tan^{-1} \sqrt{\left(\frac{R_c}{b} \right)^2 - 1}$$

$$= 2 \cos^{-1}\left(\frac{b}{R_c} \right) \qquad (b < R_c). \tag{22b}$$

The geometry [Eq. (22a)] and the dynamics [Eq. (22b)] give the same answer for the deflection as a function of the impact parameter. The function is plotted in

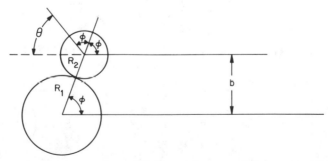

Figure 6. Geometry at the moment of collision between two hard spheres, ilustrating the quantities in Eq. (22a).

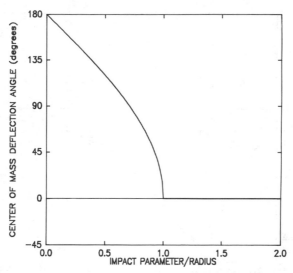

Figure 7.   Deflection function for hard spheres. For $b > R_c$, the deflection is zero.

Fig. 7. As we expected, for small impact parameters the deflection function is π (180°). As b increases, the deflection angle decreases smoothly to zero. Notice that the deflection function does not depend on the collisional energy for hard spheres. This is an unusual case, as we shall see next, in a more realistic example.

2.5   An Illustrative Example—Morse Potential

We now treat a potential that is more realistic for interatomic interactions, the Morse potential,

$$U(r) = D(e^{-2x} - 2e^{-x}), \qquad \text{where } x = \alpha(r - r_e). \qquad (23)$$

This potential has an attractive well with a minimum of $-D$ at the equilibrium separation, $r = r_e$. The parameter α controls the width of the well, the curvature at the bottom of the well, or how fast the potential rises toward zero at large r. This potential does not have the proper behavior for an ion–atom interaction at large r [cf. Eq. (7)], because it goes to zero exponentially; it is, however, a convenient functional form in the well region. For r values less than $r = r_e$, the potential rises steeply to a high value representing the repulsive interaction as two particles collide. The parameters $r_e = 1.28$ Å, $D = 4.17$ eV, and $\alpha = 1.85$ Å^{-1} give a reasonable fit to the experimentally determined potential for protons colliding with Ar atoms at low energy (3). The experimental potential for this situation is the solid curve in Fig. 8; the Morse potential obtained with the parameters given is the dashed curve.

In calculating the deflection angle θ for a given energy and impact parameter, one must first determine the turning point, r_c. The turning point is a function of impact parameter (or angular momentum, L), because of the centrifugal term in the

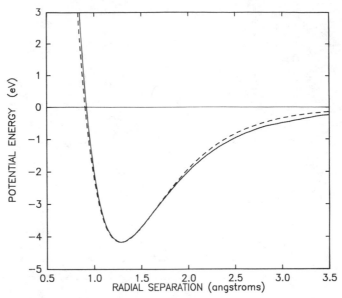

Figure 8. Comparison between an experimentally fit potential-energy function (solid line) and the Morse potential (dashed line). The Morse potential parameters are chosen to give the same equilibrium bond length, well depth, and curvature at equilibrium that the experimental curve has.

effective potential. Figure 9 shows the effective potential for $U(r)$ given by Eq. (23) and for several values of L. Each curve is labeled with the associated value of L in units of $\hbar$. If $L = 0$ in Fig. 9, we have just the Morse potential of Fig. 8. This potential has an attractive well at large distances and a repulsive wall at small distances. For sufficiently high values of L, however, the repulsive centrifugal potential completely overpowers the attractive well of the Morse potential, and the effective potential is monotonic (repulsive everywhere). The critical value of L in this case (the Morse parameters given above and the reduced mass of Ar + H$^+$) is 75.95 $\hbar$, for which there is an inflection point in the effective potential at $V = 1.153$ eV. For values of L less than 75.95 $\hbar$, the effective potential has two regions of repulsion separated by a well of attraction. The outer region of repulsion is called the centrifugal barrier; it arises because the centrifugal potential, which falls off slowly, dominates the "true" potential at large distances. Note that for sufficiently low L and E, there are three roots to $E = U_{\text{eff}}$. Classically, the turning point is the outermost root. Figure 10 shows r_c vs. b for several different energies. For large enough values of b, $r_c \approx b$, because the particles never get close enough together to feel much of the attractive well. All of the curves have one point in common. Where $V(r^*) = 0$, the energy drops out of the equation for the turning point and $r_c = b = r^*$ for all energies. For a given energy, if $b < r^*$, then $V(r_c) > 0$. Likewise, if $b > r^*$, then $V(r_c) < 0$. The value r^* at which $r_c = b$ in Fig. 10 is called the hard-sphere radius. Even though the potential is not a hard-sphere potential, for some purposes one can approximate the potential by a hard-sphere

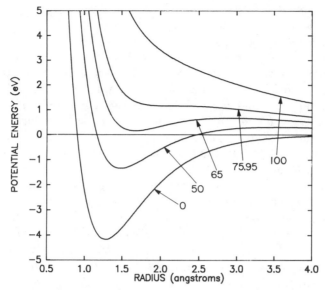

Figure 9. Effective potentials [Eq (18)] obtained for several values of L (indicated on curves in units of $\hbar$) using the Morse potential (dotted curve in Fig. 8). Note that the attractive well is canceled out for $L > 75.95\ \hbar$.

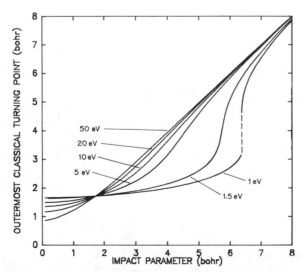

Figure 10. Outermost classical turning point as a function of impact parameter for the Morse potential of Fig. 8 and the six labeled values of the relative kinetic energy. All curves cross at $r^* = b$, where r^* is the effective hard-sphere radius.

potential with the given radius. In any case, the hard-sphere radius is a measure of size of the particles.

After evaluating r_c for a given b and E, one must evaluate the integral in Eq. (21) to obtain the deflection angle. Figure 11 shows the deflection function for the Morse potential at several different energies. The energy associated with each curve is given in electron volts. Note that for small impact parameters, the deflection angle falls from π as it did in the hard-sphere case. Large positive deflections are due to the repulsive interaction of head-on collisions. At large impact parameters, the deflection angle approaches zero from negative values. Negative deflections are due to the attractive potential at large distances. The attractive force pulls the incoming particle toward the scattering center (cf. Fig. 4). If the impact parameter is large, the turning point is so large that the incoming particle never feels the repulsive wall. In other words, the particle hits the centrifugal barrier before it can reach the repulsive wall in the interaction potential. As the impact parameter is decreased, the turning point decreases, the particle feels more and more of the attractive well, and the deflection angle becomes more and more negative. Eventually, the turning point moves to values of r that correspond to the repulsive region of the potential, and the deflection becomes positive. Consequently, there must be a maximum negative deflection angle. This maximum negative deflection is called the rainbow angle because of an analogy with light scattering from water droplets. We will explore the experimental importance of the rainbow angle below.

We noted above that for sufficiently low energy ($E < 1.153$ eV in this case) and a given nonzero value of L, there are three turning points. At any energy below 1.153 eV, there is one value of L for which two of the turning points coalesce into

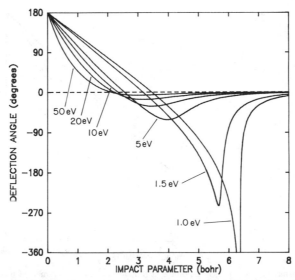

Figure 11.   The deflection function as a function of impact parameter for six different energies and the Morse potential of Fig. 8. Note that a low-energy collision can have a large negative deflection. This is called orbiting.

one at the top of the centrifugal barrier. When this position is reached for this energy and L value, the force vanishes. In the radial coordinate, the incoming particle approaches the barrier maximum and balances there. Meanwhile, in the angular coordinate, the two particles are rotating around and around. This is called orbiting, and the deflection angle increases without bound. The rainbow angle becomes infinite. A narow range of impact parameters centered at the rainbow impact parameter can be found with very large deflection angles; however, this range is usually narrow enough that only careful experiments can measure the rainbow scattering. Below, we will see that if the energy is low enough that the rainbow angle is greater than 180° (half a revolution), then one can easily recognize it in the experimental data. In summary, the rainbow angle increases in magnitude with decreasing energy until the energy is reached at which orbiting can occur. Below this energy the rainbow angle is infinite.

Many experiments begin with two beams of particles intersecting at right angles to each other. Such crossed-beam experiments measure the number of particles deflected through various angles. One can sometimes analyze these results to determine what the potential of interaction was that produced such a pattern of scattering. This process is called inversion of the scattering data (3,4).

2.6 Cross Sections

In an experiment, all magnitudes and directions of impact parameters are equally represented; a distribution of scattering angles results. If the number of particles per unit area per second approaching along the positive x-axis is I_0 (the flux), then the number per second with impact parameter between b and $b + db$ is $I_0\, 2\pi b\, db$. These particles will be scattered into angles θ to $\theta + d\theta$ (Fig. 12). The differential cross section, $I(\theta)$, is defined to be the fraction of the incident flux scattered into the angles between θ and $\theta + d\theta$:

$$I(\theta)\, d\Omega = I(\theta)2\pi \sin\theta\, d\theta = \frac{I(b)}{I_0} = 2\pi b\, db, \tag{24a}$$

where $d\Omega$ is the differential of solid angle. Note that $I(\theta)$ has units of area. Thus

$$I(\theta) = \frac{b}{\sin\theta}\left|\frac{db}{d\theta}\right| = \frac{b}{|\sin\theta\,(d\theta/db)|}. \tag{24b}$$

Figure 12. Diagram showing incoming scatterable particles with impact parameters between b and $b + db$, which are scattered into angles between θ and $\theta + d\theta$.

Here we have taken the absolute value because $I(\theta)$ must be positive. If several different impact parameters all scatter into one angle, their contributions must be added together. Quantum mechanically, we would have interferences, but classically we just add the contributions.

We can now calculate the differential cross section $I(\theta)$ for hard-sphere scattering using Eq. (22) and $b = R_c \cos(\theta/2)$:

$$I(\theta) = \frac{b(\theta)}{\sin\theta}\frac{db}{d\theta} = \frac{R_c \cos(\theta/2)}{\sin\theta}\frac{R_c}{2}\sin\left(\frac{\theta}{2}\right) = \frac{R_c^2}{4}. \tag{25}$$

Note that the hard-sphere differential cross section does not depend on θ: An equal number of particles are scattered into every angle. This is another property that is unique to the hard-sphere potential.

The differential cross section for the Morse potential that we previously used may be calculated numerically and is shown in Fig. 13. The scattering is peaked in the forward direction (small angles) for all energies. These particles are mostly the particles with large impact parameters that are only slightly deflected by the attractive well. The forward peak is enhanced by the factor of $\sin\theta$ in the denominator of Eq. (24b). For each energy, the rainbow angle shows up as a square root singularity. Note that the rainbow angle increases with decreasing energy, as we discussed above. At the rainbow angle, many impact parameters are scattering into the same angle. Mathematically, this singularity appears because the derivative $d\theta/db$ goes to zero in the denominator of Eq. (24b). In the plots for energies 20, 10,

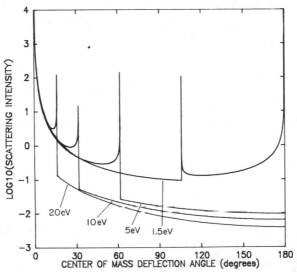

Figure 13. The classical differential cross sections corresponding to the Morse potential of Fig. 8 for four energies. Note that the rainbow angle is increasing for lower energies, but for $E = 1.5$ eV it has increased past 180° and reversed direction (bright side to large angle).

and 5 eV, the rainbow angle is less than 180°, and three different impact parameters scatter into all angles less than θ_r. Only one impact parameter scatters into angles greater than θ_r. One usually says that the "bright side" of these rainbows is at small angles. For an energy of 1.5 eV, the rainbow angle is 254°. The *experimentally* measured deflection angle is $360° - 254° = 106°$, because experimental angles of deflection must be in the range 0–180°. Note that the "bright side" of this rainbow is reversed: It is at large angles. Three impact parameters scatter into angles less than 106° and five scatter into angles greater than 106°. Note also that the scattering intensity becomes infinite at 180° (backscattering). This is due to the geometrical factor $\sin \theta$ in the denominator of Eq. (24b). Whenever one observes a backscattering peak, one knows that the particles are interacting through a sufficiently attractive force to keep them together for at least one-half of a revolution.

The total cross section, σ, is defined to be the integral of the differential cross section over all the angles:

$$\sigma = 2\pi \int_0^\pi I(\theta) \sin \theta \, d\theta. \tag{26}$$

Recall that the cross section has units of area. The geometrical significance of that area can be seen next.

For hard spheres, using the differential cross section from Eq. (22),

$$\sigma = 2\pi \int_0^\pi I(\theta) \sin \theta \, d\theta = \frac{\pi}{2} R_c^2 \int_0^\pi \sin \theta \, d\theta$$

$$= \pi R_c^2. \tag{27}$$

The total cross section is the effective area of the particles for collision, the area of a circle of radius R_c. For systems that can react or transfer energy, the total cross section is a sum of effective areas for reacting, scattering, and energy-transfer processes. Each contribution is an effective area for the respective process.

For particles that interact with potentials like Eq. (8) or (23), the cross section can be much bigger than the physical size of the particles. The attractive tail pulls passing particles in toward smaller distances to cause deflections. In fact, the classical cross section is infinite for potentials that extend to infinity: Collisions of very large impact parameter are still deflected infinitesimally. In Section 4, we will see how quantum mechanics fixes this inconvenience of classical mechanics.

3 KINETIC DATA

3.1 Rate Constants

In any experiment, the relative velocity of colliding particles takes on a large range of values. Until recently, the only observable quantities were averages over an entire distribution of velocities, $f(v)$. Historically, the average quantity of most interest was the rate constant,

$$k = \int_0^\infty v\, \sigma(v)\, f(v)\, dv. \tag{28}$$

Let us assume that $f(v)$ is a Maxwell–Boltzmann distribution and that $\sigma(v) = \sigma$ for hard spheres (i.e., independent of energy or velocity); then

$$k = \int_0^\infty v\sigma 4\pi v^2 \left(\frac{\mu}{2\pi kT}\right)^{3/2} \exp\left(\frac{-\mu v^2}{2kT}\right) dv$$

$$= \sigma \langle v_{\text{rel}} \rangle = \sigma \left(\frac{8kT}{\pi\mu}\right)^{1/2}. \tag{29}$$

We might call this the rate constant for collisions; it is only weakly dependent on temperature.

If we instead consider reactive events, we could make the assumption that no reactive collision occurs unless there is sufficient energy of activation E_a:

$$\sigma(v) = \begin{cases} 0 & \text{for } v < v_0, \\ \sigma & \text{for } v \geq v_0, \end{cases} \quad \text{where } \tfrac{1}{2}\mu v_0^2 = E_a.$$

Then

$$k = \int_{v_0}^\infty \sigma v\, f(v)\, dv$$

$$= \left(\frac{8kT}{\pi\mu}\right)^{1/2} \left(1 + \frac{E_a}{kT}\right) e^{-E_a/kT}, \tag{30}$$

which look more like the Arrhenius rate constants chemists are used to. A more accurate calculation would include additional energy dependence of σ above v_0. In any case, we see that very simple ideas about collisons can explain much of what we normally study in chemical kinetics.

We have seen so far that from a knowledge of the forces between particles, we can calculate all of the properties of collisions between the particles. We can also, in principle, calculate the macroscopic rate constants that determine both kinetic phenomena and, through competing processes, the equilibrium phenomena of chemistry. From this perspective, if we understand the collision process, we can understand all of gas-phase chemistry.

3.2 Langevin Cross Section

There is one simple cross-section calculation that has been of historical importance. Although the ideas we have presented above allow one to make accurate calculations of cross sections and rate constants given the necessary information about the form of the interaction potential, we can make useful estimates of reaction cross sections in some situations using little specific information about the interaction potential. Let us assume that the interaction potential between an ion and a molecule is given by Eq. (7). In doing this we are assuming that the repulsive potential is negligible at important values of the impact parameter. Since the attractive potential goes as r^{-4} and the centrifugal potential goes as r^{-2}, there will always be a centrifugal barrier in the effective potential. Let us denote the position of the maximum by r_m. For a given energy, there is a critical value of impact parameter, b^*, above which the effective potential barrier will be too big to get over. We can assume that these large-impact-parameter collisions do not result in reactions. However, for impact parameters smaller than this critical value, the centrifugal barrier can be surmounted. We can assume that these collisions that go on into small r values to confront the repulsive wall (which we have neglected) are reactive. The total reactive cross section at a given energy is just $\sigma = \pi b^{*2}$. We now need to calculate b^* as a function of energy. We do this by calculating the energy at the maximum in the effective potential. Using Eq. (7) and $L = b\sqrt{2\mu E}$ we obtain

$$E = \tfrac{1}{2}\mu v^2 = -\frac{\alpha Q^2}{2(4\pi\epsilon_0)^2 r_m{}^4} + \frac{\mu v^2 b^{*2}}{2r_m{}^2} \tag{31a}$$

and

$$\frac{\partial U_{\text{eff}}}{\partial r} = \frac{2\alpha Q^2}{(r\pi\epsilon_0)^2 r_m{}^5} - \frac{\mu v^2 b^{*2}}{r_m{}^3} = 0. \tag{31b}$$

Solving Eqs. (31a) and (31b) for b^* and r_m we obtain

$$b^* = 2^{1/2} r_m \quad \text{and} \quad b^{*4} = \frac{2\alpha Q^2}{(4\pi\epsilon_0)^2 E}, \tag{32}$$

$$\sigma_L = \pi \left(\frac{2\alpha Q^2}{(4\pi\epsilon_0)^2 E}\right)^{1/2} = \frac{2\pi Q}{(4\pi\epsilon_0)v} \left(\frac{\alpha}{\mu}\right)^{1/2}. \tag{33}$$

The expression is known as the Langevin cross section (5). Note that the energy dependence is in the factor v, and that the cross section has no activation energy, but increases as v decreases.

Let us go on to calculate the rate constant for such an energy-dependent cross section:

$$k_L = \int_0^\infty v \frac{2\pi Q}{v(4\pi\epsilon_0)} \left(\frac{\alpha}{\mu}\right)^{1/2} f(v)\, dv = \sigma_L v = \frac{2\pi Q}{v(4\pi\epsilon_0)} \left(\frac{\alpha}{\mu}\right)^{1/2}. \tag{34}$$

This expression is known as the Langevin rate constant (5). Note that it is independent of temperature. The large-cross-section collisons at low velocity react slower because the particles move slower. Small-cross-section collisions at high velocity react faster. The net result is that the rate constant does not depend on the average velocity. Corrections due to a permanent dipole moment and the averaging over orientations are found in Ref. 1.

4   QUANTUM THEORY OF COLLISIONS

4.1   Semiclassical Ideas—WKB Phase Shift†

The exact or correct treatment of scattering must examine the quantum mechanics of collisions. We can simplify our work by using center-of-mass coordinates just as in the classical analysis. One of the ways we can begin to develop an idea of the quantum mechanics of collisions is to think of the scattered particle as having a wavelength $\lambda = h/p = h/\mu v$. A more convenient quantity is the wavenumber $k = 2\pi/\lambda = \mu v/\hbar$. A potential affects a particle by either speeding up or slowing down the particle compared with a particle that does not feel the potential (an unscattered particle). In Fig. 14, showing two waves, the waves start out identically. One interacts with an attractive potential; the scattered wave is speeded up briefly and has its phase "shifted" as a result.‡ We will see that an understanding of these phase shifts will allow us to describe the scattering process completely. The rate of change of a travelling wave whose form is $A \cos (\mathbf{k} \cdot \mathbf{r} - \omega t)$ is governed by the *local* wavevector $\mathbf{k}$, determined by the local velocity. The total phase increment over a path is determined by integrating $\mathbf{k}(r) = \mu \, \mathbf{v}(r)/\hbar$ over the path. (We are not concerned with the phase increment $-\omega t$, since it will be the same for both scattered and unscattered particles.)

In three dimensions and with a central potential, the Schrödinger equation separates in spherical coordinates exactly as did the corresponding classical equations. The incoming particle is described by a wavefunction with a local wavenumber $k(r)$. A repulsive potential means a decrease in the velocity and wavenumber. Thus the wavefunction has fewer nodes in a given radial region and is "pushed out" relative to that for $U = 0$. A negative phase shift results. An attractive potential causes an increase in velocity and wavenumber. The wavefunction has more nodes in a given radial region and is "pulled in" relative to the $U = 0$ case. A positive phase shift results (e.g., Fig. 14).§ In a collision, the phase shift is just the difference between the phase increment for the scattered particle and that for the unscattered one. It depends on both the initial energy and impact parameter (or angular momentum):

$$\delta(E, b) = \int_{\text{actual path}} k(r) \, dr - \int_{\text{unscattered path}} k_0(r) \, dr. \qquad (35)$$

†The WKB method is due to Wentzel, Kramers, and Brillouin (1926) but several other names are associated with it, most notably Jeffreys.
‡Figure 14 does not show time, only spatial behavior. The scattered wave has a higher velocity in the interaction region than the unscattered wave; consequently its phase accumulates faster than the unscattered wave's (i.e., corresponding phases are reached at *different* times).
§This explanation is due to Professor D. R. Herschbach.

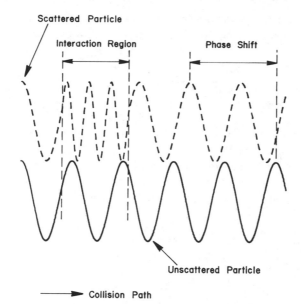

Figure 14. Diagram of two waves showing how if the upper wave is slowed down, the phase is shifted with respect to the unaffected wave.

Now let $E = \frac{1}{2}\mu v^2(r) + U(r) + L^2/2\mu r^2$, where $E = \frac{1}{2}mv^2$ and $L = b\mu v$ (v is the asymptotic velocity); then

$$v^2(r) = v^2 - \frac{2U(r)}{\mu} - \frac{b^2 v^2}{r^2}$$

$$= v^2\left(1 - \frac{U(r)}{E} - \frac{b^2}{r^2}\right), \tag{36}$$

so

$$k = \frac{\mu v}{\hbar}\left(1 - \frac{U(r)}{E} - \frac{b^2}{r^2}\right)^{1/2}. \tag{37}$$

Since the unscattered particle's motion is obtained by setting $U(r) = 0$,

$$\delta(E, b) = \frac{\mu v}{\hbar}\left[\int_{\text{actual path}}\left(1 - \frac{U(r)}{E} - \frac{b^2}{r^2}\right)^{1/2} dr\right.$$

$$\left. - \int_{\text{unscattered path}}\left(1 - \frac{b^2}{r^2}\right)^{1/2} dr\right]. \tag{38}$$

We integrate over the path from the distance of closest approach to infinity (this gives half the total difference in phase over the entire trajectory, but that is how the phase shift is defined):

$$\delta(E,\, b) = \frac{\mu v}{h} \lim_{R \to \infty} \left[\int_{r_c}^{R} \left(1 - \frac{U(r)}{E} - \frac{b^2}{r^2} \right)^{1/2} dr - \int_{b}^{R} \left(1 - \frac{b^2}{r^2} \right)^{1/2} dr \right]. \quad (39)$$

For both integrals, the lower limit of integration is the (outermost) zero of the associated integrand. The first term cannot be evaluated until we specify $U(r)$. The second term can be evaluated exactly, but first we note that both integrals in Eq. (39) are divergent, and δ depends on cancellation between the two. Let us subtract $\int_{r_c}^{R} dr$ from both terms; then the infinite integrals converge:

$$\delta(E,\, b) = \frac{\mu v}{h} \left\{ \int_{r_c}^{\infty} \left[\left(1 - \frac{V(r)}{E} - \frac{b^2}{r^2} \right)^{1/2} - 1 \right] dr + b - r_c \right.$$

$$\left. - \int_{b}^{\infty} \left[\left(1 - \frac{b^2}{r^2} \right)^{1/2} - 1 \right] dr \right.$$

$$= \frac{\mu v}{\hbar} \left[\int_{r_c}^{\infty} \left[\left(1 - \frac{V(r)}{E} - \frac{b^2}{r^2} \right)^{1/2} - 1 \right] dr - r_c + \frac{\pi b}{2}, \quad (40)$$

or

$$\delta(E,\, b) = \int_{r_c}^{\infty} [k(r) - k] \, dr - k \left(r_c - \frac{\pi b}{2} \right). \quad (41)$$

The integral term in Eq. (41) gives the phase difference between a particle under the influence of a potential and one that is not. The second term in Eq. (41) corrects for the fact that the unscattered particle does not have the same turning point as the scattered particle, and it also corrects for the effect of the centrifugal potential in $k(r)$.

It is important to note that Eq. (41) is almost precisely the expression for the WKB phase shift that can be obtained by an approximate solution of the Schrödinger equation. The only modification is that one must replace L by $\hbar \sqrt{l(l+1)}$ (6).

4.2   Relation Between Phase Shift, Delay Time, and Deflection Function

On comparing Eqs. (38) and (21), we see that the deflection function can also be written as a difference of two integrals:

$$\theta(b,\, E) = -2b \left[\int_{r_c}^{\infty} \frac{dr}{r^2} \left(1 - \frac{U(r)}{E} - \frac{b^2}{r^2} \right)^{-1/2} - \int_{b}^{\infty} \frac{dr}{r^2} \left(1 - \frac{b^2}{r^2} \right)^{-1/2} \right]. \quad (42)$$

We are interested also in the time delay in a collision: the difference between the time needed for a particle to go through a collision and that needed for a particle that does the same thing, but with $U(r)$ set equal to zero. Using Eq. (17),

$$\tau(E, b) = \int_{\text{actual path}} dt - \int_{\text{unscattered path}} dt$$

$$= \frac{2}{v} \left[\int_{r_c}^{\infty} \left(1 - \frac{U(r)}{E} - \frac{b^2}{r^2} \right)^{-1/2} dr - \int_{b}^{\infty} \left(1 - \frac{b^2}{r^2} \right)^{-1/2} dr \right]. \quad (43)$$

Comparing Eqs. (40), (42), and (43), we can see that the phase shift is closely related to the classical mechanics of deflections and trajectory times. It is not hard to show that (7):

$$\tau(E, b) = 2\hbar \left(\frac{\partial \delta(E, b)}{\partial E} \right)_b \quad (44)$$

and

$$\theta(E, b) = \frac{2\hbar}{\mu v} \left(\frac{\partial \delta(E, b)}{\partial b} \right)_E. \quad (45)$$

4.3 Quantum Scattering

We have seen how the phase shift can be understood in "semiclassical" terms. Let us now develop some simple ideas of quantum scattering. At large distances, the wavefunction for a scattered particle is a combination of an incoming plane wave and a scattered wave:

$$\psi(r, \theta) = e^{ikz} + \frac{f(\theta)}{r} e^{ikr}, \quad (46)$$

where we have now put our incoming particles in the z-direction, so θ is the usual spherical angle. The first term in Eq. (46) is a plane wave in the z-direction and the second term is an outgoing spherical wave, the flux of which must be proportional to r^{-2}; the amplitude is therefore proportional to r^{-1}. The quantity $f(\theta)$ is the scattering amplitude. This form of the wavefunction will be good as long as r is large enough that no interference terms between the incoming and outgoing waves are important. Experimentally, this corresponds to collimating the incoming particles into a beam of finite width [not included in Eq. (46)].

All of the details of scattering are contained in the scattering amplitude $f(\theta)$. If the potential depends on angles as well as on r, then f may depend of the angle ϕ as well.

The wavefunction in Eq. (46) obeys the Schrödinger equation

$$-\frac{\hbar^2}{2\mu}\nabla^2\psi + U\psi = E\psi. \tag{47}$$

We can expand the wavefunction ψ in any complete set of orthogonal functions. If we assume U depends only on r, then ϕ does not enter into the equation and we can use Legendre polynomials:

$$\psi(r,\,\theta) = \sum_{l=0}^{\infty} \frac{\psi_l(r)}{kr} P_l(\cos\theta)(2l+1)i^l, \tag{48}$$

where the extra terms are chosen for convenience in normalization. This is called expanding the solution in partial waves.

By substituting this form in Eq. (47), multiplying by $P_{l'}(\cos\theta)$, and integrating over θ we get the following equation for $\psi_l(r)$:

$$\left[\frac{d^2}{dr^2} + k^2 - \frac{l(l+1)}{r^2} - V(r)\right]\psi_l(r) = 0, \tag{49}$$

where $\hbar^2 V(r)/2\mu = U(r)$ and $\hbar^2 k^2/2\mu = E$. For large enough values of r, $V(r)$ and $l(l+1)/r^2$ become negligible [if $\lim_{r\to\infty} rV(r) = 0$] and the asymptotic equation becomes

$$\left(\frac{d^2}{dr^2} + k^2\right)\psi_l(r) = 0. \tag{50}$$

This has solutions $\psi_l(r) = a_l \sin(kr - l\pi/2 + \delta_l)$. The function ψ_l may be complicated for small r, but by looking at the form of the solution for large r, we can tell how the potential has affected ψ_l. The phase shift, δ_l, for the lth partial wave contains all the information about the potential, and the term $l\pi/2$ is separated from δ_l because that is the form of the solution with only the centrifugal potential.

Now let us compare the wavefunction with the form of Eq. (46). First we must expand Eq. (46) in partial waves; then we can compare the terms one at a time. We can expand the second term in Eq. (46) formally:

$$f(\theta) = \sum_{l=0}^{\infty} (2l+1) f_l P_l(\cos\theta). \tag{51}$$

Luckily, mathematicians have already worked out the expansion of the first term:

$$e^{ikz} = e^{ikr\cos\theta} = \sum_{l=0}^{\infty} (2l+1)i^l j_l(kr)P_l(\cos\theta), \tag{52}$$

where $j_l(kr)$, called a spherical Bessel function, is fortunately just the solution of Eq. (50) for $V(r) = 0$. For large values of k, the asymptotic form is

$$j_l(kr) = \frac{1}{kr} \sin\left(kr - \frac{\pi l}{2}\right). \tag{53}$$

Now since the functions $P_l(\cos \theta)$ are orthogonal, we can compare the lth terms of Eqs. (46) and (48):

$$\frac{i^l}{kr} \sin\left(kr - \frac{\pi l}{2}\right) + \frac{f_l e^{ikr}}{r} = i^l \frac{a_l}{kr} \sin\left(kr - \frac{l\pi}{2} + \delta_l\right). \tag{54}$$

Expressing the sine functions as complex exponentials and multiplying through by $2ikr$, we get (using $i^l = e^{i\pi l/2}$)

$$e^{ikr} - e^{-ikr}(-1)^l + 2ikf_l e^{ikr} = a_l[e^{ikr}e^{i\delta_l} - (-1)^l e^{-ikr}e^{-i\delta_l}]. \tag{55}$$

Now we require the coefficients of $e^{\pm ikr}$ to match on both sides of Eq. (55):

$$a_l = e^{i\delta_l}$$

and

$$f_l = \frac{1}{2ik}(e^{2i\delta_l} - 1) = \frac{1}{k}e^{i\delta_l} \sin \delta_l. \tag{56}$$

Now we are almost done! The scattering amplitude is

$$f(\theta) = \sum_{l=0}^{\infty} \frac{2l + 1}{k} e^{i\delta_l} \sin \delta_l \, P_l(\cos \theta), \tag{57}$$

and the differential cross section $I(\theta)$ is just the probability density for observing a particle at an angle θ, $|f(\theta)|^2$. The total cross section is obtained by integration of $I(\theta)$ over all angles:

$$\sigma = 2\pi \int_0^{\pi} \sin \theta \, |f(\theta)|^2 \, d\theta$$

$$= \frac{2\pi}{k^2} \sum_{l=0}^{\infty} (2l + 1)e^{i\delta_l} \sin \delta_l \sum_{l'=0}^{\infty} (2l' + 1)e^{-i\delta_{l'}}$$

$$\sin \delta_{l'} \int_0^{\pi} P_l(\cos \theta) P_{l'}(\cos \theta) \sin \theta \, d\theta$$

$$= \frac{4\pi}{k^2} \sum_{l=0}^{\infty} (2l + 1) \sin^2 \delta_l. \tag{58}$$

Thus, knowing the phase shifts, we can calculate both the number of scattered particles as a function of angle and the total cross section. One obtains the phase shifts by starting at $r = 0$ and integrating Eq. (49) out until $V(r)$ is negligible, then determining δ_l from the form of the solution.

One can get the total cross section by summing up the contributions of all the partial waves using Eq. (58), but there is another interesting way. For $\theta = 0$ (no deflection), $P_l(0) = 1$ for all l. Thus, using Eq. (57),

$$f(0) = \sum_{l=0} \frac{2l + 1}{k} e^{i\delta_l} \sin \delta_l,$$

$$\text{Im } f(0) = \sum_{l=0} \frac{(2l + 1)}{k} \sin^2 \delta_l. \tag{59}$$

So, comparing Eqs (59) and (58) one obtains:

$$\sigma = \frac{4\pi}{k} \text{Im } f(0). \tag{60}$$

This equation has been interpreted to mean that scattering leaves a shadow in the beam in the forward direction, $\theta = 0$. The total cross section can be calculated either by measuring the number of particles scattered into all angles except zero or by measuring the number of particles missing from the beam in the forward direction. Because of the shadow analogy, the result [Eq (60)] is called the optical theorem.

Figure 15 illustrates the differential cross section for the Morse potential [Eq. (23)] calculated using Eq. (57) and 500 phase shifts that were calculated using Eq. (41). The relative energy is 5 eV. The classical differential cross section from Fig. 13 is also shown. Note how the quantum mechanical form of $I(\theta)$ reaches a local maximum near where the classical form of $I(\theta)$ has a singularity. The additional peaks occurring at angles smaller than the rainbow angle are interferences from the three different values of the impact parameter that contribute to scattering into these angles. The quantum result tends to oscillate about the classical result, showing that classical mechanics contains most of the physics. Classical mechanics, however, cannot recover the interferences of the quantum result.

We can calculate the total elastic cross section for $p^+ + $ Ar collisions using Eq. (59). At 5 eV, we obtain 58 Å^2. This value is about what we would expect for an effective molecular area. The cross section is dependent on energy. Including the energy dependence would alter the computation we obtained in Eq. (29) or (30) and give us new temperature dependences for rate constants for processes of interest. An actual computation of these is beyond the scope of this chapter, but at least the concept should be clear.

For molecules, the potential of interaction depends on angles as well as on distance. This means that the Schrödinger equation does not separate in partial waves. Nevertheless, one can still expand the solution in partial waves, but then the individual partial-wave radial Schrödinger equations (49) will have additional

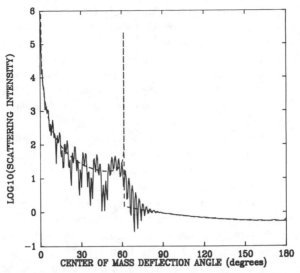

Figure 15. Quantum mechanical differential cross section (solid line) for the Morse potential of Fig. 8 at 5 eV relative kinetic energy. This was calculated using 500 phase shifts each calculated using Eq. (57). The dashed line is the corresponding classical differential cross section from Fig. 13.

terms that couple them together. There exist large computer programs that solve these coupled equations to calculate how the rotational and vibrational energies of the particles can be changed in a collision. Such theory is beyond the scope of this primer; the interested reader is referred to sources such as Ref. 8.

5 APPROXIMATE TREATMENTS AND HANDWAVING

Consider a reaction involving an atom and a diatomic molecule:

$$A + BC \rightarrow A + BC$$
$$\rightarrow AB + C$$
$$\rightarrow AC + B$$
$$\rightarrow A + B + C$$
$$\rightarrow A + BC^+ + e^-$$
$$\rightarrow A + B^+ + C + e^-$$

etc.

Collisions of A and BC can be elastic, in which case no change occurs; BC can have its internal energy state changed; or one of many possible rearrangements with a variety of possible final internal energy states of the products can occur. Even this simplest of chemical reactions is hopelessly complicated for thorough, exact calculation. A number of useful approximations and simplifications can be made, however. The most common and most conceptually fruitful approximation is to require that A, B, and C remain on a single line, and that A interact only with B, not C. This picture is called the collinear triatomic collision. There are only two coordinates: R_{AB} and R_{BC}. A contour map of the potential for this reaction can be plotted in a plane (e.g., Fig. 16), and Newton's (or Schrödinger's) equations can be solved for particles (or waves) on this surface.

Henry Eyring developed his Absolute Rate Theory (9) by thinking about reactions in these terms and then generalizing to more complicated systems.

If the potential for such a reaction has an attractive well instead of a transition state as shown in Fig. 16, then motion can be trapped for long periods in the well before it can find its way out. Such a "long-lived complex" allows many interesting processes, such as electronic excitation and vibrational radomization, to occur.

Another kind of complex formation can occur in two-body scattering at sufficiently low energies because of tunneling through the potential barrier formed by the centrifugal potential (see Fig. 9). If a particle comes into the scattering region with sufficient angular momentum that there is a barrier due to the $L^2/2\mu r^2$ term, but not enough angular momentum that the centrifugal term completely overpowers the attractive well, there is a chance that the quantum mechanical particle will tunnel through the barrier and become trapped inside the well for a long time (until it can tunnel back out again). Such quantum states have been seen experimentally and are called shape resonances. Resonances appear as jumps in the phase shift, because some partial wave suddenly gets shifted a large amount. The classical analogue of

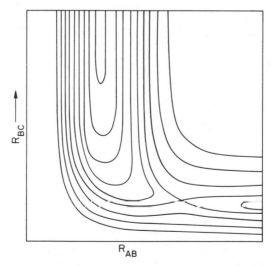

Figure 16.   Contour diagram of the collinear three-body potential energy surface for A + BC (heuristic).

these resonances is orbiting particles that have an energy and angular momentum such that they will stay on the top of the barrier for a long time while the system is rotating; the deflection angle will therefore become very large and negative (2π times the number of revolutions the pair makes before separating).

6 NEWTON DIAGRAMS

The relationships between laboratory and center-of-mass coordinates and between initial and final velocities are conveniently and clearly illustrated using Newton diagrams. The author is not sure who was the first to use this device, but it is now widely used. Consider two particles with initial velocity vectors $\mathbf{v}_1$ and $\mathbf{v}_2$ and masses m_1 and m_2. The relative velocity is $\mathbf{v} = \mathbf{v}_2 - \mathbf{v}_1$ and is the vector from $\mathbf{v}_1$ to $\mathbf{v}_2$. The center-of-mass velocity $\mathbf{V}_{cm}$ is a weighted average of $\mathbf{v}_1$ and $\mathbf{v}_2$, so it must fall along the vector $\mathbf{v}$. If $m_1 > m_2$, $\mathbf{V}_{cm}$ will be closer to $\mathbf{v}_1$; if $m_2 > m_1$, $\mathbf{V}_{cm}$ will be closer to $\mathbf{v}_2$ (see Fig. 17). Figure 17 uses $m_2 = 2m_1$.

In the center-of-mass frame, the velocities of m_1 and m_2 are $\mathbf{w}_1 = \mathbf{v}_1 - \mathbf{V}_{cm}$ and $\mathbf{w}_2 = \mathbf{v}_2 - \mathbf{V}_{cm}$ (see Fig. 18). Since the total linear momentum is zero in center-of-mass frame by definition, the velocities after collision must be in opposite directions. Therefore, in the center-of-mass frame, the deflection angles of both particles must be the same. For $\mathbf{w}'_1$ and $\mathbf{w}'_2$ in Fig. 18, we have indicated a deflection angle of θ. Note that we have drawn $|\mathbf{w}_1| = |\mathbf{w}'_1|$ and $|\mathbf{w}_2| = |\mathbf{w}'_2|$. For conservation of linear momentum, we need require only that the directions of $\mathbf{w}'_1$ and $\mathbf{w}'_2$ be opposite and that their magnitudes be in the same ratio as their respective masses. By having the magnitudes be the same before and after collision, we are conserving kinetic energy; that is, we have an elastic collision. Now note that the laboratory frame final velocities $|\mathbf{v}'_1|$ and $|\mathbf{v}'_2|$ are not the same, nor are they the same as they were before the collision. Also note that the angles of deflection in the laboratory frame, Θ_1 and Θ_2, are not equal to each other or to θ. The formulas we derived in Sections 1–5 are correct only in the center-of-mass frame. To interpret experiments

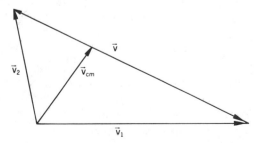

Figure 17. Newton diagram laboratory velocities $\mathbf{v}_1$ and $\mathbf{v}_2$ and the corresponding center-of-mass and relative velocities $\mathbf{V}_{cm}$ and $\mathbf{v}$.

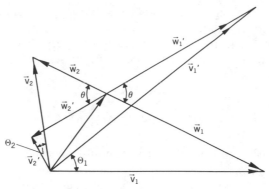

Figure 18. Newton diagram showing initial and final velocity vectors in the laboratory and center-of-mass frames. Key: **v**, laboratory vectors; **w**, center-of-mass vectors; primes indicate final vectors after an elastic collision; unprimed letters are initial vectors.

that measure laboratory deflections, we need to convert from one frame to another. Some simple trigonometry is required:

$$\tan \Theta_1 = \frac{(v_1/v_2)\sin \theta - \cos \alpha \sin \theta + (\sin \alpha)(1 - \cos \theta)}{(v_1/v_2)(m_1/m_2 + \cos \theta) + (\cos \alpha)(1 - \cos \theta) + \sin \alpha \sin \theta}, \quad (61)$$

where α is the angle between $\mathbf{v}_1$ and $\mathbf{v}_2$. If $\alpha = \pi/2$, as in many experiments, then

$$\tan \Theta_1 = \frac{(v_1/v_2)\sin \theta + 1 - \cos \theta}{(v_1/v_2)(m_1/m_2 + \cos \theta) + \sin \theta}. \quad (62)$$

If the reaction is inelastic, then $|\mathbf{w}'_1| \neq |\mathbf{w}_1|$. For example, suppose particle 2 is a diatomic molecule that is left vibrationally excited by the collision. The energy of vibrational (or possibly rotational) excitation will come out of the initial kinetic energy, and the final kinetic energy will be reduced. The occurrence and the energy of vibrational (or rotational) excitation can be inferred from a measurement of the final velocity of one of the scattered particles.

A straightforward analysis of these simple vector diagrams allows us to interpret many of experimental measurements in terms of the elementary processes taking place in a molecular collision.

7 SUGGESTIONS FOR FURTHER STUDY

We have covered only a few of the major topics in reaction dynamics, and these only sketchily. A number of excellent monographs on collision dynamics and scattering theory are available; listed below are a few references that have been useful over the years. No attempt has been made to be exhaustive. You will probably find many more in any major university library.

BIBLIOGRAPHY

M. S. Child, *Molecular Collision Theory,* Academic Press, New York (1973).

D. R. Herschbach, unpublished lecture notes, Harvard University, Cambridge, Massachusetts.

J. O. Hirschfelder, C. F. Curtiss, and R. Byron Bird, *Molecular Theory of Gases and Liquids,* John Wiley & Sons, New York (1954).

R. D. Levine and R. B. Bernstein, *Molecular Reaction Dynamics,* Oxford University Press, New York (1974).

J. D. Marion, *Classical Dynamics of Particles and Systems,* Academic Press, New York (1970).

L. I. Schiff, *Quantum Mechanics,* 3rd ed., McGraw-Hill, New York (1968), and most other quantum mechanics texts.

ACKNOWLEDGMENTS

Much of the discussion on classical collision theory contained in this chapter is derived from an excellent set of unpublished notes by Professor D. R. Herschbach. Professor J. H. Futrell also provided excellent comments.

REFERENCES

1. W. J. Chesnavich, T. Su, and M. T. Bowers, in *Kinetics of Ion Molecule Reactions,* P. Ausloos, Ed., Plenum, New York (1979), pp. 31–53.

2. J. O. Hirschfelder, C. F. Curtiss, and R. Byron Bird, *Molecular Theory of Gases and Liquids,* John Wiley & Sons, New York (1954), Chapter 13.

3. R. Klingbeil, *J. Chem. Phys.* **57,** 1066–1069 (1972).

4. U. Buck, *Rev. Mod. Phys.* **46,** 369–389 (1974).

5. P. Langevin, *Ann. Chim. Phys. Ser. 8* **5,** 245–288 (1905); English translation by F. W. McDaniel, *AFOSR Document TN-60-865* (1960), United States Air Force, Bolling AFB, D.C.

6. M. V. Berry and K. E. Mount, *Repts. Prog. Phys.* **35,** 315–397 (1972).

7. F. T. Smith, *Phys. Rev.* **118,** 349–356 (1960); *J. Chem. Phys.* **36,** 248–255 (1962); *J. Chem. Phys.* **42,** 2419–2426 (1965).

8. L. D. Thomas, M. H. Alexander, B. R. Johnson, W. A. Lester, Jr., J. C. Light, K. D. McLenithan, G. A. Parker, M. J. Redmon, T. G. Schmalz, D. Secrest, and R. B. Walker, *J. Comput. Phys.* **41,** 407–426 (1981); see also for a collection of related papers, *NRCC Proceedings #5,* L. O. Thomas, Ed., Lawrence Berkeley Laboratory, Berkeley, CA (1979).

9. H. Eyring, *J. Chem Phys.* **3,** 107–115 (1935).

PART 2

ION FORMATION
AND MASS ANALYSIS

CHAPTER 3

Electron Impact
Ionization

TILMANN D. MÄRK

Institut für Experimentalphysik
Leopold-Franzens-Universität
Innsbruck, Austria

1 INTRODUCTION

Lenard (1) appears to have been the first to bombard atoms with low-energy electrons and to report observation of the phenomenon of electron-impact-induced ionization and its dependence on energy. Since then numerous studies have been performed on this subject, which is of interest not only in its own right, but also because of the many applications of electron impact ionization (2). Despite the enormous effort put into experimental and theoretical studies, quantitative knowledge, in the form of ionization cross sections, has been an elusive goal until recently (2–6).

Electron impact ionization involves the collision of an energetic electron with a (neutral) target particle and the subsequent production of an ion (and a neutral) and the respective ejected electron(s). The term "electron impact ionization" is somewhat misleading, because an electron is quite small in molecular terms and thus would have difficulty "hitting" any part of an atomic target. It is better to think of the electron as passing close to or even through the atomic target while in quantum mechanical terms the wave of the electron interacts with and distorts the electric field of the atomic system.

In this review we will consider first the mechanism, then the outcome (i.e., the various types of ions produced), and finally the kinetics and dynamics of this ionization process.

2 BASICS OF THE INTERACTION BETWEEN AN IONIZING ELECTRON AND AN ATOMIC TARGET

2.1 Ionization Channels

Electrons accelerated through a potential of several tens of volts have a de Broglie wavelength of $\sim$0.1 nm, corresponding to short-wavelength radiation (7). In this case the wavelength and the molecular dimensions are similar and mutual quantum effects (distortions) occur. The distorted electron wave can be considered to be composed of many different sine waves, and some of these waves will be of the "correct" frequency (energy) to interact with a molecular electron, that is, to promote an electron from a lower to a higher orbital (excitation) or—if the electron energy is greater than a critical value (the *ionization energy* or *appearance energy*)—to eject an electron from the target, thus producing a positive ion (cation).

Conversely, direct attachment of the incident electron to an atomic target to give a stable anion is rather improbable. The reason for this is that the translational energy of the attaching electron and the binding energy (*electron affinity*) must be taken up (accommodated) in the emerging product (*anion*). Usually, the excess energy leads to either fragmentation of the anion or shake-off of the electron. One way to produce stable anions by direct electron attachment is to have sufficient gas pressure in the ion source for three-body reactions to occur and to remove the excess energy. However, a discussion of negative-ion formation by electrons is outside the scope of this review [e.g., see Ref. 8 for coverage of this topic].

As the electron energy is increased, the variety and abundance of the ions produced from a specific molecular target will increase, because the electron ionization process may proceed via different reaction channels, each of which gives rise to characteristic ionized and neutral products. These include the following types of ions (e.g., see Section 3): parent ion, fragment ion, multiply charged ion, excited ion, metastable ion, rearrangement ion, and ion pair (6). For the simple case of a diatomic molecule AB these reaction channels are:

$$AB + e \rightarrow$$

$$AB^+ + e_s + e_e \qquad \text{(single ionization)}, \qquad (1)$$

$$AB^{2+} + e_s + 2e_e \qquad \text{(double ionization)}, \qquad (2)$$

$$AB^{z+} + e_s + z \cdot e_e \qquad \text{(multiple ionization)}, \qquad (3)$$

$$AB^{K^+} + e_s + e_e \qquad \text{[K-shell (inner) ionization]}, \qquad (4)$$

$$AB^{**} + e_s \rightarrow AB^+ + e_s + e_e \qquad \text{(autoionization)}, \qquad (5)$$

$$AB^{+*} + e_s + e_e \rightarrow A^+ + B + e_s + e_e \qquad \text{(fragmentation)}, \qquad (6)$$

$$AB^{2+} + e_s + 2e_e \qquad \text{(autoionization)}, \qquad (7)$$

$$AB^+ + e_s + e_e + h\nu \qquad \text{(radiative ionization)}. \qquad (8)$$

$$A^+ + B + e_s + e_e \qquad\qquad \text{(dissociative ionization)}, \qquad (9)$$

$$A^+ + B^- + e_s \qquad\qquad \text{(ion pair formation)}, \qquad (10)$$

where e_s is a 'scattered' electron and e_e an 'ejected' electron. Other products may be obtained (6), especially when one is using complex atomic targets [i.e., poly-atomic molecules, clusters (9)].

2.2   Franck–Condon Principle

Inelastic collisions between electrons and molecules involve transitions between defined (electronic, vibrational, and rotational) molecular states. The energy losses of the scattered electron in the excitation of molecular vibration and rotation are relatively small compared with the energy of the electronic transition, at least if one assumes that only one vibrational quantum is excited in a single collision. For instance, the greatest separation between a ground and a first excited vibrational state of any molecular ion, namely that in H_2^+, is 0.27 eV (10), compared with the ionization potential of 15.426 eV (11). The changes in the vibrational levels from v to v' that result in ionization can be described, as is well known, in terms of the Franck–Condon principle.

Qualitatively, the Franck–Condon principle may be summarized as follows: In an electric transition no (or only negligible†) changes occur in the nuclear separation and velocity of relative nuclear motion. Due to the large ratio of nuclear to electronic mass and to the short interaction time ($\sim 10^{-17}$ sec, as compared with $\sim 10^{-13}$ sec for a bound vibration), the point on the upper potential-energy curve (corresponding to the configuration after the transition) lies directly above the starting point on the initial potential-energy curve (*vertical transition*). This leads to a number of possible electronic transitions, which depend on the relative shapes of the potential-energy curves available in a specific system. The various Franck–Condon transitions can be discussed with the help of Fig. 1, which shows some of the potential-energy curves for electronic states of H_2, H_2^+, and H_2^{2+} [see also Pichou et al. (12)].

Several cases are possible:

1.   The final accessible level lies within the region of discrete vibrational states of the upper potential-energy curve [e.g., the transition $H_2(X^1\Sigma_g^+) \rightarrow H_2(B^1\Sigma_u^+)$ in Fig. 1). The probability that the vibrational quantum number will change depends on the relative positions of the potential-energy curves.

2.   The final accessible level not only lies within the region of discrete vibrational

---

†As mentioned above, as the electron approaches the molecule, there is some distortion of the molecule during the transition to the ion; that is, ionizing transitions are not strictly vertical as in photoionization, where the wavelength of the radiation required for ionization is ~ 100 nm, which is much larger than the molecular dimensions, so the molecule is subjected only to a uniform electric field (7). Also, relative to photoionization, more spin states of the ion can be reached by electron impact ionization (due to the relaxation of the spin conservation rule).

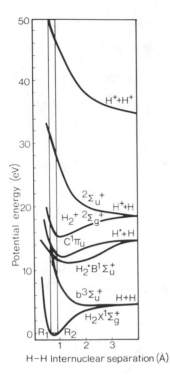

Figure 1. Schematic potential-energy diagram for the ground state of H_2 and some states of H_2^*, H_2^+, and H_2^{2+}. From Märk (6), after Massey et al. (10).

states but includes some part of the continuum [e.g., the transition $H_2(X^1\Sigma_g^+) \rightarrow H_2^+ (^2\Sigma_g^+)$]. Hence, some of the transitions will lead to dissociation.

3. The final accessible level lies within the continuum of a repulsive state and all transitions lead to dissociation (e.g., the transition $H_2(X^1\Sigma_g^+) \rightarrow H_2^+(^2\Sigma_u^+)$).

In addition, the Franck–Condon principle can be used to treat quantitatively electron ionization (fragmentation) of diatomic or pseudodiatomic molecules at low electron energy (13–15). Figure 2 shows as an example results for the H^+/H_2^+ ratio at incident energies up to 25 eV, that is, including only H^+ from the dissociation of the $H_2^+(^2\Sigma_g^+)$ state (see Fig. 1).

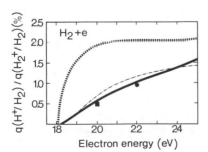

Figure 2. Partial ionization cross-section ratio for the production of H^+ and H_2^+ from H_2 as a function of incident electron energy: ■, Adamczyk et al. (16); ●, Hipple [see Crowe and McConkey (15)]; ——, electron impact data of Crowe and McConkey (15), –·–, electron impact curve predicted by Browning and Fryar (14); ■■■, photon impact data of Browning and Fryar (14). After Märk (6).

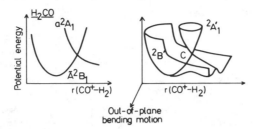

Figure 3. Potential-energy curves (left) and potential-energy surfaces (right) of the formaldehyde ion H_2CO^+. The left-hand plot shows a genuine intersection between two potential-energy curves ($^2B_1$ and $^2A_1$) of H_2CO^+ along an internuclear separation that produces $H_2 + CO^+$. This genuine curve crossing is converted into a conical intersection (right) when the out-of-plane degree of freedom is introduced as a symmetry-lowering coordinate, thus providing a low-energy, adiabatic dissociation path. After Lorquet (17).

In polyatomic molecules the two-dimensional potential-energy curves (Fig. 1) have to be replaced by n-dimensional potential-energy hypersurfaces (with n the number of atoms in the molecule). Although electron impact ionization proceeds without nuclear displacements (Franck–Condon principle), the resulting (excited) polyatomic ion can undergo further internal transitions (e.g., a raditionless transfer of energy can occur; see, e.g., Fig. 3), possibly leading to subsequent unimolecular decompositions (see below). Therefore, it is impossible to interpret the ionization process fully in terms of a detailed knowledge of the hypersurfaces; it is necessary to use statistical methods. This is discussed further in Section 3.2.

In addition, rotational effects in electron impact ionization of molecules have been the subject of some recent investigations. It was found that the electric dipole selection rule $|\Delta J| = 1$ applies at high electron energies (>1000 eV), whereas at lower energies this rule breaks down and large ΔJ transitions are significant (18,19).

2.3   Ionization Mechanisms

Most of the ionization reactions summarized in the previous sections [e.g., processes (1)–(4), (9), and (10)] can be classified as *direct* ionizations, in which the ejected and the scattered electron leave the ion within 10^{-16} sec of each other (20,21). Conversely, there exists an alternate ionization channel (competing with direct ionization) in which the electrons are ejected one after the other (see Fig. 4 for a schematic representation of possible ionization mechanisms). This *autoionization* event [e.g., processes (5) and (7)] can be described as a two-step reaction: First, a neutral molecule (or atom) is raised to a superexcited state, which can exist for some finite time. Then, radiationless transition into the continuum occurs. For molecules, the upper autoionization rate (and hence the ionization cross section) is limited by the characteristic energy-storage mode frequency (see Fig. 4 and Refs. 7, 20, 21). In addition, if predissociation (into two neutrals) is faster than autoionization the latter will not occur at an appreciable rate.

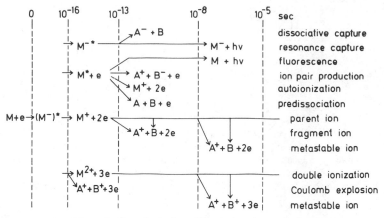

Figure 4. Schematic time chart of possible electron impact ionization processes.

Moreover, autoionization is a resonance process and this will complicate the respective ionization cross-section function at low electron energy (giving it a broadened threshold; e.g., see Fig. 5 and Ref. 6), but also at higher energies (e.g., see the partial ionization cross-section function of Ar shown in Fig. 6). The top curve in Fig. 6 shows the variation of the cross-section function obtained by summing over all possible ionization mechanisms. The middle curves illustrate the variation of the strengths of the $3d$ and $4p$ autoionization processes [e.g., Ar $+ e \rightarrow$ Ar$(3s^2 3p^5 3d) + e \rightarrow$ Ar$^+(3s^2 3p^5) + 2e$]. The bottom curve shows the behavior of the direct ionization mechanism (24).

Quite similarly, multiply charged ions can be formed in a two-step autoionization

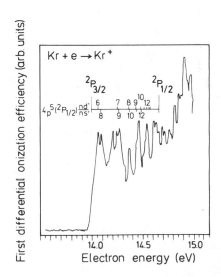

Figure 5. First-differential electron impact ionization efficiency curve of Kr$^+$ close to threshold. The curve shows structure that can be correlated with known autoionizing levels lying between the $^2P_{1/2}$ and $^2P_{3/2}$ states, although other structures exist between 14.10 and 14.25 eV that can be attributed to negative-ion resonances of Kr$^-$ (22). After Johnson et al. (23).

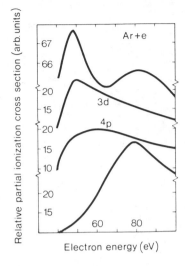

Figure 6. Variation of electron impact ionization efficiency curve with electron energy for Ar⁺. Top curve shows total Ar⁺ ionization efficiency. Middle curves illustrate the variation of the strengths of the $3d$ and $4p$ autoionization features. Bottom curve shows the efficiency of production of ions via direct ionization processes. After Crowe et al. (24).

process. First, a singly charged ion is produced by the ejection of an electron from an inner shell [*inner-shell ionization process* (4)]. This internally ionized atom (or molecule) may then be transformed into a multiply charged ion by a series of radiationless transitions (*Auger effect*). Figure 7 shows as an example the ionization cross-section curve of N_2^{2+}, which demonstrates the occurrence and effect of this Auger autoionization process at ~430 eV (25,26).

2.4   Electron–Electron Correlation

In concluding this section on the basic mechanisms of ionization, it is interesting to consider kinematic aspects of the electron ionization process (which are shown by single-, double-, and triple-differential cross sections), particularly the long-range electron–electron correlations present in the near-threshold region (27,28). Figure 8 is a schematic diagram of the kinematics of an ionizing electron collision.

   Introducing hyperspherical coordinates and assuming the ion to be a point charge,

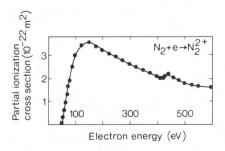

Figure 7. Partial electron impact ionization cross-section function for the production of N_2^{2+}. After Halas and Adamczyk (25).

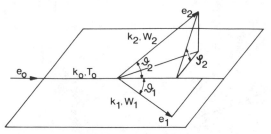

Figure 8. Schematic view of the kinematics of an electron impact ionization process: e_0, incident electron with kinetic energy T_0 and momentum k_0; e_1 and e_2, scattered and ejected electrons with kinetic energies W_1 and W_2 and momenta k_1 and k_2, respectively. After Märk (6).

the potential energy of the system of the scattered and ejected electrons and a positive ion is given by

$$V = \frac{C(\alpha,\theta)}{2R}, \tag{11}$$

with

$$R = (r_1^2 + r_2^2)^{1/2}, \qquad \alpha = \tan^{-1}\left(\frac{r_2}{r_1}\right), \qquad \theta = \cos^{-1}(\mathbf{r}_1,\mathbf{r}_2). \tag{12}$$

Figure 9 shows this functional dependence $C = C(\alpha,\theta)$ (29). The two ditches (at $\alpha = 0$ and $\pi/2$) correspond to attractive forces between the electrons and the ion, whereas the spike (at $\alpha = \pi/4$ and $\cos\theta = 1$) corresponds to the repulsive force between the two electrons.

This so-called *relief plot* can be used to explain the escape process of the three-particle system after the ionizing collision without any knowledge of the ionization process in the reaction zone itself (30). It follows that ionization after the collision occurs if the three particles start in the neighborhood of the *Wannier point* ($\alpha = \pi/4$, $\cos\theta = -1$) and remain there until R is greater than a critical distance R_c at which $|V|$ equals the total excess kinetic energy. In this case $|V|$ is able to continue to decrease with increasing R, with the result that both electrons eventually become free. In other words, actual ionization occurs when the distances r_1 and r_2 of the two electrons remain approximately equal [radial correlation; $\alpha = \tan^{-1}(r_2/r_1) = \tan^{-1}1 = \pi/4$]. If one electron lags behind ($\alpha \neq \pi/4$) it is decelerated in the attractive field of the ion and may end up in a bound state (the ditches of the relief plot) (31). In addition, the repulsive Coulomb interaction forces the electrons to move in nearly opposite directions (angular correlation; $\cos\theta = -1$, $\theta = \pi$). The lower the value of the total excess kinetic energy E, the closer the three particles must pass through the Wannier point to allow double escape (single ionization) of the electrons. This is due to the fact that for low E the system has a longer time to move from the unstable Wannier saddle region into one of the ditches. Owing to

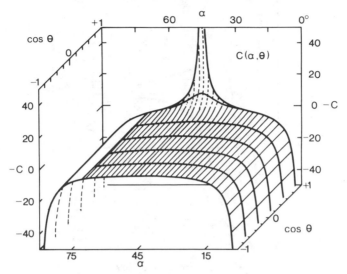

Figure 9. Functional dependence (relief plot) given in Eq. (11): $C = C(\alpha,\theta)$. The ordinates represent a potential surface at $R = 1$. After Lin (29).

these correlation effects, the ionization cross section starts at the threshold at a less than linear slope (*threshold law;* for more detail see Ref. 2).

3   TYPES OF IONS PRODUCED

Electron ionization of *atoms* results in the production of singly and multiply charged atomic ions. On the other hand, ionization of *molecules* gives rise to a number of different ions. The properties of some of these ions will be discussed in the following. For more details, see Märk (6).

3.1   Parent Ions

Parent ions are positively charged ions produced by reaction (1) through removal of one electron from the neutral precursor. The production of these parent ions relative to that of other ions originating from the same neutral precursor (molecule) depends on the electron energy and on the properties of the neutral molecule. At and just above the ionization potential, only singly charged (parent) ions are produced, but at higher electron energies, other ions (see below) are observed. The relative importance of the various ions as a function of electron energy can be demonstrated with help of clastograms (e.g., see Fig. 10).

In general, for small molecules the parent ion is the dominant ion at all electron energies (e.g., Fig. 11), although there are exceptions, such as CCl_4, CF_4 (34,35), and IF_7 (36). Conversely, for large molecules the parent-ion intensity usually de-

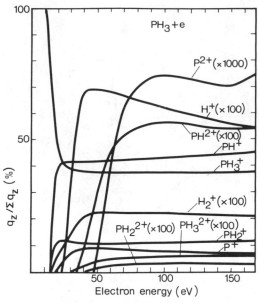

Figure 10.   Clastogram for PH_3, plotting the partial ionization cross sections (32) divided by the counting ionization cross sections (see Section 4.1) as a function of electron energy. After Märk (6).

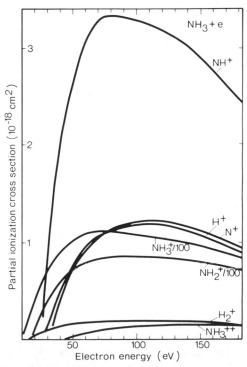

Figure 11.   Partial electron impact ionization cross-section functions for singly and doubly charged ions obtained from NH_3. After Märk et al. (33).

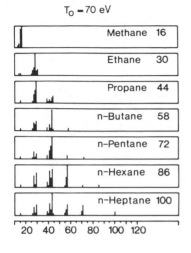

$T_0 = 70$ eV

Methane 16

Ethane 30

Propane 44

n–Butane 58

n–Pentane 72

n–Hexane 86

n–Heptane 100

20   40   60   80   100   120

Reduced mass m/z

Figure 12. Schematic mass spectra of some light paraffins at an electron energy of 70 eV. The numbers give the reduced mass of the parent ion. From Märk (6), after Brunnee and Voshage (37).

creases with increasing molecular weight (Fig. 12) and increasing electron energy (Fig. 13).

In this context it is interesting to point out that the reason for running mass spectra at relatively high electron energies (50–100 eV) is that fragmentation patterns do not vary very much with electron energy in this energy range (e.g., see Fig. 10) and that the partial ionization cross-section functions (and thus the detection efficiency) have their respective maxima in this energy range (e.g., see Fig. 11). On the other hand, so much energy can be transferred to the molecular ion in this energy range that extensive fragmentation results (see below), sometimes making it difficult to identify the parent ion. Because of this, electron impact ionization is considered a "nonsoft" ionization technique as compared with photoionization or chemical ionization. However, if similar energies are used for photoionization and electron impact ionization very similar mass spectra are often observed (e.g., see Figs. 2 and 14).

The parent-ion intensity depends also on the temperature of the molecular gas target (41). At constant ion-source gas pressure a decrease in all ion intensities is noted with increasing temperature, while an increase in vibrational energies of the molecular ion (due to an increase in the vibrational energies of the neutral precursors) may lead to an appreciable decrease in the relative parent (polyatomic) ion intensity (42,43) (e.g., see Fig. 13). For polyatomic molecules this latter effect can be explained in terms of statistical theories (see Section 3.2).

3.2 Fragment Ions

If the energy of the incident electron is increased above the ionization potential of the molecule, fragment ions appear (according to cases 2 and 3 in Section 2.2). In the case of polyatomic molecules a wide range of fragment channels are available;

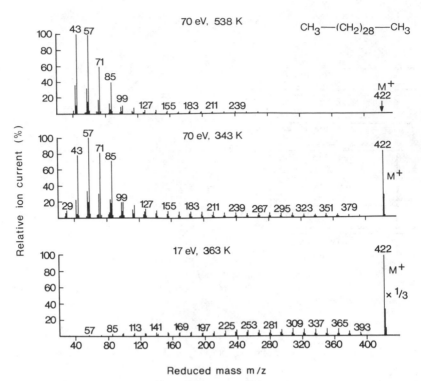

Figure 13. Schematic mass spectra of triacontane at different electron energies and target gas temperatures. From Märk (6), after Remberg et al. (38).

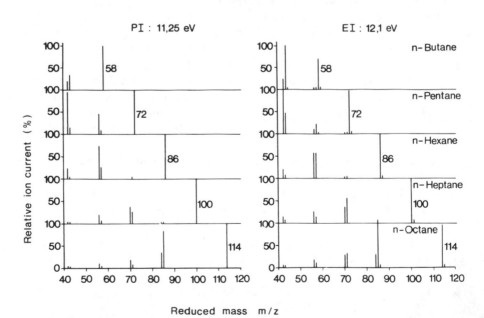

Figure 14. Schematic mass spectra of the *n*-alkanes obtained by photoionization (PI) with photons of 11.25 eV energy [after Steiner et al. (39)] and by electron impact ionization (EI) at 12.1 eV [after Maccoll (40)]. From Märk (6).

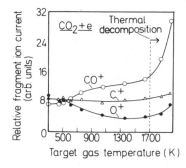

Figure 15. Relative fragment-ion current (fragment-ion current divided by parent-ion current) as a function of target gas temperature. All curves are renormalized for the target gas at room temperature. After Jackson et al. (43).

that is, the molecular ion can immediately decay into a fragment ion with an even or odd number of electrons (*primary fragment ion*). These primary fragment ions may also be produced in (unstable) excited states and immediately decay into further fragments.

The number of fragment ions and their relative cross-section functions are characteristic of the corresponding parent molecule. In the case of diatomic parent molecules, the fragmentation can be treated quantitatively in terms of the Franck–Condon principle (see Fig. 1) (13–15).

For the dissociation of small polyatomic molecules spectroscopic and quantum theoretical ideas (correlation rules) can be used to determine the dissociation path and predict the resulting electronic states (43–46). Figure 15 shows as an example the temperature dependence of the production of fragment ions of CO_2. The enhanced fragmentation at higher temperature results from the thermal population of certain vibrationally excited states of the neutral molecule (43). With the advent of *ab initio* CI (configuration interaction) calculations detailed information on the potential-energy surfaces of the different states and on their interactions is becoming available (45,46). However, to deal with large polyatomic molecules statistical theories have to be used. The so-called quasiequilibrium theory provides the most successful theoretical approach for the discussion of dissociative ionization of large molecules (see Chapter 1) (17,47).

In general, the relative abundance of any fragment ion is related to its rate of formation and its rate of dissociation by unimolecular decomposition. Hence, the measured fragmentation pattern of a molecule is a record in time of the "quasi-equilibrium" balance of these rates. In other words, because the fragmentation pattern is a slice of the three-dimensional plot of ion current as a function of electron energy and mass-to-charge ratio, the respective partial ionization cross-section functions will depend on the time after formation of the primary ion. Trapped-ion mass spectrometry (48) has been used to investigate this phenomenon for electron impact ionization. Typical results for short and long delay times are shown for $C_6H_6^+$, $C_6H_5^+$, and $C_6H_4^+$ in Fig. 16 (50). The increased fragmentation of $C_6H_6^+$ at long delay times is obvious (metastable ions, see Section 3.3).

Finally, it should be pointed out that dissociative ionization yields fragments with small to moderate kinetic energies. To describe completely the dissociative ionization process, these kinematic properties must be known. The translational energy with

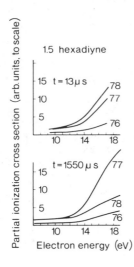

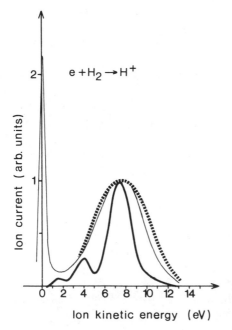

Figure 16. Partial ionization cross-section functions for ions of 1.5-hexadiyne (a linear isomer of benzene) with m/ze = 76, 77, and 78 at two different times following electron impact ionization. From Märk (49), after Lifshitz and Gefen (50).

which fragment ions are formed and the dependence of the appearance energy on this kinetic energy have been measured with a number of different methods (6). This kinetic energy depends on the effective Franck–Condon region (12); one important application of the Franck–Condon principle is for the prediction of kinetic energies of fragment ions. Detailed measurements and calculations have been made more frequently on the hydrogen molecule than on any other molecule, and Fig. 17 shows evidence of the complex nature of the proton kinetic-energy distribution. In some cases translational-kinetic-energy distribution measurements have been sup-

Figure 17. Ion kinetic-energy distribution for H^+ produced by electron impact ionization of H_2. Data are as follows: broken curve, Van Brunt and Kieffer (51) for T_0 = 160 eV, ϑ = 23°; heavy curve, Crowe and McConkey (15) for T_0 = 100 eV, ϑ = 27°; light curve, Köllmann (52) for T_0 = 150 eV, ϑ = 35°. T_0 is the electron kinetic energy and ϑ the detection angle. From Märk (6).

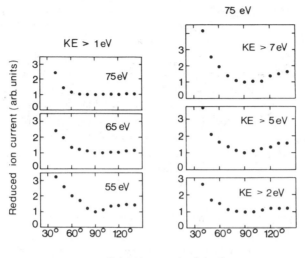

Detection angle (deg)

Figure 18. Angular distributions of H^+ produced by electron impact ionization of H_2 at different incident electron energies and for different ion kinetic energies (KE) after ionization. From Märk (6), after Johnson and Franklin (54).

plemented with data on angular distributions. As expected from theoretical considerations [Dunn's selection rule (53)], angular distributions of fragment ions in the dissociative ionization of small molecules were found to be quite anisotropic (e.g., see Fig. 18). For more information, see Märk (6) and Teubner (55).

3.3 Metastable Ions

Ions produced by electron impact with lifetimes longer than those of excited ions that decompose in the source ($\sim 10^{-8}$ sec) are called metastable ions (47,56). The existence of metastable ions can be explained by different mechanisms depending on the size and property of the precursor ion, namely *electronic predissociation, tunneling through a barrier, vibrational (statistical) predissociation,* and *rearrangement transitions* (6,57,58).

If the metastable decomposition occurs in flight before the ion reaches the analyzer of a magnetic-type instrument, a typical metastable peak is observed in the mass spectrum. Hipple and Condon (59) were the first to interpret the occurrence of these peaks correctly. They showed that the resultant daughter ion of mass m_2 appears at a position on the mass scale designated as the apparent mass m^* (60) and given by

$$m^* = \frac{(m_2/z_2)^2}{m_1/z_1} \left(1 + \frac{m_1 - m_2}{m_2} \frac{U_A - U}{U_A} \right), \tag{13}$$

where U_A is the accelerating voltage, U is the potential difference through which the parent (precursor) ion of mass m_1 falls before decomposition, and z_1 and z_2 are the respective charges. It follows from this equation that metastable peaks need not and in general do not occur at an integer m/ze value. A considerable part of today's knowledge about metastable ions is based on this technique. (For reviews, see Refs. 6, 56, 58, and 60–62.)

The intensity of metastable peaks in a mass spectrum is usually less than 1% of that of the base peak. For certain metastable transitions, such as $CF_4^{+*} \rightarrow CF_3^+$ and $CCl_4^{+*} \rightarrow CCl_3^+$ (34,35,63), it is not possible to detect any precursor-ion signal. In addition, it has been found for certain ions that the ratio between (metastable) product ion and precursor ion is a strong function of the temperature of the neutral precursor molecule; the ratio may even exceed 100% (64,65).

3.4 Multiply Charged Ions

Multiply charged atomic and molecular ions were observed and identified as early as 1912 (66). Subsequent observation of numerous doubly charged ions followed (6). Triply charged molecular ions have been detected in low abundances in the mass spectra of some species (e.g., CO_2, CS_2, C_2N_2, and aromatic compounds) (67). Moreover, electron impact ionization of free van der Waals clusters can lead to multiply charged ions with up to five elementary charges (68–70). Conversely, with certain molecules [e.g., H_2O and CH_4 (71,72)], it is not possible to detect any doubly charged ions by electron impact ionization.

Most of the doubly charged diatomic ions AB^{2+} observed satisfy the relationship

$$IP(A^+) + IP(B^+) < IP(A^{2+}) \leq IP(B^{2+}), \tag{14}$$

where IP denotes the respective single- and double-ionization potentials for the single atoms A and B. In this case the repulsive Coulomb state arising from $A^+ + B^+$ lies energetically below weakly bound states arising from $A^{2+} + B$ at

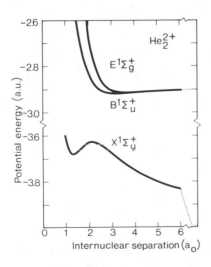

Figure 19. Schematic potential-energy diagram for He_2^{2+} (a.u., atomic units). From Märk (6), after Cohen and Bardsley (73).

all internuclear distances. Metastable ions AB^{2+} exist when short-range chemical forces impose a sufficiently strong attractive well in the repulsive Coulomb interaction. Figure 19 gives as an example the potential curves of He_2^{2+} (see also Ref. 74).

Conversely, as has been pointed out recently by Helm et al. and Stephan et al. (75), a second category of doubly charged diatomic ions exists satisfying the relationship

$$IP(A^+) + IP(B^+) > IP(AB^{2+}). \tag{15}$$

Here the repulsive Coulomb state arising from A^+ and B^+ lies above the weakly bound state arising from $A^{2+} + B$ (see Fig. 20), allowing the formation of stable AB^{2+} ions.

Unstable or metastable doubly charged molecules can decompose in two different ways, depending on how the two charges are distributed in the reaction, namely

$$AB^{2+} \rightarrow A^+ + B^+, \tag{16}$$

$$AB^{2+} \rightarrow A^{2+} + B. \tag{17}$$

In general, doubly charged ions produced by electron impact decay by process (16) rather than by process (17) (76,77). The product ions A^+ and B^+ of reaction (16) are identical with those formed in the fragmentation of the singly ionized parent species AB^+, and hence a decomposition of this type is usually established only if the corresponding metastable peak is detected. This "charge-separation" process (78) is sometimes termed *Coulomb explosion* (79–82).

The partial ionization cross sections for the production of multiply charged atoms and molecular ions rarely exceed 1–5% of that of the dominant singly charged ion

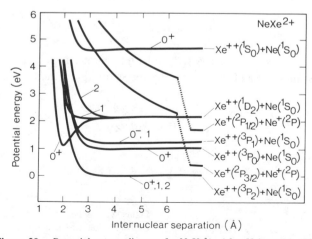

Figure 20. Potential-energy diagram for $NeXe^{2+}$. After Helm et al. (74).

(2–6,83,84). Certain atoms and compounds, however, have been found to possess increased ability to sustain two positive charges, namely aromatic, heteroaromatic, and polyfluor compounds; aminoboranes; and organometallic compounds (85–87). In some cases doubly charged parent ions have been found to be more abundant than the corresponding singly charged ion [e.g., 1-0-nitrophenylanthra-9,10-quinone (88)].

4 IONIZATION EFFICIENCY: PARTIAL ELECTRON IMPACT IONIZATION CROSS SECTIONS

4.1 Definitions

Consider, as shown in Fig. 21, a parallel, homogeneous, and monoenergetic beam of electrons crossing a semiinfinite medium containing N_t target particles per cubic centimeter at rest. If $n(0)_e$ represents the initial intensity of the incident electrons per square centimeter per second, the density of the electron beam at depth x is given by the exponential absorption law

$$n(x)_e = n(0)_e \exp(-N_t q x). \tag{18}$$

If $N_t q x \ll 1$ (single-collision condition), the number of ions generated per second along the collision interaction path $x = L$ (over which the ions are collected and analyzed) is

$$n(L)_i = n(0)_e N_t q_c L, \tag{19}$$

where q_c is the *counting ionization cross section* in square centimeters. The total positive-ion current i_t produced in this interaction volume is given by

$$i_t = n(0)_e e N_t q_t L, \tag{20}$$

where q_t is the *total ionization cross section*. If the produced ions are analyzed with respect to their mass m and charge ze, the respective individual ion currents are given by

$$i_{ms} = n(0)_e e N_t q_{zi} L, \tag{21}$$

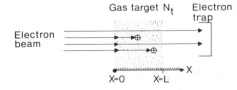

Figure 21. Schematic view of an electron impact ionization experiment. L, interaction length. After Märk (49).

where q_{zi} is the *partial ionization cross section* for the production of a specific ion i with charge ze. If electron impact is used as a convenient means of ionizing (and detecting) molecules and producing specific ions (whose properties are to be studied with other methods), it is this partial cross section that must be known for quantitative studies.

Total and counting ionization cross sections of a specific target system are the weighted and the simple sum of the various single and multiple partial cross sections, respectively:

$$q_t = \Sigma q_{zi} z \quad \text{and} \quad q_c = \Sigma q_{zi}. \tag{22}$$

Sometimes the *macroscopic cross section* $s = qN_t$, which represents the total effective cross-sectional area for ionization of all target molecules in 1 cm^3 of the target medium, is used. This quantity is numerically equal to the ionization efficiency, usually defined as the number of charges (or ion pairs) produced per centimeter of path at 1 torr and 0°C.

4.2 Experimental Considerations

The first mass spectrometric studies of partial ionization cross-section functions date back to the 1930s (2–6,49). Only a few of these early studies have been repeated; until recently (3), large differences in magnitude and shape existed among experimentally determined partial ionization cross-section functions, even for the simplest gases. The main condition required for measuring accurate functions is constant and/or complete ion source–mass spectrometer collection efficiency, independent of the mass to charge ratio m/ze, the electron energy T_0, and the initial ion kinetic energy. As has been pointed out by a large number of authors (e.g., see 2–6,49), large discrimination effects frequently occur at the ion-source exit and mass spectrometer slits. Moreover, a common problem (never solved satisfactorily) is the absolute calibration of ionization cross sections. Closely related to this is the fact that discrimination may occur at the multiplier–detector. Recent experimental approaches that largely overcome these difficulties include the following:

1. Fast-ion-beam charge-exchange technique (89–93).
2. Cycloidal mass spectrometry with high ion-extraction efficiency (16,25,94–99).
3. Field-free diffusive extraction (15,24,100–103).
4. Improved crossed-beam apparatus (104–109).
5. Magnetic sector field mass spectrometer with large acceptance (quadrupole lenses) (110,111).
6. Trapped-ion mass spectrometry (48,49,112,113).
7. Penetrating field extraction and deflection method with a Nier-type ion source and a sector field mass spectrometer (4–6,34,35,49,114–122).

The last method was developed and recently improved in our laboratory and is discussed in detail below. The widespread use of the Nier-type ion source in most modern mass spectrometers has already been discussed (see Refs. 4, 49).

The extraction of ions from the ionization region in a Nier-type ion source, depicted schematically in Fig. 22, depends under usual experimental conditions on various parameters: the initial energy of the ions, the mass-to-charge ratio, the electron collimation magnetic field, the electron beam space charge, and the applied extraction field. Usually ions are extracted from the ionization region (which contains a crossed electric and magnetic field) by a weak electric field applied between the collision chamber exit slit (L_1 in Fig. 22) and an electrode, the pusher (P), opposite the exit slit. This extraction, however, is not complete and results in discrimination for ions with different m/ze (e.g., see discussions in Refs. 2–6 and 49). In an alternate approach (123–130) a penetrating field from external electrodes (L_2) may be used. In this approach, all electrodes confining the collision chamber (P, C, and L_1 in Fig. 22) are kept at the same potential and ions are drawn out of the collision chamber through the slit in L_1 under the action of an electric field applied to L_2. It has been shown that this *penetrating field extraction* assures saturation of the ion current; that is, complete ion collection is achieved. Ions extracted in this manner are then centered by elements L_3 and L_4. The earth slit L_5 marks the end of the acceleration region. Stephan et al. (114) additionally introduced in front of the mass spectrometer entrance-slit deflection plates ($L_{6,7}$ and $L_{8,9}$), which sweep the ion beam across the mass spectrometer entrance-slit S_1 in the y-direction (perpendicular to S_1)

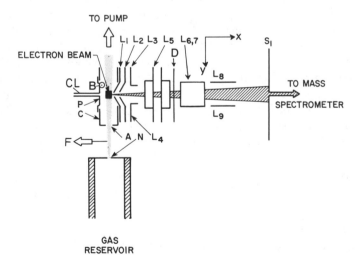

Figure 22.   Schematic view of three-electrode Nier-type ion source. A, aperture; B, guiding magnetic field; C, collision chamber; P, pusher (repeller); CL, capillary leak for stagnant gas target; D, defining aperture; F, molecular beam flag; L_1, extraction electrodes (direct extraction); L_2, extraction electrodes (penetrating field extraction); L_3, L_4, beam-centering half-plates; L_5, earth slit; $L_{6,7}$, L_8, L_9, deflection plates; N, nozzle (10–50 μm) for molecular gas target; S_1, adjustable mass spectrometer entrance slit. After Märk (49).

and z-direction (parallel to S_1) (131,132). This allows the recording and/or integration of the ion beam profile, so that discrimination at S_1 can be corrected. This technique has recently been extended to the quantitative detection of fragment ions (34,35,122).

In conjunction with this it is interesting to point out that the y and z deflection method is a useful technique for distinguishing ions produced in the static target gas (capillary leak gas inlet) from ions [e.g., cluster ions (9)] produced in the neutral beam (molecular beam gas inlet through nozzle expansion; see Fig. 22) (e.g., see Section 4.3.6).

4.3 Some Characteristic Examples

It is outside the scope of this review to give a detailed account of all available results on partial ionization cross-section functions. Kieffer and Dunn (3) published in 1966 a comprehensive discussion and review of experimentation in this field. Studies between 1966 and 1983 have been summarized in later compilations and reviews (4–6,49). The reader is also referred to the respective chapters in a recent book on electron impact ionization (2) and the appropriate references given there. In the present review a representative group of targets is presented illustrating the different shapes and magnitudes of the partial ionization cross-section functions.

4.3.1 Radical Atoms

Figure 23 shows partial ionization cross-section functions for the process $H + e \rightarrow H^+ + 2e$. The normalization procedures used and the associated problems are discussed by Kieffer and Dunn (3). Figure 24 shows for comparison partial ionization cross-section functions for the process $H(2s) + e \rightarrow H^+ + 2e$. These results are especially interesting, because Dolder (142) points out that classical scaling suggests that the partial cross section ratio $q(H^+/H(2s)/q(H^+/H(1s))$ should be ~ 16 when the incident electron energy is expressed in units of the corresponding threshold energy. The measured ratio, however, is between 10 and 15 (Figs. 23 and 24). Experimental results exist also for C, N, and O (49).

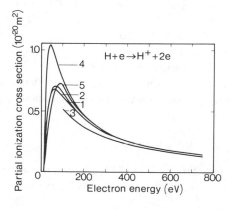

Figure 23. Partial ionization cross sections for the process $H + e \rightarrow H^+ + 2e$ as a function of electron energy. Curves as follows: 1, experimental results of Fite and Brackmann (133); 2, experimental results of Boyd and Boksenberg (134); 3, experimental results of Rothe et al. (135); 4, Born approximation (136); 5, impulse approximation (137). From Märk (49).

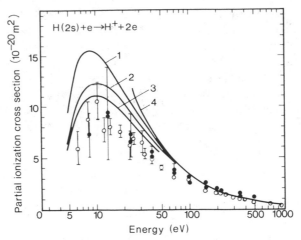

Figure 24. Partial ionization cross-section functions for the process $H(2s) + e \rightarrow H^+ + 2e$. Filled circles, experimental results of Dixon et al. (90); open circles, experimental results of Defrance et al. (138); curve 1, Born-A results (139); curve 2, Born-B results (140); curve 3, Born exchange results (140); curve 4, Born–Bethe approximations (141). From Märk (49).

4.3.2 Rare Gases

Since rare gases are inert and act as ideal gases at the pressures used to study electron impact ionization, they were the first substances, and remain the substances most frequently investigated quantitatively. Hence, partial ionization cross-section data for the rare gases appear to be the most reliable data available. Märk (4) has recently reviewed these measurements and given a set of recommended rare-gas partial ionization cross sections (see Table 1). Partial ionization cross-section functions for the processes $Ar + e \rightarrow Ar^+$, Ar^{2+}, Ar^{3+}, and Ar^{+m} are given in Fig. 25a and b, respectively.

Excitation into a specific spin-orbit state is expected to be governed by the statistical weights at electron energies well above the threshold. For example, consider the ionization process $Xe + e \rightarrow Xe^{2+} + 3e$, which has three states close to threshold—the ^3P ground state and the $^1D_2$ and $^1S_0$ excited states (with threshold energies of 33.34, 35.46, and 37.98 eV, respectively). The triplet ground-state configurations are $^3P_2$, $^3P_1$, and $^3P_0$ and their proportions are expected to be 5:3:1. Moreover, statistical weight considerations would predict the $^3P/^1D_2/^1S_0$ proportions to be 9:5:1. Observed proportions (148) differ significantly from these expected proportions, presumably because the Xe^{2+} ions may be formed initially in unstable higher-energy excited states that can radiatively cascade to lower-energy metastable states (see Fig. 26).

4.3.3 Alkalis, Alkaline Earths, and Metals

McDowell (149) has recently reviewed (see also Refs. 150 and 151) electron impact ionization of the alkalis and tabulated the results on single ionization for Li through

TABLE 1 Set of Recommended Absolute Partial Ionization Cross Sections for the Rare Gases[a]

Electron Energy (keV)	$\sigma(10^{-20}\ m^2)$						
	He^+/He	He^{2+}/He	Ne^+/Ne	Ne^{2+}/Ne	Ne^{3+}/Ne	Ar^+/Ar	Ar^{2+}/Ar
0.05	0.245	—	0.338	—	—	2.50	0.016
0.10	0.366	0.00014	0.655	0.0059	—	2.51	0.17
0.15	0.367	0.00073	0.730	0.021	0.00004	2.34	0.17
0.50	0.221	0.0013	0.534	0.025	0.0015	1.29	0.071
1.0	0.140	0.00067	0.355	0.014	0.00087	0.805	0.046
2.0	0.0790	0.00032	0.215	0.0075	0.00047	0.471	0.024
3.0	0.0568	0.00022	0.159	0.0051	0.00032	0.344	0.018
4.0	0.0452	0.00016	0.126	0.0039	0.00026	0.269	0.014
5.0	0.0372	0.00013	0.107	0.0033	0.00023	0.226	0.011
6.0	0.0324	0.00010	0.0932	0.0026	0.00015	0.200	0.011
8.0	0.0254	0.000071	0.0739	0.0020	0.00012	0.157	0.0085
10.0	0.0211	0.000060	0.0624	0.0016	0.000095	0.132	0.0072
12.0	0.0182	0.000050	0.0541	0.0013	0.000081	0.117	0.0065
14.0	0.0163	0.000045	0.0476	0.0012	0.000068	0.101	0.0055

Electron Energy (keV)	$\sigma(10^{-20}\ m^2)$						
	Ar^{3+}/Ar	Kr^+/Kr	Kr^{2+}/Kr	Kr^{3+}/Kr	Xe^+/Xe	Xe^{2+}/Xe	Xe^{3+}/Xe
0.05	—	3.64	0.10	—	4.87	0.27	—
0.10	0.00075	3.56	0.31	0.0064	4.91	0.53	0.072
0.15	0.0044	3.20	0.28	0.021	4.34	0.49	0.18
0.50	0.0074	1.66	0.086	0.033	2.11	0.23	0.11
1.0	0.0064	1.10	0.069	0.034	1.25	0.16	0.073
2.0	0.0054	0.638	0.049	0.027	0.732	0.11	0.046
3.0	0.0040	0.458	0.036	0.021	0.533	0.084	0.036
4.0	0.0033	0.355	0.029	0.017	0.409	0.065	0.027
5.0	0.0030	0.299	0.025	0.015	0.336	0.054	0.023
6.0	0.0027	0.250	0.021	0.013	0.289	0.046	0.018
8.0	0.0022	0.197	0.017	0.011	0.227	0.037	0.015
10.0	0.0019	0.161	0.015	0.0094	0.186	0.032	0.012
12.0	0.0017	0.137	0.013	0.0085	0.159	0.027	0.011
14.0	0.0014	0.123	0.012	0.0072	0.140	0.024	0.0096

[a]After Märk (4,49), and based on data from various experimental determinations of high precision (111,114,115,121,143–146).

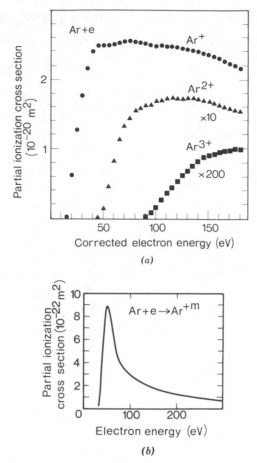

Figure 25. (*a*) Absolute partial ionization cross-section functions for the processes Ar + *e* → Ar⁺ + 2*e*, Ar + *e* → Ar²⁺ + 3*e*, and Ar + *e* → Ar³⁺ + 4*e*. After Stephan et al. (114,115). See also Fig. 6. (*b*) Absolute partial ionization cross-section function for the production of singly charged metastable Ar⁺ ions. After Varga et al. (147).

Cs. Figure 27 gives as an example the partial ionization cross-section functions for the production of Ba⁺ and Ba²⁺ via electron impact on Ba. Note the strong influence of autoionization and inner-shell ionization (see Section 2.3) on the shape of the ionization cross-section function. Experimental results exist also for Mg, Ca, Sr, Cu, Zn, Ag, Hg, and Pb (see Ref. 49).

4.3.4 Diatomic Molecules

In the case of diatomic target species, dissociative ions may be produced by electron impact, and as mentioned earlier, these ions have excess kinetic energy and are difficult to collect quantitatively in a mass spectrometer. Therefore, only a few studies have yielded reliable partial ionization cross-section functions. In addition,

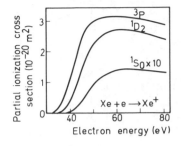

Figure 26. Partial ionization cross sections of the Xe^{2+} metastable excited states ($^1S_0$, $^1D_2$) and the ground state (^3P) as a function of electron energy using production rates measured by Adams et al. (148) in the electron impact ion source of a selected ion flow tube (SIFT) apparatus and calibrated with the help of recent data of Stephan and Märk (121).

Rapp et al. (155) have measured without mass analysis the percentage of total ion current due to energetic (dissociative) ions with kinetic energies in excess of 0.25 eV. It ranged from ~7% in H_2 to ~35% in N_2O, with other gases intermediate. Figure 28 shows as an example partial ionization cross-section functions for N_2. Experimental results exist also for H_2 (see also Fig. 2), O_2, CO, NO, and HCl (see Ref. 49).

4.3.5 Polyatomic Molecules

This subject has recently been reviewed by Märk (49), who provides bibliographies on the following target species: H_2O, CO_2, N_2O, NO_2, O_3, SO_2, $HgBr_2$, C_2H_2, C_2N_2,

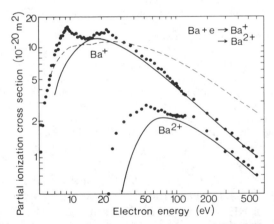

Figure 27. Partial ionization cross-section curves for Ba^+ and Ba^{2+}: ●, data of Dettmann and Karstensen (108); ——, scaled Born approximation by McGuire (152); – – –, Lotz semiempirical formula (153). Dettmann and Karstensen interpret the maximum of the Ba^+ at ~9 eV to be caused by direct ionization of a $6s$ electron, whereas the maximum at ~22 eV is caused, according to these authors, by the autoionization process $Ba(5p^66s^2) + e \rightarrow Ba(5p^56s^25d) + e \rightarrow Ba(5p^66s) + 2e$. Ziesel and Abouaf (154) find the threshold for this autoionization process to be ~12.5 eV. The maxima of the Ba^{2+} curve at ~55 and ~110 eV are interpreted by Dettmann and Karstensen as resulting from contribution from direct inner-shell ionization, whereas a break at ~20 eV (not shown in this figure) is ascribed to an Auger process occurring after the removal of one $5p$ electron and resulting in the removal of two $6s$ electrons.

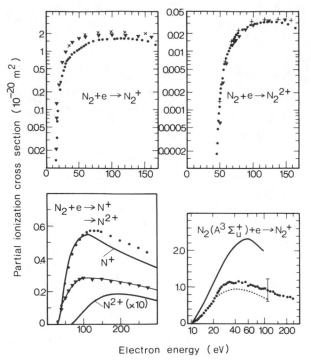

Figure 28. Absolute partial ionization cross-section functions as given by Märk (49) for the following processes: $N_2 + e \rightarrow N_2^+$ [filled circles, experimental results of Märk (156); filled triangles, data derived from experimental results of Rapp and Englander-Golden (143) and Rapp et al. (155); crosses, experimental results of Halas and Adamczyk (25)]; $N_2 + e \rightarrow N_2^{2+}$ [filled circles, experimental results of Märk (156); filled triangles, experimental results of Halas and Adamczyk (25); crosses, experimental results of Daly and Powell (157)] (see also Fig. 7); $N_2 + e \rightarrow N^+$ [filled circles, data derived from experimental results of Rapp and Englander-Golden (143) and Rapp et al. (155); solid line, experimental results of Crowe and McConkey (101); line plus triangles, experimental results of Halas and Adamczyk (25)]; $N_2 + e \rightarrow N^{2+}$ [solid line, experimental results of Crowe and McConkey (101)]; $N_2(A^3\Sigma u^+) + e \rightarrow N_2^+$ [filled circles, experimental results of Armentrout et al. with error bar (93); solid line, binary encounter approximation (158); dashed line, semiclassical calculation (159)].

NH_3, PH_3, SO_3, UO_3, CH_4, CF_4, CCl_4, CF_2Cl_2, SiH_4, $TiCl_4$, $TaCl_5$, $POCl_5$, SF_6, $CrCO_6$, C_2H_6, and SiH_6. Partial ionization cross sections are shown in Fig. 10 for PH_3, in Fig. 11 for NH_3, in Fig. 29 for N_2O, in Fig. 30 for CF_4, in Fig. 31 for CCl_4, and in Fig. 32 for SF_6. Note the absence of any parent ions for CF_4 and CCl_4.

4.3.6 Clusters

Recently, there has been growing interest in a new category of molecules, the so-called (van der Waals) *clusters* (9). Neutral atomic and/or molecular clusters are produced in free jet nozzle expansions, and most experiments use electron impact ionization in combination with mass spectrometry for the detection of these weakly

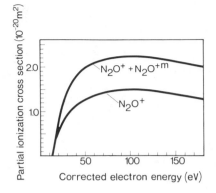

Figure 29. Absolute partial ionization cross-section functions for N_2O. After Märk et al. (120).

bound species. However, little quantitative information is known yet in terms of partial ionization cross sections. This is due mainly to the fact that it is not possible to produce beams of neutral clusters of known density and specific cluster size. To obtain absolute cross sections (or detect clusters quantitatively by mass spectrometry) some investigators have assumed that encounters between an electron and a cluster, whether monomer, dimer, or polymer, have equal probabilities of producing the corresponding ion. Since a dimer is roughly twice the size of a monomer, the probability that a particular dimer will collide with an electron should be twice the probability that a particular monomer will collide with an electron. This additivity rule (6) predicts a cross-section ratio of 2. This working rule, however, includes

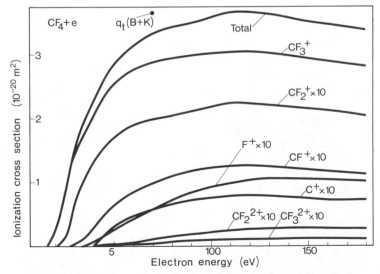

Figure 30. Absolute partial and total ionization cross-section functions for ions of CF_4 as measured by Stephan et al. (35). Filled circle is total ionization cross section at 70 eV determined by Beran and Kevan (160) recalibrated to the Ar value of Rapp and Englander-Golden (143).

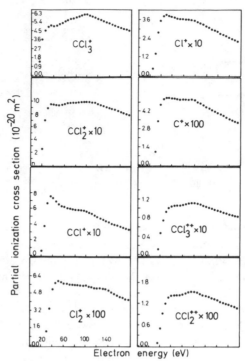

Figure 31. Absolute partial ionization cross-section functions for ions of CCl$_4$ as measured by Leiter et al. (34).

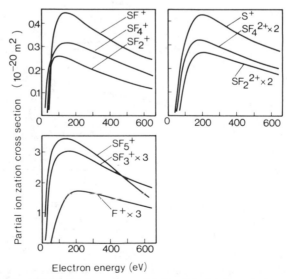

Figure 32. Absolute partial ionization cross-section functions for fragment ions of SF$_6$ as measured by Stanski and Adamczyk (99).

the assumption that there is no fragmentation of dimers (or polymers) during the ionization process; unfortunately, this assumption has recently been shown to be incorrect (9). Märk (4,49) has recently advanced a modified additivity rule that takes into account the dissociative ionization channel (161), and has thus obtained estimated partial ionization cross sections for Ar_2, Kr_2, Xe_2, ArKr, and KrXe. For more detail on electron–cluster interaction, see Chapter 12 and Refs. 9 and 82.

ACKNOWLEDGMENTS

This work was partially supported by the Österreichischer Fonds zur Förderung der Wissenschaftlichen Forschung.

REFERENCES

1. P. Lenard, *Ann. d. Phys.* **8**, 149 (1902).
2. T. D. Märk and G. H. Dunn, Eds., *Electron Impact Ionization*, Springer-Verlag, Vienna (1985).
3. L. J. Kieffer and G. H. Dunn, *Rev. Mod. Phys.* **38**, 1 (1966).
4. T. D. Märk, *Beitr. Plasmaphys.* **22**, 257 (1982).
5. T. D. Märk, *Int. J. Mass Spectrom. Ion Phys.* **45**, 125 (1982).
6. T. D. Märk, "Ionization of Molecules by Electron Impact," in *Electron–Molecule Interactions and Their Applications,* Vol. 1, L. G. Christophorou, Ed., Academic Press, New York (1984), pp. 251–334.
7. M. E. Rose and R. A. W. Johnstone, *Mass Spectrometry for Chemists and Biochemists,* Cambridge University Press, Cambridge, England (1982).
8. L. G. Christophorou, Ed., *Electron-Molecule Interactions and Their Applications*, Vol. 1, Academic Press, New York (1984) and references given therein.
9. T. D. Märk and A. W. Castleman, Jr., *Adv. Atom. Molec. Phys.* **20**, 65 (1985).
10. H. S. W. Massey, E. H. S. Burhop, and H. B. Gilbody, *Electronic and Ionic Impact Phenomena*, Clarendon Press, Oxford (1969).
11. H. M. Rosenstock, K. Draxl, B. W. Steiner, and J. T. Herron, *J. Phys. Chem. Ref. Data* **6**, Suppl. 1 (1977).
12. F. Pichou, R. I. Hall, M. Landau, and C. Schermann, *J. Phys.* **B16**, 2445 (1983).
13. G. H. Dunn, *J. Chem. Phys.* **44**, 2592 (1966).
14. R. Browning and J. Fryar, *J. Phys.* **B6**, 364 (1973).
15. A. Crowe and J. W. McConkey, *J. Phys.* **B6**, 2088 (1973).
16. B. Adamczyk, A. J. Boerboom, B. L. Schram, and J. Kistemaker, *J. Chem. Phys.* **44**, 1 (1966).
17. J. C. Lorquet, *Org. Mass Spectrom.* **16**, 469 (1981).
18. S. P. Hernandez, P. J. Dagdigian, and J. P. Doering, *J. Chem. Phys.* **77**, 6021 (1982).
19. P. J. Dagdigian and J. P. Doering, *J. Chem. Phys.* **78**, 1846 (1983) and references cited therein.
20. C. E. Berry, *Adv. Mass Spectrom.* **6**, 1 (1974).
21. L. P. Johnson and J. D. Morrison, *Int. J. Mass Spectrom. Ion Phys.* **18**, 355 (1975).
22. P. Marmet, E. Bolduc, and J. C. Quemener, *J. Chem. Phys.* **56**, 3463 (1972).
23. L. P. Johnson, J. D. Morrison, and A. L. Wahrhaftig, *Int. J. Mass Spectrom. Ion Phys.* **26**, 1 (1978).

24. A. Crowe, J. A. Preston, and J. W. McConkey, *J. Chem. Phys.* **57**, 1620 (1972).

25. S. Halas and B. Adamczyk, *Int. J. Mass Spectrom. Ion Phys.* **10**, 157 (1972/73).

26. R. K. Nesbet, *J. Chem. Phys.* **40**, 3619 (1964).

27. F. W. Read, in *The Physics of Ionized Gases*, M. Malic, Ed., Boris Kidric Institute, Beograd (1980), p. 11.

28. H. Klar, *Z. Phys.* **A307**, 75 (1982).

29. U. Fano and C. D. Lin, *Atom. Phys.* **4**, 47 (1975); C. D. Lin, *Phys. Rev.* **A10**, 1986 (1974).

30. G. H. Wannier, *Phys. Rev.* **90**, 817 (1953).

31. E. van de Water and H. G. M. Heideman, in *Proceedings of the 12th International Conference of the Physics of Electron and Ionic Collisions (ICPEAC)*, *Gatlinburg*, S. Datz, Ed. (1981), pp. 270–271.

32. T. D. Märk and F. Egger, *J. Chem. Phys.* **67**, 2629 (1977).

33. T. D. Märk, F. Egger, and M. Cheret, *J. Chem. Phys.* **67**, 3795 (1977).

34. K. Leiter, K. Stephan, E. Märk, and T. D. Märk, *Plasma Chem. Plasma Proc.* **4**, 235 (1984).

35. K. Stephan, H. Deutsch, and T. D. Märk, *J. Chem. Phys.* **83**, 5712 (1985).

36. C. J. Schack, D. Pilipovich, S. N. Cohz, and D. F. Sheehan, *J. Phys. Chem.* **72**, 4697 (1968).

37. C. Brunnee and H. Voshage, *Massenspektrometrie,* Thiemig, Munich (1964).

38. G. Remberg, E. Remberg, M. Spiteller-Friedman, and G. Spiteller, *Org. Mass Spectrom.* **1**, 87 (1968).

39. B. Steiner, C. F. Giese, and M. G. Inghram, *J. Chem. Phys.* **34**, 189 (1961).

40. A. Maccoll, *Org. Mass Spectrom.* **17**, 1 (1982).

41. F. H. Field and J. L. Franklin, *Electron Impact Phenomena,* Academic Press, New York (1970).

42. H. Erhardt and O. Osberghaus, *Z. Naturforsch.* **13a**, 16 (1958).

43. W. M. Jackson, R. T. Brackmann, and W. L. Fite, *Int. J. Mass Spectrom. Ion Phys.* **13**, 237 (1974).

44. J. Momigny, H. Wankenne, and C. Krier, *Int. J. Mass Spectrom. Ion Phys.* **35**, 151 (1980).

45. J. A. Popple, *Int. J. Mass Spectrom. Ion Phys.* **19**, 89 (1975).

46. J. C. Lorquet, *Adv. Mass Spectrom.* **8A**, 3 (1980); J. C. Lorquet and B. Leyh, in *Ionic Processes in the Gas Phase,* M. A. Almoster-Ferreira, Ed., Reidel, Dordrecht (1984), pp. 1–6.

47. C. Lifshitz, *Adv. Mass Spectrom.* **7A**, 3 (1978).

48. C. Lifshitz and P. E. Eaton, *Int. J. Mass Spectrom. Ion Phys.* **49**, 337 (1983) and references therein.

49. T. D. Märk, in *Electron Impact Ionization,* T. D. Märk and G. H. Dunn, Eds., Springer-Verlag, Vienna (1985), pp. 137–197.

50. C. Lifshitz and S. Gefen, *Int. J. Mass Spectrom. Ion Phys.* **35**, 31 (1980).

51. R. J. Van Brunt and L. J. Kieffer, *Phys. Rev.* **A2**, 1293 (1970).

52. K. Köllmann, *Int. J. Mass Spectrom. Ion Phys.* **17**, 261 (1975).

53. G. H. Dunn, *Phys. Rev. Lett.* **8**, 62 (1962).

54. J. P. Johnson and J. L. Franklin, *Int. J. Mass Spectrom. Ion Phys.* **33**, 393 (1980).

55. L. Teubner, in *Electron Impact Ionization,* T. D. Märk and G. H. Dunn, Eds., Springer-Verlag, Vienna (1985), pp. 89–136.

56. A. G. Brenton, R. P. Morgan, and J. H. Beynon, *Ann. Rev. Phys. Chem.* **30**, 51 (1979).

57. G. Herzberg, *Molecular Spectra and Molecular Structure III,* Van Nostrand, Princeton (1967).

58. R. G. Cooks, J. H. Beynon, R. M. Caprioli, and G. R. Lester, *Metastable Ions,* Elsevier, Amsterdam (1973).

59. J. A. Hipple and E. U. Condon, *Phys. Rev.* **68**, 54 (1945).

60. J. H. Beynon, R. A. Saunders, and A. E. Williams, *Table of Metastable Transitions for Use in Mass Spectrometry*, Elsevier, Amsterdam (1965).

61. J. H. Beynon, *Adv. Mass Spectrom.* **4**, 123 (1968).

62. J. H. Beynon and R. G. Cooks, *Int. J. Mass Spectrom. Ion Phys.* **19**, 107 (1976).

63. H. Deutsch, K. Leiter, and T. D. Märk, *Int. J. Mass Spectrom. Ion Phys.* **67**, 191 (1985).

64. K. Stephan and T. D. Märk, *Chem. Phys. Lett.* **90**, 51 (1982).

65. K. Stephan and T. D. Märk, *Int. J. Mass Spectrom. Ion Phys.* **47**, 195 (1983).

66. J. J. Thomson, *Phil. Mag.* **24**, 668 (1912).

67. T. D. Märk, *Int. J. Mass Spectrom. Ion Proc.* **55**, 325 (1983/84).

68. M. Dole, L. L. Mack, R. L. Hines, R. C. Mobley, L. D. Ferguson, and M. B. Alice, *J. Chem. Phys.* **49**, 2240 (1968).

69. J. Gspann and K. Körting, *J. Chem. Phys.* **59**, 4726 (1973).

70. O. Echt, K. Sattler, and E. Recknagel, *Phys. Lett.* **90A**, 185 (1976).

71. T. D. Märk and F. Egger, *Int. J. Mass Spectrom. Ion Phys.* **20**, 89 (1976); W. Koch, N. Heinrich, H. Schwarz, F. Maquin, and D. Stahl, *Int. J. Mass Spectrom. Ion Proc.* **67**, 305 (1985).

72. A. W. Hanner and T. F. Moran, *Org. Mass Spectrom.* **16**, 512 (1981); P. G. Fournier, J. Fournier, F. Salama, P. J. Richardson, and J. H. D. Eland, *J. Chem. Phys.* **83**, 241 (1985).

73. J. S. Cohen and J. N. Bardsley, *Phys. Rev.* **A18**, 1004 (1978).

74. M. Guilhaus, A. G. Brenton, J. H. Beynon, M. Rabrenovic, and P. v. R. Schleyer, *J. Phys.* **B17**, L605 (1984).

75. H. Helm, K. Stephan, T. D. Märk, and D. L. Huestis, *J. Chem. Phys.* **74**, 3844 (1981); K. Stephan, T. D. Märk, and H. Helm, *Phys. Rev.* **A26**, 2981 (1982).

76. B. Brehm and G. De Frenes, *Int. J. Mass Spectrom. Ion Phys.* **26**, 251 (1978).

77. B. Brehm and G. De Frenes, *Adv. Mass Spectrom.* **8A**, 138 (1980).

78. T. Ast, *Adv. Mass Spectrom.* **8A**, 555 (1980).

79. K. Sattler, J. Mühlbach, O. Echt, P. Pfau, and E. Recknagel, *Phys. Rev. Lett.* **47**, 160 (1981).

80. T. Jentsch, W. Drachsel and J. H. Block, *Chem. Phys. Lett.* **93**, 144 (1982).

81. P. Pfau, K. Sattler, R. Pflaum, and E. Recknagel, *Phys. Lett.* **104A**, 262 (1984).

82. T. D. Märk, in *Proceedings of the 10th International Mass Spectrometry Conference*, J. F. J. Todd, Ed., Wiley, New York (1986).

83. J. H. Beynon, *Mass Spectrometry and Its Application to Organic Chemistry*, Elsevier, Amsterdam (1960).

84. K. Biemann, *Mass Spectrometry: Organic Chemical Application*, McGraw-Hill, New York (1962).

85. H. Kienitz, *Massenspektrometrie*, Verlag Chemie, Weinheim (1968).

86. M. J. S. Dewar and P. Rona, *J. Am. Chem. Soc.* **87**, 5510 (1965).

87. R. B. King, *Can. J. Chem.* **47**, 559 (1969).

88. E. S. Waight, *Chem. Commun.*, 1258 (1969).

89. C. J. Cook and J. R. Peterson, *Phys. Rev. Lett.* **9**, 164 (1976).

90. A. J. Dixon, M. F. A. Harrison, and A. C. H. Smith, *J. Phys.* **B9**, 2617 (1976) and references therein.

91. E. Brook, M. F. A. Harrison, and A. C. H. Smith, *J. Phys.* **B11**, 3115 (1978).

92. T. D. Märk and F. J. de Heer, *J. Phys.* **B12**, L429 (1979).

93. P. B. Armentrout, S. M. Tarr, A. Dori, and R. S. Freund, *J. Chem. Phys.* **75**, 2786 (1981).

94. B. L. Schram, B. Adamczyk, and A. J. H. Boerboom, *Rev. Sci. Instrum.* **43**, 638 (1966).

95. J. Schutten, F. J. de Heer, H. R. Moustafa, A. J. H. Boerboom, and J. Kistemaker, *J. Chem. Phys.* **44**, 3924 (1966).

96. B. Adamczyk, *Ann. Univ. Curie-Sklodowska* **24**, 141 (1970).

97. B. Adamczyk, A. J. H. Boerboom, and M. Lukasiewicz, *Int. J. Mass Spectrom. Ion Phys.* **9**, 407 (1972).

98. K. Bederski, L. Wojcik, and B. Adamczyk, *Int. J. Mass Spectrom. Ion Phys.* **35**, 171 (1980).

99. T. Stanski and B. Adamczyk, *Int. J. Mass Spectrom. Ion Phys.* **46**, 31 (1983).

100. J. W. McConkey, A. Crowe, and M. A. Hender, *Phys. Rev. Lett.* **29**, 1 (1972).

101. A. Crowe and J. W. McConkey, *J. Phys.* **B6**, 2108 (1973).

102. A. Crowe and J. W. McConkey, *J. Phys.* **B7**, 349 (1974).

103. A. Crowe and J. W. McConkey, *Int. J. Mass Spectrom. Ion Phys.* **24**, 181 (1977).

104. F. Karstensen and H. Köster, *Astron. Astrophys.* **13**, 116 (1971).

105. M. Schneider, *J. Phys.* **D7**, L83 (1974).

106. F. Karstensen and M. Schneider, *Z. Phys.* **A273**, 321 (1975).

107. F. Karstensen and M. Schneider, *J. Phys.* **B11**, 167 (1978).

108. J. M. Dettmann and F. Karstensen, *J. Phys.* **B15**, 287 (1982).

109. U. T. Raheja, C. Badrinathan, and D. Mathur, *Ind. J. Phys.* **57B**, 27 (1983).

110. V. Schmidt, N. Sandner, and H. Kuntzmüller, *Phys. Rev.* **A13**, 1743 (1976).

111. P. Nagy, A. Skutlartz, and V. Schmidt, *J. Phys.* **B13**, 1249 (1980).

112. C. Lifshitz and M. Weiss, *Int. J. Mass Spectrom. Ion Phys.* **49**, 337 (1983).

113. C. Lifshitz, P. Gotchiguian, and R. Roller, *Chem. Phys. Lett.* **95**, 106 (1983).

114. K. Stephan, H. Helm, and T. D. Märk, *J. Chem. Phys.* **73**, 3763 (1980).

115. K. Stephan, H. Helm, and T. D. Märk, *Adv. Mass Spectrom.* **8A**, 122 (1980).

116. T. D. Märk and A. W. Castleman, Jr., *J. Phys.* **E13**, 1121 (1980).

117. K. Stephan, H. Helm, Y. B. Kim, G. Sejkora, J. Ramler, M. Grössl, E. Märk, and T. D. Märk, *J. Chem. Phys.* **73**, 303 (1980).

118. T. D. Märk, E. Märk, and K. Stephan, *J. Chem. Phys.* **74**, 3633 (1981).

119. Y. B. Kim, K. Stephan, E. Märk, and T. D. Märk, *J. Chem. Phys.* **74**, 6771 (1981).

120. E. Märk, T. D. Märk, Y. B. Kim, and K. Stephan, *J. Chem. Phys.* **75**, 4446 (1981).

121. K. Stephan and T. D. Märk, *J. Chem. Phys.* **81**, 3116 (1984).

122. K. Leiter, K. Stephan, and T. D. Märk, in *Proceedings of the 10th International Mass Spectrometry Conference*, J. F. J. Todd, Ed., Wiley, New York (1986).

123. A. O. Nier, *Rev. Sci. Instrum.* **18**, 398 (1947).

124. W. Paul, *Z. Phys.* **124**, 244 (1948).

125. H. D. Hagstrum, *Rev. Mod. Phys.* **23**, 185 (1951).

126. H. D. Hagstrum, *Rev. Sci. Instrum.* **24**, 1122 (1953).

127. F. Fiquet-Fayard and M. Lahmani, *J. Chim. Phys.* **59**, 1050 (1962).

128. H. E. Stanton and J. E. Monahan, *J. Chem. Phys.* **41**, 3694 (1964).

129. R. Locht and J. Momigny, *Int. J. Mass Spectrom. Ion Phys.* **2**, 425 (1969).

130. H. J. Drewitz, *PTB Bericht, PTB-APh6*, Braunschweig (1974), p. 14.

131. R. M. Reese and J. A. Hipple, *Phys. Rev.* **75**, 1332 (1949).

132. C. E. Berry, *Phys. Rev.* **78**, 597 (1950).

133. W. L. Fite and R. T. Brackmann, *Phys. Rev.* **112**, 1141 (1958).

134. R. L. F. Boyd and A. Boksenberg, in *Proceedings of the 4th International Conference on the Physics of Ionized Gases*, (ICPIG), *Uppsala*, Vol. I (1959), p. 529.

135. E. W. Rothe, L. L. Marino, R. H. Neynaber, and S. M. Trujillo, *Phys. Rev.* **125**, 582 (1962).

136. G. Peach, *Proc. Phys. Soc.* **85**, 709 (1965).

137. R. Akerib and S. Borowitz, *Phys. Rev.* **122,** 1177 (1961).

138. P. Defrance, W. Claeys, A. Cornet, and G. Poulaert, *J. Phys.* **B14,** 111 (1981).

139. B. Piraux and C. Joachain, private communication (1980), cited in Ref. 138.

140. S. S. Prasad, *Proc. Phys. Soc.* **87,** 393 (1966).

141. L. Vriens and T. F. M. Bonsen, *J. Phys.* **B1,** 1123 (1968).

142. K. T. Dolder, in *Atomic and Molecular Processes in Controlled Thermonuclear Fusion,* M. R. C. Mc Dowell and A. M. Ferendec, Eds., Plenum, New York (1980), pp. 313–349.

143. D. Rapp and P. Englander-Golden, *J. Chem. Phys.* **43,** 1464 (1965).

144. B. L. Schram, *Physica* **32,** 197 (1965).

145. B. L. Schram, F. J. de Heer, M. J. van der Wiel, and J. Kistemaker, *Physica* **31,** 94 (1965).

146. B. L. Schram, A. J. H. Boerboom, and J. Kistemaker, *Physica* **32,** 185 (1966); B. L. Schram, H. R. Moustafa, J. Schutten, and F. J. de Heer, *Physica* **32,** 734 (1966).

147. P. Varga, W. Hofer and H. Winter, *J. Phys.* **B14,** 1341 (1981).

148. N. G. Adams, D. Smith, and D. Grief, *J. Phys.* **B12,** 791 (1979).

149. M. R. C. McDowell, in *Case Studies in Atomic Collision Physics,* Vol. 1, E. W. McDaniel and M. R. C. McDowell, Eds., North Holland, Amsterdam (1969), pp. 47–97.

150. R. H. McFarland, *Phys. Rev.* **159,** 20 (1967).

151. P. Laborie, J. M. Rocard, and J. A. Rees, *Electronic Cross Sections and Macroscopic Coefficients. 2. Metallic Vapours and Molecular Gases,* Dunod, Paris (1971).

152. E. J. McGuire, *Phys. Rev.* **A20,** 445 (1979).

153. W. Lotz, *Z. Phys.* **232,** 101 (1970).

154. J. P. Ziesel and R. Abouaf, *J. Chim. Phys.* **64,** 702 (1967).

155. D. Rapp, P. Englander-Golden, and D. D. Briglia, *J. Chem. Phys.* **42,** 4081 (1965).

156. T. D. Märk, *J. Chem. Phys.* **63,** 3731 (1975).

157. R. N. Daly and R. E. Powell, *Proc. Phys. Soc.* **89,** 273 (1966).

158. D. Ton-That and M. R. Flannery, *Phys. Rev.* **A15,** 517 (1977).

159. P. D. Tannen, M. S. Thesis, Air Force Institute of Technology, Order No. 74-14940, University Microfilms, Ann Arbor, Michigan (1973).

160. J. A. Beran and L. Kevan, *J. Phys. Chem.* **73,** 3866 (1969).

161. H. Helm, K. Stephan, and T. D. Märk, *Phys. Rev.* **A19,** 2154 (1979).

CHAPTER 4

Photoionization and Multiphoton Ionization

JAMES D. MORRISON

Physical Chemistry Department
La Trobe University
Melbourne, Australia

1 THEORY

Photoionization as a technique for making ions possesses some special theoretical advantages as compared with electron impact, but it also suffers from some severe practical limitations. To examine the first of these claims, it is necessary to consider in detail the mechanisms of the energy-transfer processes involved in ionization. To create a positive ion from a neutral molecule, enough energy must be transferred to excite one of the electrons from a bound to an unquantized orbital, the minimum energy needed for this being the ionization potential. In the case of electron impact, this process can be written as

$$M + e \rightarrow (M^*)^- \rightarrow M^+ + 2e.$$

Experimentally, the efficiency or probability of such a process as a function of the electron energy rises from zero at the ionization potential more or less linearly over some tens of electron volts, reaches a maximum at an electron energy in the region of 100–150 eV, and then decreases monotonically over some thousands of electron volts.

For helium, in which the first excited state of the positive ion He^+ is some 35 eV above the ground state of the ion, the probability curve is very closely linear above the threshold at 24.5 eV, apart from an initially curved region of 0.5 eV. This initial curvature can be shown to be due almost entirely to the fact that the source of the ionizing electron beam is an incandescent rhenium wire, and the electrons possess a quasi-Maxwellian energy distribution of 0.35–0.5 eV [full width at half maximum (fwhm)]. There is little change in the ionization efficiency curve as the first or higher excited states of the ion are reached. In the case of more complex atoms, or molecules, although many states of the ion are potentially accessible within the first few electron volts above the first ionization potential, again this hardly shows in the ionization efficiency curves.

The key to explaining the energy-transfer process in ionization by electron impact was provided by Wigner (1), who offered two simple postulates. The first assumed that the ionization process can be divided into three events: (i) approach of an electron and a molecule in some electronic state, (ii) formation of a collision complex, and (iii) dissociation of this complex into a stable positive ion and two free electrons. The second proposed that the ionization probability does not to a first approximation depend on what went on in steps (i) and (ii), but is critically dependent on the likelihood of success of the separation step (iii). In this model, separation depends on the probability of disposing of the energy excess over the minimum required for the process, and since the electrons are very much lighter than the positive ion, the bulk of this excess energy must appear as translation.

Wannier (2), using a very simple phase-space argument, showed that this probability depended on the number of degrees of freedom n for sharing the energy excess between the electrons. The ionization probability as a function of energy excess above the minimum E_c is given by

$$P(E - E_c) = k(E - E_c)^n$$

For emission of two electrons from the collision complex, $n = 1$. For single ionization of molecules, the gradient of the linear increase in P, the coefficient k, is determined by the electronic transition probability, which is proportional to the overlap of the vibronic wavefunctions for the neutral ground state and the state of the ion in question. Extension of this argument suggested that for n-fold ionization in a given state induced by electron impact, the ionization probability curve should have the form $P(E - E_c) = k_n(E - E_c)^n$. Later, detailed calculation by Wannier indicated that these power laws should all be increased by 0.127 (3). The time required for these direct ionization processes is short, of the order of 10^{-15}–10^{-16} sec, so all transitions are "vertical."

In photoionization, the basic reaction is

$$M + h\nu \rightarrow (M^*) \rightarrow M^+ + e$$

The way in which the probability of this process will behave as a function of the photon energy can be predicted in a simplistic fashion using the previous arguments of Wigner and Wannier. If one neglects the approach region and the collision complex and examines the degrees of freedom for removal of the excess energy, for n-fold photoionization this probability will have n less by 1 than the value for the electron-impact-induced process. For single ionization, this is zero, suggesting that direct photoionization should obey a step-function cross-section law. That is, the ionization probability as a function of photon energy will be zero until the ionization potential is reached, at which point the probability will rise immediately to a value determined by the electronic transition probability for the process.

More recent and sophisticated treatments of this problem (4) have not significantly modified these conclusions, and the experimental data bear them out. The simple theory gives no prediction of the energy range above threshold for which the assumptions on which it is based will hold. In practice the threshold laws are modified, in that the observed ionization probability is better represented by multiplying the threshold law curve by an exponential decrease, the exponent being such that the probability for single photoionization falls to 10% of its maximum value at 30–50 eV above threshold (5).

It is customary to assume that there is a separate ionization probability for the transitions giving rise to each state of the ion, and that to a first approximation these probabilities are additive. That is, at any given energy of the ionizing electrons or photons, the total ionization efficiency, which is closely related to the ionization cross section, is the sum of the separate ionization probabilities for transition to every state that is accessible at that energy. This certainly appears to be the case for the individual levels within a given electronic state. In electron impact, there are few limitations on the states of the ion that may be excited, since optical selection rules do not apply, but for photons some restrictions are imposed.

For many atoms and molecules, a significant contribution to the total ionization can arise by the indirect processes of autoionization, in which the primary step of energy transfer is excitation to levels of a neutral state above the lowest ionization potential. These levels undergo internal conversion of their electrons, ejecting one electron after a time delay ranging from 10^{-14}–10^{-6} sec and depending on the coupling between the excited discrete and continuum states. The threshold law for such processes is assumed to be that for the primary step of excitation; that is, it will have a power law one less than that for direct ionization. Geltman (6) has shown theoretically that this simple assumption is less satisfactory for excitation processes than for ionization and predicted that for electron impact at threshold such a process should have a probability closer to $(E - E_c)^{1/2}$. Experimentally the probability of autoionization by electron impact is found to be approximately a step function (7), while for photon impact it is a sharp peak at the energy of the autoionizing state (8).

However, studies of autoionization cross sections with highly monochromatic photon beams indicate that the simple notion of the additivity of cross sections is seldom true. It is applicable in some cases, but in others complex interactions take place between the bound excited states above the lowest ionization potential and the underlying continuum states, so that the cross sections for the autoionizing state either add to or subtract from the underlying ionization continuum. In extreme cases, this interaction is so strong that the ionization continuum is reduced almost to zero, for example in the so-called window resonances (9). These interactions have been accounted for very satisfactorily in terms of configuration interaction by Fano (10).

For the above reasons of differing threshold law behavior, in electron impact the populations of the various states of the ion immediately after ion formation depend less on the individual transition probabilities for these states, and more on the actual ionizing electron energy and the threshold value for each state. Once formed, of course, molecular ions in some excited states are unstable and undergo dissociation to form fragment ions before they reach the ion current detector. At all impact energies, the ground state of the molecular ion predominates, but the relative populations of all states depend on the impact energy (Fig. 1, top).

With photon impact, at any impact energy above the ionization potential, the population of each state of the ion lying below this energy depends to a first approximation on the relative electronic transition probabilities for these states and not on the energy excess of the photon beam (Fig. 1, bottom). Further, autoionization does not contribute to the total photoionization current unless an autoionizing level happens to coincide with the photon energy. These considerations are implicit in the use of photoelectron spectra to determine electronic transition probabilities.

Apart from these theoretical factors, a number of practical considerations must be taken into account when choosing between electron and photon impact for making ions. Abundant ionization can be produced by electron impact, since it is possible to obtain ionizing electron beams of several hundred milliamperes, even in quite a small source: By operation at electron energies of around 100 eV, intense beams of ions can be produced. The ground state will predominate, but there will be considerable rotational and vibrational excitation.

However, it is very difficult to obtain electrons with accurately known energy. Not only does the thermal energy spread already referred to affect matters, but the energy of the electrons in the ionizing region depends on the potential at every point

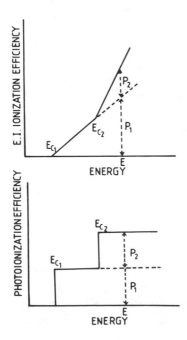

Figure 1. Idealized ionization efficiency curves for electron impact (top) and photon impact (bottom), showing contributions P_1 and P_2 to the total curve at energy E for ions formed in two states with ionization energies E_{c_1} and E_{c_2}. For electron impact at energies above E_{c_2} the ratio P_1/P_2 depends on $E - E_{c_2}$; for photon impact this ratio is constant.

along their path. To extract the ions that are formed, a potential gradient has to be maintained in the source, and this will perturb the electron energy. Due to their different methods of construction, a contact potential difference exists between the cathode surface and the body of the ion chamber, and this can change by a variable amount depending on the nature and pressure of the gas sample present.

The ionizing electron beam creates a space charge, which modifies the potential at different places within the electron beam. When a gas is present, ions are trapped in the electron beam by this space charge after formation, and partly neutralize it, but the extent of this depends on gas pressure, electron beam density, and electron energy. It is possible to reduce the electron-energy spread by approximately 50% by the use of oxide cathodes or, to a more considerable extent, by the use of energy analyzers, but the latter dramatically reduce electron currents available (11). The energy of electrons can very readily be varied over a wide range, up to thousands of electron volts, but unless great care is taken, the acceleration process will broaden the energy spread. In all electron impact work where the electron energy has to be known with accuracy, it is essential to include a calibrating gas whose ionization potential is known spectroscopically, such as Ar, along with the sample, and to record ionization curves for both.

In photoionization the energy of the photons can be determined to very high accuracy. The energy is not affected by any applied potentials along the path of the beam, and ion sources can be designed to use optimum ion drawout potentials. Some care must be exercised to prevent the photon beam from striking metal surfaces in the vicinity of the ion chamber; otherwise copious quantities of photoelectrons may be produced, accelerated by the ion drawout potentials, and create spurious ions by electron impact. The use of grating monochromators permits beams of photons of very narrow energy spread (1 meV or less fwhm) to be selected; changing the wavelength is simply a matter of rotating the grating. However, the range of energies available for photons is much more restricted than that for electrons. A complication in the use of photons is that at the wavelengths required to produce ionization, 150 nm or less, it is not possible to use windows in the path of the photon beam. Even the best samples of LiF, used as windows for ultraviolet (UV) light, become opaque at wavelengths below 115 nm. Accordingly, the light source, which usually involves a gas discharge of some kind at pressures up to 200 torr; the ionization region, which may be operated at up to 1 torr; and any monochromator employed must be coupled together inside a high-vacuum chamber, and differential pumping must be employed to prevent cross contamination.

2 RADIATION SOURCES

2.1 Monochromatic Line Sources

Monochromatic UV radiation of good intensity and fixed wavelength can be obtained by exciting the resonance lines of the elements. The rare gases are most commonly used for this; their wavelengths are listed in Table 1 (12). This excitation can be

TABLE 1   Properties of Resonance Lines of the Rare Gases

	HeI	NeI	ArI	KrI	XeI	H(I)
Wavelength (nm)	58.4334	74.3718	106.6659	123.5838	146.961	121.567
Energy (eV)	21.218	16.671	11.623	10.032	8.436	10.198

achieved by electrical discharge, either continuous or pulsed, between electrodes in the gas at pressures of 10^{-2}–10^{-3} torr, or by coupling microwave energy to the gas in a resonant cavity. Photon beams with a flux of 10^{14} photons/sec, which is comparable in intensity to electron beams, can be achieved in this way. Helium discharges producing the HeI and HeII lines are widely used in photoelectron spectroscopy (13). At very short wavelengths, the characteristic x-ray lines produced by electron bombardment of metal anodes can also be used (14).

2.2   Continuum Sources

To study photoionization efficiencies requires sources of radiation that are both as monochromatic as possible and variable in energy. These goals can be achieved by using a continuum source that emits radiation over a wide band of wavelengths, coupled with a monochromator to select one wavelength at a time.

A simple and widely used source of good intensity is the hydrogen many-line spectrum produced by a capillary discharge in H_2 gas at pressures between 0.2 and 10 torr. A typical lamp design is given in Fig. 2 (15). This produces useful energy over the range 165–105 nm. The radiation produced is not a true continuum, but a multitude of lines. Strictly speaking, measurements should be made only at the maximum of each individual line; otherwise spurious structure may appear in ion-

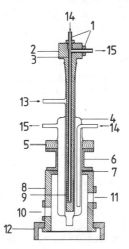

Figure 2. Discharge light source for producing H_2 many line spectrum. Key: 1, cooling-water inlet fittings; 2, aluminum anode with internal cooling; 3, stainless steel tube; 4, quartz tube with integral cooling; 5, split clamp, Perspex; 6, centering cup, aluminum; 7, Viton 0 ring; 8, main body, aluminum; 9, platinum sleeve, 0.05 mm thick; 10, viewing port, aluminum; 11, pumping port; 12, captive nut, brass; 13, gas inlet; 14, cooling water in; 15, cooling water out. Reproduced with permission from Ref. 15.

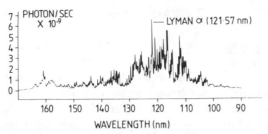

Figure 3. Hydrogen many-line spectrum obtained using 1200 line/mm holographic grating and sodium salicylate/EMI 9502S photomultiplier. Reproduced with permission from Ref. 15.

ization cross sections measured using this source (Fig. 3). A mixture of D_2 and H_2 in the discharge gives a set of even more closely spaced lines. Hydrogen does produce a true continuum at longer wavelengths (16).

The Lyman radiation source, and later developments of it by many others (17) generate pseudocontinua when a spark discharge in low-pressure gas takes place through a restricting orifice. These sources give useful radiation from the visible range down to wavelengths of 30 nm. They are used widely in absorption spectroscopy, but have not found much application in ionization work, partly because of their lack of reproducibility between successive electrical pulses, and also because of the electrical noise they generate.

The rare-gas continua produced by discharges in these gases at pressures of 70–150 torr (Fig. 4) are true continua, and cover the range from 180 to 55 nm, the (7–22 eV) Ar and He continua being the most useful for ionization work (19).

A powerful source of UV radiation is provided by a synchrotron or storage ring, which gives continuous radiation over a very wide range of wavelengths at high intensity (20). These facilities are not widely available, which has limited their use to scientists with convenient access to a center having one.

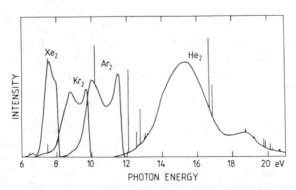

Figure 4. Intensity distribution as a function of photon energy in the continuum emission spectra of xenon, krypton, and argon excited by microwaves and of helium excited by a condensed discharge. The lines superposed on the continua are due to emission from hydrogen, oxygen, nitrogen, and carbon present as impurities, and provide internal calibration points. Curves are replotted from the data of Huffman and Wilkinson given in Ref. 18.

2.3 Monochromators

To obtain monochromatic radiation from a continuum source, some kind of grating monochromator is essential (21). Ideally, one would like to have a monochromator with a constant-energy bandwidth that could be scanned in a linear fashion over its energy range by a simple motor drive. The intensity per unit bandwidth of radiation sources in the UV range is usually low, and the reflectivity of all surfaces decreases dramatically at normal incidence below 100 nm. To keep the number of reflections (and hence loss of light) to a minimum, concave gratings are employed, and wavelength change is achieved by moving either the entrance or the exit slit on the Rowland circle. At wavelengths above 60 nm, the excellent focusing qualities and very low astigmatism of a normal-incidence mounting more than counterbalance the poor reflectivity, and the wavelength drive can be made almost linear over short wavelength regions very easily. Most work at high resolution (0.012 nm) is done with monochromators using this configuration. The Seya Namioka mounting is convenient, in that wavelength change is achieved by simple rotation of the grating (22) (Fig. 5). The image is curved and shows some astigmatism, so it is suitable only for intermediate resolutions of 0.05 nm. At wavelengths below 50 nm, it becomes almost essential to use grazing-incidence spectrometers (24). These maintain reasonable reflectivity but have very high astigmatism, so they are of low efficiency.

Excellent concave gratings with 1200 lines/mm are produced nowadays by holographic techniques, and give adequate dispersion for most photoionization work. Gratings can be obtained blazed to give maximum intensity in a given wavelength region using a given order. It is claimed that platinum and gold coatings are superior to aluminum as regards reflectivity in the vacuum UV (25) in the region of 100 nm, but in practice the difference in performance is not marked.

The use of a grating monochromator imposes one restriction on the radiation source: The continuum emitted by the source should not extend over a wavelength range the ratio of whose end points is greater than 2:1; otherwise problems may be encountered with overlapping orders. With synchrotron sources, a preliminary monochromator is often used to give a long-wavelength cutoff; in this case there is enough energy to spare to allow for the losses due to this additional reflection.

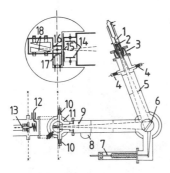

Figure 5.   Seya Namioka UV monochromator used in photoionization studies. Key: 1, discharge lamp; 2, entrance slit; 3, bellows for adjusting lamp; 4, micrometers for adjusting and centering entrance slit; 5, light baffles; 6, grating; 7, grating drive monochromator; 8, pumping lead for monochromator; 9, light shutter; 10, micrometers for adjusting exit slits; 11, ionization-source chamber; 12, sample-gas inlet; 13, entrance slit of mass analyzer; 14, photoelectron suppressors; 15, exit slit of monochromator; 16, ionization source; 17, filament for producing electrons; 18, light-monitoring plates. Adapted with permission from figures in Ref. 23.

2.4   Tunable Laser Sources

Until very recently, it has not been possible to produce usable tunable radiation in the vacuum UV range with a laser. High-intensity infrared radiation produces ionization and fragmentation in molecules, the extent of which is determined more by the laser power than by the wavelength in multiphoton absorption processes.

Multiphoton ionization (MPI) occurs by a number of mechanisms. Resonance-enhanced ionization is believed to begin with coherent n-photon excitation of ground-state molecules to an excited Rydberg state. This state then absorbs additional photons to excite it into the ionization continuum. The photons need not all be of the same wavelength: Separate lasers operating at different wavelengths can be used, providing a most useful method for probing the spectroscopy of excited states.

By variation of the laser power, multiphoton ionization can be made to range from "soft" ionization, giving only the parent ion, to total atomization (26). (Fig. 6). In the latter case, it is believed that the first step is formation of the parent

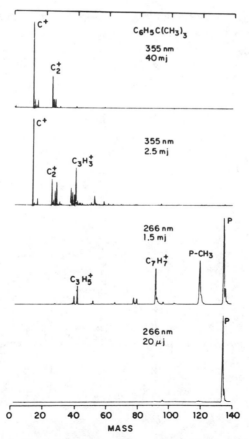

Figure 6.   Multiphoton ionization fragmentation patterns of t-butyl benzene observed at different combinations of laser power density and pulse length. Reproduced with permission from Ref. 26.

molecular ion, and that this is then photofragmented by further photons. MPI is unusual in that an apparently rather indiscriminate form of ionization gives very different mass spectra for isomeric molecules, whereas with controlled electron impact the mass spectra are almost indistinguishable.

A tunable pumped dye laser of high power has been frequency doubled, then used to generate third-harmonic radiation in the vacuum UV range in a pulsed argon jet. This radiation is tunable over the range 97–102 nm (27). (Fig. 7). This system opens up exciting possibilities for the extension of photoionization studies.

2.5   Wavelength Calibration

Calibration of the wavelength of a radiation source in the vacuum UV presents some problems. Most of the commonly used continua produce a number of superposed atomic lines over their useful output, and the wavelengths of the resonance series of the rare gases have been tabulated and are extensively used as calibration points (29).

When lasers are used for multiphoton ionization, if their output lies in the visible region, part of the light may be passed through a cell containing iodine vapor and the intensity recorded. The iodine absorption spectrum gives a multitude of rovibrational lines, which have been tabulated (30) and allow calibration of the energy scale to within 10 μeV or better (31) (Fig. 8).

2.6   Detectors

The experimental measurement of ionization efficiencies or cross sections requires monitoring of the radiation intensity. With the low power outputs available from

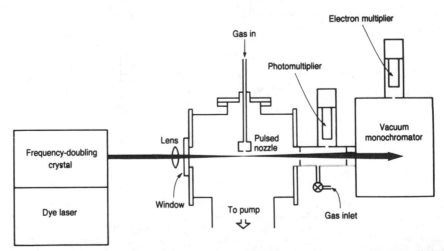

Figure 7.   Elements of a recently developed laser source producing coherent tunable UV radiation. Reproduced with permission from Ref. 28.

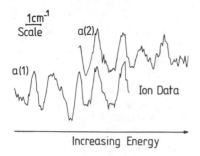

$\dfrac{1cm^{-1}}{Scale}$ a(2)

a(1)

Ion Data

Increasing Energy

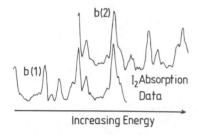

b(2)

b(1)

I_2 Absorption Data

Increasing Energy

Figure 8. The use of a simultaneously recorded I_2 absorption spectrum for energy-scale and intrascan calibration of ion photodissociation (IPD) data. Consecutive spectra are recorded at 0.3 cm^{-1} (37 μeV) resolution. Curves a(1) and a(2) form the IPD spectrum, and b(1) and b(2) make up the I_2 absorption spectrum. The superposition of a(1) on a(2) and b(1) on b(2) to match details of features provides the necessary intrascan calibration. Reproduced with permission from Ref. 31.

most sources, absolute measurements of radiation flux are not possible. The following are three representative means of detection:

1. A sodium salicylate coating on the end of a light pipe, with a photomultiplier mounted outside the vacuum chamber, provides a very simple method of light detection. The salicylate coating has to be renewed fairly frequently, since it deteriorates in a vacuum. This detector has a fairly constant response from 160 to 50 nm (32).

2. The radiation can be allowed to strike a metal plate, and the emitted photoelectrons may be either collected on a positively charged electrode and measured as a current directly or amplified with an electron multiplier, allowing direct photon counting. The output of the metal-plate detector rises smoothly with decreasing wavelength, so it requires calibration, but it has the advantage that it gives clean and reproducible results (23).

3. The pyroelectric detector (33) is very satisfactory for measurement of pulsed radiation. This device gives a signal that is proportional to the differential of the light intensity and requires integration to give the light intensity itself. It has a fast response, and is claimed to have a uniform response over the wavelength range 1000–10 nm. The author has not used it in the vacuum UV range, but it certainly works well with 10-nsec laser pulses in the visible and near UV regions.

REFERENCES

1. E. P. Wigner, *Phys. Rev.* **73**, 1002 (1968).
2. G. H. Wannier, *Phys. Rev.* **90**, 817 (1953).

3. G. H. Wannier, *Phys. Rev.* **100**, 1180 (1956).

4. J. Berkowitz, *Photoabsorption, Photoionization and Photoelectron Spectroscopy*, Academic Press, New York (1979), Chapter 4.

5. J. D. Morrison, "Transport d'Energie dans les Gaz," in *Proceedings of the Solvay Conference*, R. Stoops, Ed., Interscience, New York (1962), p. 397.

6. S. Geltman, *Phys. Rev.* **102**, 171 (1956); **112**, 176 (1958).

7. J. D. Morrison, "Ionization and Appearance Potentials," in *MTP International Review of Science, Physical Chemistry Series One*, Vol. 5, A. D. Buckingham and A. Macoll, Eds., Butterworths, London (1972).

8. D. M. Mintz and T. Baer, *Int. J. Mass Spectrom. Ion Phys.* **25**, 39 (1977).

9. R. P. Madden, D. L. Ederer, and K. Codling, *Phys. Rev.* **177**, 136 (1969).

10. U. Fano, *Phys. Rev.* **124**, 1866 (1961).

11. L. P. Johnson, J. D. Morrison, and A. L. Wahrhaftig, *Int. J. Mass Spectrom. Ion Phys.* **26**, 1 (1978).

12. C. E. Moore, *Atomic Energy Levels Circular 467*, Vols. I–III, National Bureau of Standards, Washington, D.C. (1949, 1952, 1958).

13. J. W. Rabelais, *Principles of UV Photoelectron Spectroscopy*, John Wiley & Sons, New York (1977).

14. U. Gielius, E. Basilier, S. Svensson, T. Bergmark, and K. Siegbahn, *J. Electron Spectrosc. Relat. Phenom.* **2**, 405 (1974).

15. R. G. McLoughlin, Ph.D. thesis, La Trobe University, Melbourne, Australia (1980).

16. J. A. R. Samson, *Techniques of Vacuum UV Spectroscopy*, John Wiley & Sons, New York (1967), pp. 95–98.

17. W. R. S. Garton, *J. Sci. Instrum.* **36**, 11 (1959); see also Ref. 16, pp. 163–171.

18. Ref. 16, pp. 101 and 105.

19. Y. Tanaka, A. S. Jursa, and F. J. LeBlanc, *J. Opt. Soc. Am.* **48**, 304 (1958).

20. K. Codling, *Rep. Prog. Phys.* **36**, 544 (1973).

21. For a detailed account of grating monochromators, see Ref. 16, Chapters 2 and 3.

22. T. Namioka, *J. Opt. Soc. Am.* **49**, 951 (1959).

23. H. Hurzeler, M. G. Inghram, and J. D. Morrison, *J. Chem. Phys.* **28**, 76 (1958).

24. M. Salle and B. Vodar, *Compt. Rend.* **230**, 380 (1950).

25. Ref. 16, pp. 34–40.

26. R. B. Bernstein, *J. Phys. Chem.* **86**, 1178 (1982).

27. C. T. Rettner, E. E. Marinero, R. N. Zare, and A. H. Kung, in *Excimer Lasers—1983*, C. K. Rhodes, H. Egger, and H. Pummer, Eds., American Institute of Physics, New York (1983), p. 345.

28. *Physics Today,* **36**(11), 19 (1983).

29. Ref. 16, p. 323.

30. S. Gerstenkorn and P. Luc, *Atlas de Spectre d'Absorption de la Molecule d'Iode*, Centre National de la Recherche Scientifique, Paris (1978).

31. D. L. Smith, Ph.D. thesis, La Trobe University, Melbourne, Australia (1982).

32. J. A. R. Samson and G. N. Haddad, *J. Opt. Soc. Am.* **64**, 1346 (1974).

33. C. A. Hamilton, R. J. Phelan, and G. W. Day, *Opt. Spectra* **9**, 37 (1975).

CHAPTER 5

Ion Focusing, Mass Analysis, and Detection

JAMES D. MORRISON

Physical Chemistry Department
La Trobe University
Melbourne, Australia

The energy-transfer processes that take place in an ionization source give rise to a wide variety of ions. One may wish to identify the ionic species present (e.g., by measurement of their m/z ratio), to record their relative abundances, or to select a beam of ions of given m/z ratio to carry out some further collision experiment. Each of these applications requires a mass analyzer, and one is faced with a very wide range of possibilities. Not only the molecular weight of the ions of highest mass likely to be present and the resolution in mass believed to be necessary, but also the conditions under which the ions are created, are important issues in choosing an analyzer.

All mass spectrometers operate as flow systems. Sample is continually supplied to the ionization source and some fraction of this, usually less than 0.1%, is ionized. The remaining sample and any ions or radicals formed in the ionization process diffuse rapidly to the walls of the ionization chamber, where the ions are neutralized and all volatile constituents are pumped away as quickly as possible.

Because of the Coulomb repulsion of like charges, it is difficult to produce a dense flux of ions; that is, ion sources are of low luminosity, with ion densities usually very much less than 10^{16} ions/cm^3. Space-charge neutralization, such as that occurring in a plasma, can allow higher luminosities to be obtained, but such plasma sources allow little control over the formation of ions from complex molecules.

1 REQUIREMENTS OF A MASS ANALYZER

1. Clearly a mass analyzer should be efficient; that is, it should be able to make use of as many of the ions formed in the ionization source as are needed to give good sensitivity of detection.

2. At the same time, extraction of ions from the ion source for mass analysis should not unduly perturb the conditions of their formation. Sometimes this is difficult to achieve, as, for example, in sampling from a plasma or a flame.

3. The ionization region may be at pressures ranging from 10^{-10} torr to above atmospheric pressure. High pressures require differential pumping between the ionization region and the mass analyzer. A few mass analyzers can tolerate pressures as high as 10^{-3} torr, but most require a vacuum of 10^{-6} torr or better to operate efficiently.

4. In some types of ionization sources, the ions are not formed at rest, but are generated with a wide range of translational energies. Not all types of mass analyzers can operate satisfactorily with such energy spreads.

5. Once extracted from the ionization region, ions are usually formed into a beam for projection into the mass analyzer. With ion sources of very low luminosity or in which ions are formed in an extremely small volume (as, e.g., at the focus of a laser), it may be advantageous to have a mass analyzer that will accept beams with a large solid angle. In other cases, ions at low luminosity can be formed over a wide area, and ability to accept a broad beam is a requirement.

6. A short analysis time may be imposed by the necessity to respond to rapid changes in intensity of a given ion, as in kinetic studies of fast reactions involving ions, or in state-selected experiments, where ion intensities may be pulsed or modulated.

7. The operation of the mass analyzer should not perturb the ion source; this precludes the use of magnetic analyzers in some cases.

8. If required to produce complete mass spectra, the mass analyzer should not suffer unduly from mass discrimination; that is, it should reproduce faithfully the abundances of all ions at all masses present in the source.

9. Finally, there may be practical limitations on the size, weight, or cost of the mass analyzer.

It is not always possible to satisfy all of these requirements, but it is desirable to achieve the best compromise attainable using the very wide range of mass analyzers available.

2 PRINCIPLES OF MASS ANALYSIS

The basic principles of mass analysis (1) and the ways in which these are applied to the construction of practical mass analyzers are summarized briefly in this section. All ion mass analyzers operate by virtue of the behavior of charged particles in electric and magnetic fields, and we shall consider briefly these interactions prior to considering actual mass analyzers.

2.1 Motion in a Uniform Electric Field

If a potential difference V is established between two electrodes and a positive ion of charge ze is formed at rest adjacent to the more positive electrode, it will strike the negative plate, having acquired just before impact a kinetic energy of $zeV = \frac{1}{2}Mv^2$, where M is mass and v velocity. If there is an aperture in the negative plate, the ion may pass through it, and if the space beyond is at the same potential, this ion will continue on its path with the same energy.

An ion with mass M atomic mass units and charge ze ($e = 1.6 \times 10^{-19}$ coulomb) accelerated through V volts will reach a final velocity of

$$v = 1.39 \times 10^4 \, (Vz/M)^{1/2} \text{ m/sec.} \tag{1}$$

Table 1 uses this equation to calculate the velocity of selected ions.

2.2 Motion in a Uniform Magnetic Field

A charged particle of charge ze and velocity v injected at some angle θ into a uniform magnetic field B experiences a force

$$F = Bzev \sin \theta \tag{2}$$

perpendicular to the field direction.

If $\theta = 0°$, ions travel parallel to the magnetic field, and no force is exerted upon them. At $\theta = 90°$ the ion experiences the force $Bzev$ perpendicular to both the field and the direction of motion of the ion. These orthogonally injected ions will follow a circular path for which the centripetal and centrifugal forces are equal; hence,

$$\frac{Mv^2}{R} = Bzev \tag{3}$$

defines the radius R of the circular path.

TABLE 1   Variation of Ion Velocities (in m/sec) with Energy

Ion	Energy (eV)			
	1	10	100	1000
$^1H_2^+$	9.83×10^3	3.109×10^4	9.83×10^4	3.109×10^5
$^{14}N_2^+$	2.63×10^3	8.317×10^3	2.63×10^4	8.317×10^4
$^{204}Hg^+$	9.73×10^2	3.077×10^3	9.73×10^3	3.077×10^4
$^{204}Hg^{2+}$	1.38×10^3	4.363×10^3	1.38×10^4	4.363×10^4

TABLE 2   Larmor Frequencies and Radii for Some Ions at 100 eV and 1000 eV

Ion	Radius R(m)		Larmor Frequency
	100 eV	1000 eV	
$^1H_2^+$	2.036×10^{-3}	6.438×10^{-3}	7.815 MHz
$^{14}N_2^+$	7.620×10^{-3}	2.410×10^{-2}	0.558 MHz
$^{204}Hg^+$	2.057×10^{-2}	6.505×10^{-2}	76.6 kHz
$^{204}Hg^{2+}$	1.454×10^{-2}	4.598×10^{-2}	153.2 kHz

If all ions enter the magnetic field with the same energy $zeV = \frac{1}{2}Mv^2$, Eq. (3) rearranges to

$$R = \frac{Mv}{zeB} = \frac{1}{B}\left(\frac{2MV}{ze}\right)^{1/2} \tag{4}$$

Thus, for constant B and V, ions travel on circular paths with radii proportional to $(M/z)^{1/2}$.

Expressing B in teslas (1 Tesla = 10,000 gauss), R in meters, V in volts, M in atomic mass units, and z in units of electric charge e, the radius is given by

$$R = \frac{1.44 \times 10^{-4}}{B}\left(\frac{MV}{z}\right)^{1/2}. \tag{5}$$

The time to travel 360° on a circular path of length $s = 2\pi R$ is given by

$$T = \frac{s}{v} = \frac{6.398 \times 10^{-4}}{B}\frac{M}{Z}. \tag{6}$$

The reciprocal of the characteristic time, $f = 1/T$, is independent of the energy of the charged particles; this frequency is called the Larmor frequency. Characteristic Larmor frequencies for the ions in Table 1 are given in Table 2 along with the radii of their paths at 100 and 1000 eV.

3   PRACTICAL MASS ANALYZERS

A wide variety of combinations of electric and magnetic fields can be used for mass analysis. Mass analysis can be either dynamic or static, depending on whether or not advantage is taken of the time dependence of the equations of motion. The list of examples that follows is not intended to be exhaustive; more detailed accounts of this topic are given by McDowell (2), White (3), and Roboz (4).

The most useful combinations of fields are those that focus the ions to some degree. The general focusing properties of uniform magnetic and electric fields will therefore be examined and some special cases considered.

3.1 Direction Focusing of a Uniform Magnetic Field

Ions formed and accelerated at a point inside a uniform magnetic field travel in circular paths, and all return to this initial point, regardless of their initial direction or energy, after traveling through 360°. This property gives perfect refocusing but no mass analysis. If the ions are all given the same energy, first-order focusing is obtained after 180° of travel, with ions of different M/z ratio being separated (Fig. 1).

For a half-angle of injection α, the width of the focused image at the 180° position is $R\alpha^2$; that is, for an analyzer of $R = 10$ cm and $\alpha = 3°$, the image at 180° is broadened by 0.25 mm. This sets an obvious limit to the resolution attainable. Moreover, because of the need to immerse both the ion source and the ion current detector in the magnetic field, very little use is made today of 180° analyzers. Their main application is as leak detectors and residual-gas analyzers.

The 180° focusing for total immersion in a magnetic field is a special case of general first-order magnetic focusing in any sector field that ions enter and leave normal to the field boundary. The object and image focus points lie on a straight line passing through the apex of the sector (Fig. 2). This is true only for a perfectly defined field boundary, however, and is not attainable in practice. With iron magnets there is always a fringe field, which causes defocusing. Experimentally this is corrected by setting the field boundaries back from the theoretical position by an amount equal to the magnet gap. Sector mass analyzers are usually operated with $\alpha = 1°$ or $2°$ and a sector angle Φ of 60° or 90°. The smaller the sector angle, the less iron is required in the magnet for a given ion radius and the greater are the distances of the source and collector from the magnetic field; however, fringe-field defocusing effects become more pronounced with smaller sector angles. Asymmetric configurations can also be employed to obtain a magnifying or reducing lens action at the focus.

Sector mass analyzers are easy to construct, have fairly good ion transmission, and, provided ion energies above 1 kV are used, produce well-collimated ion beams.

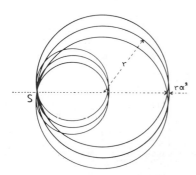

Figure 1. Trajectories of ions when totally immersed in a magnetic field, showing first-order direction focusing at 180° position. S denotes the ion source point (object focus point).

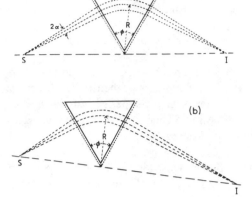

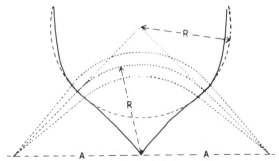

Figure 2. General first-order focusing of ions in a sector magnetic field in (*a*) symmetric case and (*b*) asymmetric case. S = source (object focus point), I = image focus point, ϕ = sector angle, R = ion path radius, α = half-angle of injection.

Being momentum filters, they require initial ion energy spreads of less than 0.2 eV to obtain reasonable resolution.

3.2 Perfect Direction Focusing

Kerwin (5) showed that monoenergetic ions traveling with radius R in a uniform magnetic field will return to a direction focus at a symmetrical point $(x, y) = (A, 0)$ after starting at a point $(x, y) = (-A, 0)$ outside the field for a wide range of initial angles of direction, provided the magnetic-field boundary is defined by $y = x(A - x)/(R^2 - x^2)^{1/2}$ (Fig. 3). Practical values of α can be up to 20°. For reasonably monoenergetic ions (energy spreads not greater than 0.2 eV), this gives mass analyzers of very high transmission and excellent direction focusing. Surprisingly few instruments of this type have been built, perhaps because of the difficulties of shaping the magnet pole piece boundaries. The availability of computer-controlled milling machines now makes this much easier to achieve. The Kerwin solution can be approximated by circular magnet pole pieces and a sector angle of 90° (3), and this has often been used in small mass analyzers.

Figure 3. Field boundaries for perfect magnetic direction focusing (solid line) and circular approximation (dashed line) for $A = \sqrt{2}R$.

These machines have the same advantages and disadvantages as the sector machines discussed above, but their high luminosity makes them particularly suited for the analysis of ions formed in a very small volume, as at the focus of a laser.

Direction focusing machines work satisfactorily when the initial energy spread ΔV of the ions extracted from the ion source is small, which minimizes the ratio $V/\Delta V$, where V is the ion energy used in the analyzer. This ratio determines the resolution. Raising V is one way of increasing resolution, but has an obvious limitation. The lower limit to the ion-accelerating voltage is set by the need to keep the mean free path of the ions very much greater than the size of the apparatus, and mass analyzers of this type will not work reliably with ion energies much less than 100 eV. The higher the ion-accelerating voltage, the better the performance of the ion optics, but high ion energies both require high magnetic fields and introduce other problems, such as corona discharge and unwanted field ionization. The maximum value that can be conveniently used is of the order of 20–30 keV.

3.3 Energy-Focusing Properties of an Electric Field

To attain higher resolution or to analyze ions formed with a wide spread of initial energies, it is necessary to reduce the energy spread of the ions by means of an energy analyzer. Ions injected tangentially between concentric cylindrical plates are energy deflected and dispersed, and Hughes and Rojansky (6) determined that first-order energy refocusing occurred at every $\pi/2$ radians along the approximately circular path. Energy-selected beams of ions can be obtained using defining slits; the exit slit of the energy analyzer can serve as the entrance slit of the mass analyzer. This combination can give very high mass resolution, but necessarily exhibits low luminosity.

3.4 Double-Focusing Mass Analyzers

Mattauch and Herzog (7) pointed out that for ions of a given M/z ratio and for a suitable choice of angles of an electric sector and a magnetic sector, the slit between the two sectors could be greatly widened. The principle employed was to counteract the velocity dispersion of the electric sector by means of the direction dispersion of the magnetic one. This instrument, giving focusing in both velocity and direction, refocused ions of all M/z values along an image plane, allowing photographic detection. Other instruments of this general type, but with electrical detection, were designed and built by Bainbridge and Jordan (8) and Nier and Roberts (9) (Fig. 4).

Hintenberger and König (10) have calculated numerous solutions to these double-focusing conditions; for certain angles of the two sectors focusing of higher order may be achieved. Double-focusing machines of this type have very good sensitivity, tolerate ions with large initial energy spreads, and achieve excellent resolution in mass, along with their good transmission characteristics (11).

Numerous machines of this type have been built, from the tiny instruments used in the Mars probes to machines with a magnetic sector radius of 240 cm that are capable of resolutions $M/\Delta M$ of $\sim 10^6$. They are widely available commercially.

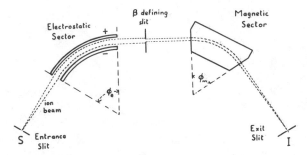

Figure 4.   Components of double-focusing mass analyzer (not to scale).

Not only does the electric sector allow ion sources with wide energy spreads, but it also allows analysis of ion energy distributions when it is used in the reverse geometry. The emergent analyzed ion beams also have excellent collimation, making them very suitable for beam experiments.

3.5   Perfect Double Focusing

Tandem electrostatic and magnetic analyzers can give very good focusing to second order in both velocity and direction, but some aberrations are still present. If both fields are made coincident, perfect double focusing can theoretically be achieved.

If ions are injected at a point into an electric field E uniform along the y-axis, and at right angles to a magnetic field B uniform along the z-axis, they will return to a focus at a second point displaced at right angles to the electric field along the x-axis by an amount

$$a = 2\pi n \frac{E}{B^2} \frac{M}{ze} \quad (n = \text{integer}) \tag{7}$$

and the time required to reach this point will be

$$t = \frac{2\pi n}{B} \frac{M}{ze}. \tag{8}$$

These expressions involve neither the initial velocity nor the initial direction of motion of the ions. The ions trace out either cycloidal or trochoidal paths, and the M/z scale is linear with ion-accelerating energy. In spite of these apparent advantages, commercial machines of this type (12,13) were not as successful scientifically as they deserved to be, and surprisingly few machines of this type have been built.

They require a relatively large magnet for their intended mass range, and it is difficult to achieve a truly linear electrostatic field in them, even with the use of corrector plates. Probably the main objection to their use is that both the ion source and the ion detector have to be immersed in the magnetic field, limiting ion-source design and making it difficult to use electron multipliers for ion detection.

The magnet is the most costly item in magnetic mass analyzers. It is heavy, and bulky power supplies are necessary to achieve stable high fields. For small, low-resolution machines, permanent magnets give excellent stability, but make it necessary to use voltage scanning to produce a mass spectrum. All magnetic mass analyzers are limited in the rate at which the mass scale may be scanned by the inductance of the field coils and eddy currents in the magnet yoke. This can be overcome to some extent by laminated construction of the magnet, which also helps in attaining field uniformity; in the author's experience it is much easier to achieve uniform magnetic properties in rolled soft iron plate than in cast ingots. When mass scanning is achieved by varying the energy of the ions, it can be very fast, but it usually introduces an unacceptable degree of mass discrimination.

3.6 Wien Filter

This much neglected type of mass analyzer is obtained when ions are injected at right angles to crossed uniform electrostatic and magnetic fields. The electric field is adjusted until the electric deflection just balances the magnetic deflection for a given M/z value, and ions of this value pass through undeflected. Although capable of focusing, the instrument is usually employed to accept parallel ion beams of large cross-sectional area, giving excellent transmission but low resolution.

In general, magnetic mass analyzers give good ion beam collimation, allow a very wide choice of ion-source design, and are very reliable, in that they can tolerate a moderate amount of contamination of electrode surfaces. They require pressures much below 10^{-5} torr to avoid scattering of ions in their relatively long flight paths.

3.7 Dynamic Mass Analyzers

The time-dependent properties of the equations of motion for ions have been used to develop a wide range of mass analyzers. With few exceptions, these do not employ focusing in energy or direction. Depending on the principle of operation, Blauth (14) distinguishes four main classes of these instruments:

1. Time of flight.
2. Energy balance.
3. Path stability.
4. Characteristic-frequency generator.

Each of these may make use of different types of ion motion:

a. Linear direct motion.
b. Linear periodic motion.
c. Circular periodic motion.

There are so many variations of these that reference is made below to only a few, commonly used examples (15).

3.7.1　Time of Flight (TOF) Instruments (Class 1a)

In TOF instruments, a pulse of ions is extracted from the ion source and accelerated to some energy eV; these ions travel down a field-free tube to a detector. They acquire velocities

$$v = \left(\frac{2zeV}{M}\right)^{1/2} = 1.39 \times 10^4 \left(\frac{Vz}{M}\right)^{1/2} \text{ m/sec}$$

and for a drift tube L meters long, the difference in transit time between ions of $M/z = M_1$ and M_2 is

$$\Delta t = 7.19 \times 10^{-5} \frac{L(M_1^{1/2} - M_2^{1/2})}{(Vz)^{1/2}} \text{ sec.} \tag{9}$$

For good resolution, the pulse of ions must be short, the path length L long, and the accelerating potential V low. Once all the ions in a pulse have reached the detector, the pulse can be repeated. Typical values used in the Bendix TOF instrument (16) are $L = 1$ m, pulse length $= 0.3$ μsec, and repetition rate $= 10$ kHz, giving a mass resolution of 300. Time of flight analyzers require an ion source with a low spread in ion initial energies, but can make effective use of ions formed in a diffuse region over a wide area. They are currently widely used for pulsed laser experiments because of this property. Revisions to the optics that provide a degree of energy focusing yield somewhat better mass resolution than that given by the basic instrument described here.

3.7.2　Energy Balance Machines (Class 2a)

A beam of ions traveling down a drift tube can be selected in M/z by causing them to interact with a set of grids or apertures that are switched periodically to either positive or negative polarity using either pulses (as in the Bennett mass analyzer) (17) or sine waves (as in the topatron) (18). Only ions of one M/z value are able to continue through to the detector without loss of energy. Because of the similarity in principle these are sometimes termed "traffic light" mass analyzers (Fig. 5). They are of low resolution and have only moderate transmission, but can be made

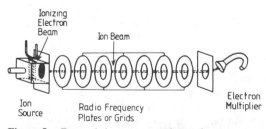

Figure 5.　Energy balance mass analyzer of topatron type.

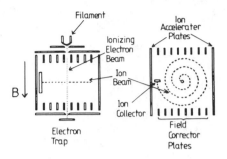

Figure 6. Horizontal and vertical cross sections of an omegatron, showing idealized electron and ion trajectories.

in extremely small sizes (e.g., 2–3 cm long). Spurious peaks may be generated at harmonic masses, but these can be suppressed by the use of retarding grids. These analyzers have the same source requirements as the TOF machines. Neither these nor the TOF machines are well suited for the production of collimated beams.

Two other important mass analyzers using the energy balance principle are the omegatron (19) (Class 2c), which uses a spiral path in crossed radio frequency (RF) electric and magnetic fields (Fig. 6), and the ion cyclotron resonance (ICR) mass analyzer (20) (Class 2c), which uses ion trajectories very similar to those of the omegatron, but with a superposed drift potential. Both of these have the drawback that the ionization region is either in or very close to the region where mass analysis takes place. Since they require very low pressures (10^{-7} torr) to operate successfully at high mass resolution, this imposes rather strict limitations on the sample pressures in the ionizing region. The ICR machine has long dwell times for ions of a selected mass, making it suitable for studying ion dissociation kinetics and energy-absorption processes of low probability. When used with a superconducting magnet and operated in the Fourier transform mode (21), these instruments have attained mass resolutions of up to 10^7. They also are very efficient, in that all ions formed are accepted for mass analysis. See Chapter 6 for a more detailed discussion of the operating principles of these spectrometers.

3.7.3 Path Stability Machines

These instruments use only RF fields to maintain ions of only one mass in a stable trajectory. They include the Farvitron (22) (Class 3b) and the quadrupole and monopole devices. The Farvitron is capable of only low mass resolution and is seldom used.

The quadrupole mass analyzer (23) (Class 3b) employs a combination of alternating and constant transverse electrostatic hyperbolic fields to achieve path stability for ions with only one value of M/z (Fig. 7). Mass scanning is achieved by varying either the frequency of the applied RF potential or the amplitudes of both RF and DC potentials. Hyperbolic electrodes are claimed to give the best performance, but satisfactory mass analysis can be achieved using cylindrical rods, or even rectangular rods or wires (24).

The quadrupole mass analyzer is relatively insensitive to initial ion energy; spreads of up to 30 eV can easily be accommodated. By insertion of an additional electrode

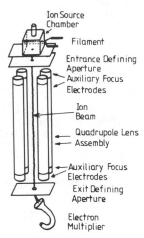

Ion Source
Chamber

Filament

Entrance Defining
Aperture

Auxiliary Focus
Electrodes

Ion
Beam

Quadrupole Lens
Assembly

Auxiliary Focus
Electrodes

Exit Defining
Aperture

Electron
Multiplier

Figure 7. Relative positions of components in a quadrupole mass analyzer.

at the exit of the quadrupole, ions from a Penning discharge ion source can be mass analyzed (25). Because of the strong focusing properties of the quadrupole lens, quadrupole mass analyzers can be operated at pressures up to 10^{-3} torr.

An excellent property of the quadrupole is that by a simple change in the ratio between the RF and DC potentials, resolution and transmission may be traded off against each other. Even a crudely constructed quadrupole can give a resolution of 150, but to obtain resolution up to 2000 considerable care has to be taken, including precise construction of the rod assembly and careful control of the applied potentials. It has also been claimed that the power-supply requirements are simplified if the quadrupole rods are operated with square waves (26).

The quadrupole gives poor collimation of the emergent ion beam, but can be operated with ions of very low energy (down to 0.5 eV). At these emergent energies it is essential to use auxiliary electrodes at the entrance and exit of the quadrupole lens to reduce the defocusing effect of the fringe electric fields (27). These electrodes also considerably reduce the mass discrimination effect evident in many quadrupoles and improve ion peak shapes.

The monopole mass analyzer (28) (Class 3b) acts as a one-quarter quadrupole. It does exhibit focusing, and has been claimed to be capable of resolutions up to 500. It is difficult to adjust, however, and is little used.

The quistor (29) (Class 3c) uses a three-dimensional hyperbolic field and acts as a most efficient trap for ions of given M/z value. Trapping times of up to 1 sec have beem claimed. It has recently found application as an extremely compact mass analyzer for use in gas chromatography (30).

4 MATCHING MASS ANALYZER TO ION SOURCE

In selecting a mass analyzer for a particular task, the duty cycle of the mass analyzer should be matched to that of the ion source, if possible. For example, it may be required to monitor the abundance of only one ion continuously, or to monitor the whole mass range periodically. There is no choice in TOF instruments: The whole mass range has to be scanned repetitively. Hence ions of a given M/z value are sampled in pulses occurring perhaps only 0.1% of the time. On the other hand, this instrument is ideal when used with a pulsed laser photoionization source, in which all the ions are created in bursts. In mass spectrographs a 100% duty cycle is achieved; until recently this was counterbalanced by the fact that the necessary photographic detection was a thousandfold less sensitive than electron multipliers are. Channel plates and semiconductor area detectors may remedy this situation in the future.

Magnetic mass analyzers give less discrimination against ions of higher mass than do quadrupoles and other TOF analyzers, and are better suited to work with ions of high molecular mass. These are also the instruments of choice for obtaining representative mass spectra without the need for sensitivity calibration of the mass scale.

It is also important to consider whether it is desired to use the mass-analyzed ions in some subsequent part of the apparatus where collimation or controlled low energy is essential, as in a beam experiment.

5 DESIGN OF ION OPTICAL SYSTEMS

Using the methods of electron and ion optics it is possible to design electrode configurations that allow retardation, acceleration, change of direction, or changes in cross-sectional area of ion beams. To achieve retardation without beam spreading requires a uniform field gradient. Focusing can be obtained with simple combinations of plane electrodes, cylinder lenses, or quadrupoles. It is preferable not to intercept the ion beam at any point with mesh electrodes, to avoid production of secondary electrons.

To obtain maximum interaction with a laser beam, or to correct for image aberrations in a mass analyzer, it may be necessary to modify the ion beam profile. The former objective can be achieved with a double LDC quadrupole lens, while hexapoles and octopoles serve the latter purpose when placed between the electric and magnetic sectors of double-focusing analyzers.

The AC-only quadrupole, with its strong focusing properties for a broad range of masses, can be used to "pipe" even the divergent beams of ions emerging from a quadrupole mass analyzer over distances of 30 cm or more. By constructing such a quadrupole lens, using fine tungsten wire to delineate the hyperbolic electrode surfaces, an almost transparent region can be obtained for ion fluorescence studies. Multipole structures (up to 12 poles have been used by the author) can also be used as ion pipes; these have the advantage that the radial potential variation within the assembly near the axis is much less than for the quadrupole.

Sector mass analyzers produce well-collimated ion beams, but with relatively high energies, and may require the use of retardation lenses to give ion energies suitable for collision experiments. Space-charge defocusing is aggravated as the ion energy is reduced, so that output ion beams are somewhat limited. See Chapter 8 for a discussion of this application.

6 CHARGED-PARTICLE TRAJECTORY CALCULATION

In designing and testing electron and ion optical systems, it is very desirable to determine the optimum operating potentials, to examine where the charged particles travel, and to study the effects of space charge. The properties of each element in an ion optical system can be described by a transfer matrix, and the matrix product of all such matrices, taken in the proper order, gives the transfer matrix for the whole system. This method has been used to design lens systems for quadrupole mass analyzers (31).

A suite of computer programs entitled SIMION has been developed in the author's laboratory (32) that enables a wide range of calculations concerning charged-particle trajectories to be carried out. To determine an ion trajectory it is necessary first to calculate the potential field created in the electrode system. Electrode structures, potentials, and field boundaries must first be delineated, after which the solution for the potential is obtained by an iterative method to any desired degree of accuracy.

To calculate trajectories, the charge, mass, initial velocity components (if any), and point of origin of a charged particle placed in the field are specified. The program then solves its equations of motion, and traces out the trajectory as a function of real or extended time on a display unit. A check is made at each unit of time of the energy and position of the particle and the potential of the field at that point. Any significant difference indicates that errors are accumulating in the calculation.

The program allows calculation of three dimensions, and can take advantage of any axial symmetry, mirror planes, and so on in the array to reduce calculation time. A later variant of this program allows trajectory calculations to be done for an electrode system immersed within a uniform magnetic field. Time-dependent

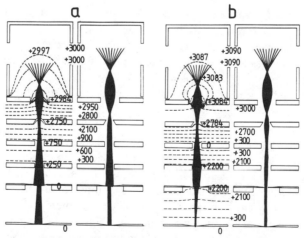

Figure 8. Use of trajectory program to determine potentials for optimum focusing in an ion gun of ions formed in different regions of the ion chamber: (*a*) first trial; (*b*) optimum potential distribution. Reproduced with permission from Ref. 39.

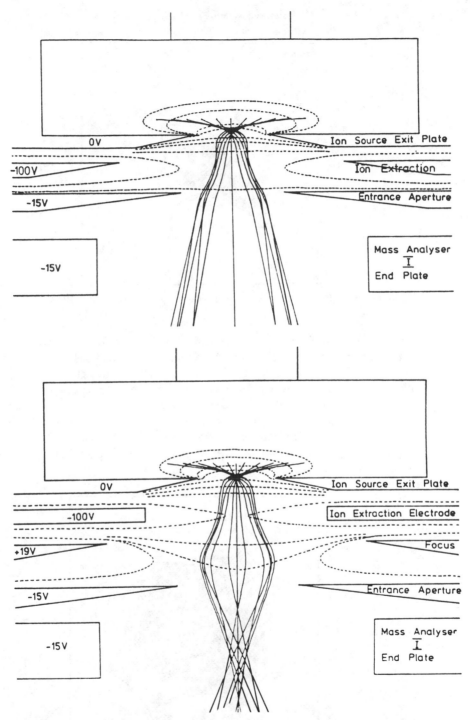

Figure 9. Improvement in focusing at entrance to quadrupole before (top) and after (bottom) addition of a retarding focus electrode.

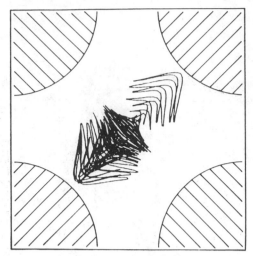

Figure 10. A stable ion trajectory in a quadrupole mass analyzer operating near the apex of the stability diagram.

variation of the potential field is permitted, allowing paths in an AC/DC quadrupole to be calculated. Some examples of these trajectory calculations are shown in Figs. 8–11 (33). The program can also calculate many adjacent trajectories automatically, allowing the charge density at various points in a beam to be calculated (Fig. 12) (34). These densities can then be used to modify the original potential to make some allowance for space charge. One interesting result of the application of this program was the discovery that the most likely fate of any ions formed is striking the cathode rather than entering the mass analyzer for many of the electron impact ion sources commonly in use (36).

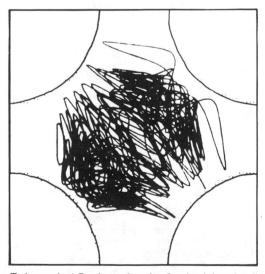

Figure 11. Trajectory in AC-only quadrupole of an ion injected well off the axis.

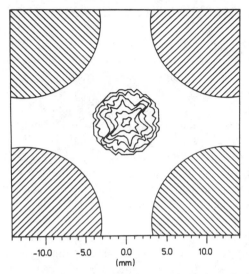

Figure 12.   Cross section of ion density in AC-only quadrupole determined by scanning repeated trajectories from the ion source. Reproduced with permission from Ref. 34.

It is the author's experience that ion optical systems work best if kept as simple as possible. Elaborate arrays of electrodes may seem to have theoretical advantages, but in practice these are usually lost.

7   ION-DETECTION SYSTEM

While not strictly part of the mass analyzer, the ion-detection system is to some extent limited by the choice of mass analyzer. Provided the detectors can be placed outside any magnetic fields present, electron multipliers of either discrete or continuous dynode types are generally used. The threshold for detection of ions with a good electron multiplier is of the order of one ion every 20 minutes, provided a pulse discriminator is used to eliminate thermal electrons generated in the multiplier. Where ion currents are relatively high, integration of the current output and amplification by a DC amplifier is best carried out with a discrete dynode multiplier. Channeltrons are well suited to particle counting for ion beams of $1-10^5$ ions/sec. They suffer from fatigue effects at higher count rates, with the gain sometimes falling irreversibly after large current bursts. Discrete dynode multipliers with easily replaceable dynodes offer gain and rise times almost as high as those of channeltrons, with much lower fatigue effects (35). When the detector has to be in a magnetic field, cither a magnetically focused multiplier of the Bendix type must be used (37) or the ions can be allowed to create secondary electrons, which then strike a phosphor and produce pulses of light that are detected with a photomultiplier (38,39). The complexity of this excellent system has discouraged most researchers from trying it.

8 MATERIALS FOR CONSTRUCTION

In building mass analyzers, it is essential to use metals that are nonmagnetic and insulating materials (such as synthetic ruby or machinable ceramics) that both are dimensionally stable and do not outgas. Stainless steels of types 302, 304, and 347 are suitable for both vacuum systems and electrodes. All joints should be inert-gas welded. Instruments are best constructed in segments, for easy disassembly and cleaning, and demountable joints should be sealed with Viton or, preferably, gold wire gaskets. Free-machining stainless steel 303 is convenient for fabricating ion chambers, but this material cannot be welded. Inconel is superior to all of the stainless steels, as it is easily welded and possesses superior dimensional stability. It also has the advantage that both the vacuum chamber walls and all electrodes can be electrolytically polished, greatly decreasing their effective surface area and giving excellent high-vacuum characteristics.

It is customary to achieve the vacua needed to operate mass analyzers by means of well-baffled diffusion pumps backed by rotary pumps; pressures are monitored by ionization, cold-cathode Penning and thermocouple, or Pirani gauges. The working fluid in the diffusion pumps should be of polyphenyl ether type. Silicone oils should not be used under any circumstances; residual vapors decompose in electron or ion beams, depositing insulating films that are difficult to remove on electrodes, which can then charge up.

When gas loads on the system are high, as in systems using molecular beams, or where light sources are connected directly to the mass analyzer, differential pumping using booster stages is almost mandatory. Pulsed gas inlet valves can give transient high sample-gas pressures while lessening speed requirements for the pumping system. As their cost decreases, turbomolecular pumps are becoming an increasingly attractive solution in situations where it is desired to minimize hydrocarbon background. Where gas loads are light, ion pumps backed at start up by sorption pumps give very clean background, but can sometimes suffer from annoying regurgitation effects when pumping argon.

REFERENCES AND NOTES

1. An excellent account is given of these principles by W. L. Fite, in *Proceedings of the 31st Annual Conference on Mass Spectrometry and Allied Topics*, American Society for Mass Spectrometry, Boston, MA (1983), pp. 19–41; also in *Applied Charged Particle Optics*, (Advances in Electronics and Electron Physics, Supplements 13A and 13B), A. Septier, Ed., Academic Press, New York (1980).

2. C. A. McDowell, Ed., *Mass Spectrometry*, McGraw-Hill, New York (1963).

3. F. A. White, *Mass Spectrometry in Science and Technology*, John Wiley & Sons, New York (1968).

4. J. Roboz, *Introduction to Mass Spectrometry*, Wiley-Interscience, New York (1968).

5. J. L. Kerwin, in *Mass Spectrometry*, C. A. McDowell, Ed., McGraw-Hill, New York (1963), Chapter 5.

6. A. L. Hughes and V. Rojansky, *Phys. Rev.* **34**, 284 (1929).

7. J. Mattauch and R. Herzog, *Z. Phys.* **89**, 786 (1934).

8. K. T. Bainbridge and E. B. Jordan, *Phys. Rev.* **50**, 282 (1936).

9. A. D. Nier and T. R. Roberts, *Phys. Rev.* **81**, 507 (1981).

10. H. Hintenberger and L. A. König, *Z. Naturforsch. A* **12**, 443 (1957).

11. R. P. Morgan, J. H. Beynon, R. H. Bateman, and B. N. Green, *Int. J. Mass Spectrom. Ion Phys.* **28**, 171 (1978).

12. Consolidated Electrodynamics Corporation mass spectrometer type CEC 61.620A; Varian Corporation mass spectrometer type M 66.

13. D. I. Ioanoviciu and D. Cuna, *Int. J. Mass Spectrom. Ion Phys.* **25**, 117 (1977).

14. E. W. Blauth, *Dynamic Mass Spectrometers*, Elsevier, Amsterdam (1966); see also Ref. 4.

15. Reviews of these instruments and their applications are to be found in the series *Dynamic Mass Spectrometry*, Heyden & Son, London; Vol. 1 (1970), D. Price and J. E. Williams, Eds.; Vols. 2 (1971) and 3 (1972), D. Price, Ed.; Vols. 4 (1975), 5 (1978), and 6 (1981), D. Price and J. F. J. Todd, Eds.

16. R. W. Kiser, *Introduction to Mass Spectrometry and Its Applications*, Prentice-Hall, Englewood Cliffs, New Jersey (1965), p. 68.

17. W. H. Bennett, *J. Appl. Phys.* **21**, 143 (1950).

18. P. F. Varadi, L. G. Sebestyen, and E. Rieger, *Vakuum Technik* **7**, 13, 46 (1958).

19. H. Sommer, H. A. Thomas, and J. A. Hipple, *Phys. Rev.* **82**, 697 (1951).

20. J. H. Futrell, in Ref. 15, Vol. 2, p. 97 (1971).

21. M. B. Comisarow and A. G. Marshall, *J. Chem. Phys.* **62**, 293 (1975); R. T. McIver, *Am. Lab.* **12**, 18 (1980).

22. W. Tretner, *Z. Angew. Phys.* **11**, 395 (1959).

23. W. Paul and H. Steinwedel, *Z. Naturforsch.* **8A**, 448 (1953); W. Paul and M. Raether, *Z. Phys.* **140**, 262 (1955); F. von Busch and W. Paul, *Z. Phys.* **164**, 581 (1961); D. R. Denison, *J. Vac. Sci. Technol.* **8**, 266 (1971); P. H. Dawson, Ed., *Quadrupole Mass Spectrometry and Its Applications*, Elsevier, Amsterdam (1976).

24. C. G. Pearce and D. Hasall, *Int. J. Mass Spectrom. Ion Phys.* **27**, 31 (1978); T. Hayashi and N. Sakudo, in *Proceedings of the International Conference on Mass Spectroscopy Kyoto, Japan* K. Ogata and T. Hayakawa, Eds. (1969), p. 263.

25. D. L. Swingler, *Int J. Mass Spectrom. Ion Phys.* **33**, 57 (1980); **54**, 341 (1983).

26. J. A. Richards, R. M. Huey, and J. Hiller, *Int. J. Mass Spectrom. Ion Phys.* **12**, 317 (1973); J. A. Richards, *Int. J. Mass Spectrom. Ion Phys.* **24**, 219 (1977).

27. W. H. Brubaker, *Adv. Mass Spectrom.* **4**, 293 (1968); W. H. Brubaker, *J. Vac. Sci. Technol.* **10**, 291 (1973); W. L. Fite, *Rev. Sci. Instrum.* **47**, 326 (1976); P. H. Dawson, *Int. J. Mass Spectrom. Ion Phys.* **25**, 375 (1977).

28. U. von Zahn, *Rev. Sci. Instrum.* **34**, 1 (1963).

29. R. E. Mather, R. M. Waldren, and J. F. J. Todd, in Ref. 15, Vol. 5, p. 71 (1978).

30. Finnigan MAT Corporation ITD (Ion Trap Detector).

31. M. Baril and A. Septier, *Rev. Phys. Appl.* **9**, 525 (1974).

32. D. C. McGilvery, Ph.D. thesis, La Trobe University, Melbourne, Australia (1978).

33. J. E. Delmore, *Int. J. Mass Spectrom. Ion Proc.* **56**, 151 (1984).

34. D. C. McGilvery and J. D. Morrison, *Int. J. Mass Spectrom. Ion Phys.* **28**, 81 (1978).

35. D. L. Swingler, *Int. J. Mass Spectrom. Ion Phys.* **27**, 367 (1978).

36. L. P. Johnson, J. D. Morrison, and A. L. Wahrhaftig, *Int. J. Mass Spectrom. Ion Phys.* **26**, 1 (1978).

37. W. C. Wiley and I. H. McLaren, *Rev. Sci. Instrum.* **26**, 1150 (1955).

38. N. R. Daly, *Rev. Sci. Instrum.* **31**, 264 (1960).

39. R. G. McLoughlin, Ph.D. thesis, La Trobe University, Melbourne, Australia (1980).

CHAPTER 6

Ion Cyclotron Resonance

JEAN H. FUTRELL
Department of Chemistry
University of Utah
Salt Lake City, Utah

The fundamental principle underlying ion cyclotron resonance (ICR) mass spectrometry was described in 1930 by Ernest O. Lawrence, the inventor of the cyclotron particle accelerator (1). It was well known then that charged particles are constrained to move in circular orbits with a constant period of revolution in a uniform magnetic field. Lawrence noticed this and pointed out that charged particles could be accelerated by being exposed to an oscillatory electric field having their cyclotron frequency.

The principle of ICR was first applied in mass spectroscopy by Hipple, Sommer, and Thomas in 1949 (2). Their instrument, called the omegatron, was designed to achieve the common objectives of early mass spectrometer development, namely, high resolving power and high abundance sensitivity. Although both these objectives were attained to a rather high degree in the omegatron, the several problems associated with the stringent stability requirements for the electronics and for very high vacuum (arising from the extremely long ion paths associated with this instrument) prevented its commercial development as a general-purpose mass spectrometer. However, it was developed into a high-sensitivity, low-resolution residual-gas analyzer.

The operating principles of the cyclotron resonance spectrometer are illustrated by Figs. 1–4. In Fig. 1, the motion of a charged particle in a magnetic field (in the z-direction) is illustrated. An ion possessing a velocity vector, $\mathbf{v}$, is continuously accelerated in a direction perpendicular to the magnetic-field lines by the Lorentz force

$$F = q\mathbf{v} \times \mathbf{B}, \qquad (1)$$

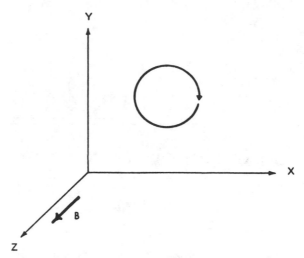

Figure 1. Motion of positively charged particles in a uniform magnetic field B. The radius is a function of ion velocity, but the frequency of circulation, the cyclotron resonance frequency, $\omega_c = 2\pi f_c = qB/m$, is not.

where q is the charge of the ion and $\mathbf{B}$ the magnetic-field intensity. Hence the ion will follow a circular trajectory, as illustrated in the figure, with a natural frequency of revolution, the cyclotron resonance frequency,

$$\omega_c = 2\pi f_c = \frac{qB}{m} \tag{2}$$

characteristic of the ion mass m. It is important to note from this expression that ω_c does not depend on ion velocity. Hence, although the radii of their paths will differ, ions with different kinetic energies will have the same cyclotron resonance frequency. The ICR spectrometer is used primarily for the study of low-energy ion–molecule reactions. For thermal-velocity ions, the radii are quite small (an argon ion, for example, in a 1.6-T magnetic field has a resonance frequency $f_c = \omega_c/2\pi$ of 614 kHz and an orbital radius of 0.01 cm when it is moving perpendicular to the magnetic field with 25 meV kinetic energy).

If a radio frequency (RF) field oscillating at the cyclotron frequency is introduced in the x–y plane (Fig. 2), the ion will absorb energy continuously from the field. In the absence of collisions the ion will execute the Archimedes sprial shown in Fig. 2, absorbing energy continuously from the irradiating field as it does so. If the irradiating oscillator is so constructed that it is extremely sensitive to changes in load, a small impedance change resulting from the absorption of energy by the resonant ion can readily be detected. The amplitude-limited oscillator detector originally developed for nuclear magnetic resonance applications is well suited to this application. This setup provides a useful alternative to the charge-collection method used in the omegatron.

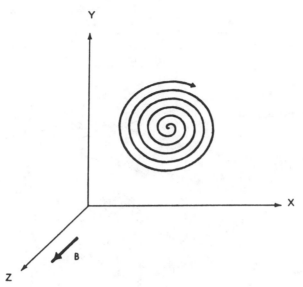

Figure 2. Motion of positively charged particles in a uniform magnetic field excited by a radio frequency field oscillating at their cyclotron resonance frequency.

This approach to ion detection and other important contributions to ICR technology were made by D. Wobschall and co-workers in the mid-1950s (3). Their apparatus used a 0.15-T solenoidal magnet to confine ions inside a long, cylindrical measurement chamber. Ions were produced by an electron gun that fired a pencil-thin beam of electrons along the axis of the cylinder. As the ions orbited about the axial magnetic-field lines, they were accelerated at their cyclotron frequency by a RF signal applied to a pair of electrodes on opposite sides of the cylinder. Power absorbed by the ions was detected as an imbalance signal in the RF bridge circuit. Wobschall reported being able to detect an average concentration of 1–10 ions/cm^3 (4).

1   DRIFT-CELL ICR SPECTROMETERS

Figure 3 illustrates schematically the ICR cell developed by Dr. Peter Llewellyn of Varian Associates in collaboration with Professor John Baldeschwieler at Stanford University in the late 1960s (5). A significant advantage of this design is the spatial separation of the ion source and resonance regions. Ions are created by electron impact as indicated by the dashed line leading from the filament to the electron collector, and are drifted through the reaction–analyzer region. The acceleration of an ion by the applied electric field interacts with the perpendicular acceleration of the magnetic field to cause the ion to execute cycloidal motion as illustrated in Fig. 4. All types of cycloids—prolate, oblate, and even straight lines—are executed by ions in ICR devices, depending on circumstances. The ion migrates at right angles

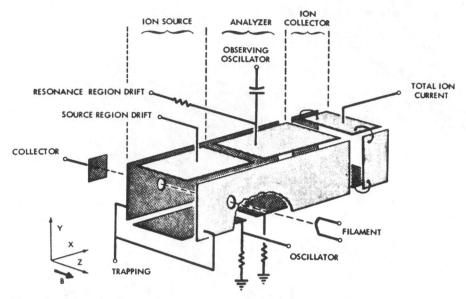

Figure 3. Schematic of Varian Syrotron ICR analyzer cell, which uses the types of motion described in Figs. 1, 2, and 4 for controlled studies of ion chemistry. Ions are created by electron impact in the source region, drifted by **E** × **B** motion (Fig. 4) into the reaction region, reacted and detected there by a marginal oscillator, and monitored by an electrometer attached to the total ion current plates (where the drift field is zero).

to both fields with a net drift velocity given by

$$v_D = \frac{E_s}{\mathbf{B}}, \tag{3}$$

where E_s is the static electric-field strength and **B** is the magnetic-field strength. Coupling this third class of ICR motion with those shown in Figs. 1 and 2 created an extremely versatile spectrometer.

One problem with the three-section drift cell is that the residence times of the ions in each region of the cell are difficult to estimate, largely because space-charge, ion-trapping, and field inhomogeneities significantly perturb ion trajectories from the simplified two-dimensional description given here. A second problem with this cell is that the marginal oscillator signal for a reactive ion is an average power absorption for a distribution of ions along the length of the analyzer region. For a simple $A^+ \longrightarrow B^+$ conversion reaction, the number of A^+ ions entering the analyzer region is greater than the number leaving. The opposite situation is true for the product ion B^+. This situation leads to complex integrals even for simple ion–molecule reaction schemes (6). A third problem is that while the ions are reacting in the analyzer region their translational energy is perturbed by the marginal oscillator. This is an undesirable situation, especially for measurement of thermal-energy rate constants.

Many of these problems were circumvented by the development of a four-region ICR drift cell by Futrell and co-workers (Fig. 5a) (7,8). Measurement of ion–molecule reaction rate constants is facilitated in this cell by the addition of a reaction region between the source region and the resonance region. Reactant ions are removed rapidly from the source region and drifted slowly through the reaction region. The residence time of ions in the reaction region is readily measured by applying an ion-ejection pulse to the reaction-region trapping plates at various time delays after ion formation. For time delays of the ejection pulse shorter than the source residence time, the total ion current (TIC) signal is unaffected. When the first ions arrive in the reaction region the TIC signal decreases, falling smoothly to zero in a time interval ΔT corresponding approximately to the ionizing pulse width. The TIC remains at zero until the ions have transited the reaction analyzer region, at which point it climbs back to the original value. The slope of the TIC rise and fall curves provides evidence for any spreading out in time of the ion beam that may result from stray potentials, space charge, and the like. Experiments conducted first with the same drift potentials in the reaction and analyzer regions and then with differing settings permit one to estimate the reaction and analyzer residence times quite accurately.

A modification of this constant-current operation of the four-section cell shown in Fig. 5a has been used to study the effect of ion kinetic energy on rates and mechanisms. A gating pulse on the electron gun creates monoenergetic ion bunches, which are then accelerated to a specific terminal ion energy by a RF oscillator connected to the source drift plates. The pulse sequence is illustrated schematically in Fig. 5b. Typically, the electron gun pulse lasts for 100 μsec, and is followed by activation of the irradiating RF oscillator at the cyclotron resonance frequency of the ion to be energized as soon as the electron energy pulse turns off. As indicated in Fig. 5b a RF pulse duration of 250 μsec is a convenient timing sequence for

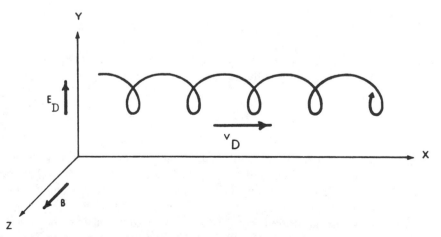

Figure 4. Motion of charged particles under the influence of crossed uniform electric and magnetic fields $E_D \times B$. This motion is used to control both ion migration and reaction time.

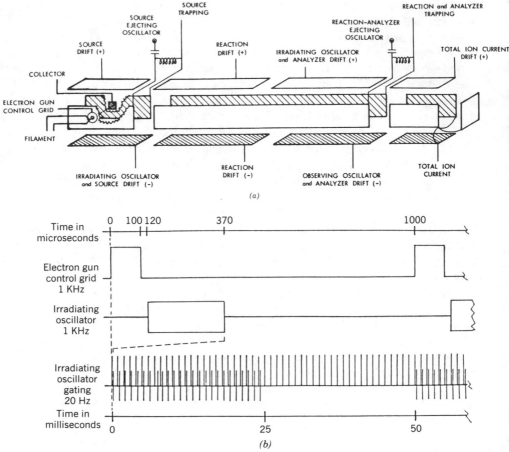

Figure 5. (*a*) Schematic of the four-section ICR cell developed at the University of Utah. For description, see text. (*b*) Pulse sequence used to generate ions of fixed translational energy.

these translational-energy experiments; the amplitude of the RF is varied to establish the terminal energy according to the equation

$$E_{\text{terminal}} = \frac{\varepsilon^2 q^2 t^2}{8m} + \frac{1}{2}(q\varepsilon v_0 t)\cos\gamma + \frac{1}{2}mv_0^2, \tag{4}$$

where ε is the amplitude of the RF field strength, q is the ion charge, t is the duration of the applied RF, m is the mass of the ion irradiated, v_0 is the initial velocity of the ion, and γ is the phase angle by which the ion leads $\varepsilon(t)$. The second and third terms, involving initial ion energy, are generally much smaller than the first term. The accelerated ions are drifted through the reaction zone and into the analyzer section as a nearly monochromatic group.

A variation of this technique can be used to display directly the effect of ion kinetic energy on reaction rates (8). As before, the basic pulse sequence of ion creation, excitation, reaction, and analysis at a frequency of the order of 1 kHz is followed. However, instead of detection of the ions by magnetic-field modulation, a second oscillator and pulse generator are used to modulate a diode switch at a lower frequency (typically 25 Hz) to which the phase-sensitive detector is referenced. The output of the irradiating RF oscillator is coupled to the ICR cell via this diode switch. Thus the RF excitation that energizes reactant ions is ''on'' only for alternate cycles. The limited oscillator detector output is fed to the phase-sensitive detector; since it is phase-locked to the lower ''on–off'' modulation frequency it amplifies the difference in signal level for selected product ions produced alternately by thermal-energy and translationally excited ions. Only the change in signal level is amplified and this technique is a very sensitive indicator of the dependence of rate constants on ion kinetic energy.

2   TRAPPED-CELL ICR SPECTROMETERS

In 1970 a one-region trapped-ion analyzer cell using a pulsed mode of operation was introduced by McIver (9). This cell is shown schematically in Fig. 6. The sequential pulse schemes described above for drift cells were all accommodated in a one-region cell. The experiment is initiated when a control grid pulse admits the electron beam to the cell and ions are produced. Either positive or negative ions are trapped in the cell, depending on the polarity of the DC trapping voltages. Ions move perpendicular to the electric-field lines and drift slowly back and forth within the cell (10); the end plates complete the electrostatic trap and prevent leakage of ions out of the cell. Ions are stored in the cell for a variable reaction time. The detect pulse is then triggered; a RF pulse from the marginal oscillator accelerates ions of a selected mass-to-charge ratio and measures the number of these ions in the cell. Finally, a pulse is applied that inverts the polarity of the DC voltage on one of the trapping plates and sweeps all ions out of the cell. The sequence is repeated at progressively longer delay times to trace out the abundances of various ions as a function of reaction time. For experiments that are not seriously limited by detection sensitivity this method is a convenient, direct method for measuring rate coefficients. Moreover, long time delays often permit the relaxation of translationally or internally excited ions prior to reaction.

3   FOURIER TRANSFORM ICR (FT-ICR)

The first FT-ICR experiments were performed in 1974 by Comisarow and Marshall (11). In this technique ions are stored inside a one-region trapped-ion analyzer cell and are constrained by the magnetic field to move in circular cyclotron orbits. Cyclotron resonance is induced by exposing the ions to an oscillating electric field just as in the ICR techniques already described. However, in FT-ICR the *image current* induced by the cyclotron motion of ions stored in the analyzer cell is detected.

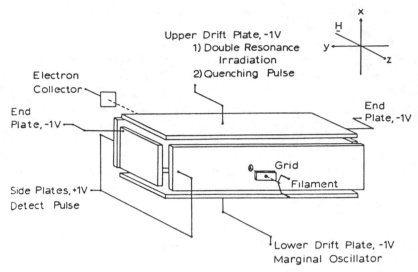

Figure 6. Schematic drawing of trapped-ion analyzer cell used in pulsed ion cyclotron double-resonance experiments. The DC potentails applied to each plate of the cell are suitable for trapping positive ions. Ions migrate through the cell, tracing out the equipotential lines on which they are formed. Reproduced from R. T. McIver, Jr., and R. C. Dunbar, *Int. J. Mass Spectrom. Ion Phys.* **7**, 471 (1971), with the kind permission of the Elsevier Publishing Company.

A resistor connected between the two electrodes corresponding to the drift plates in Fig. 6 senses an alternating current whose frequency is the same as the cyclotron frequency of the ions; its amplitude is proportional to the number of ions in the analyzer cell. The AC voltage developed across the resistor is amplified and detected. For a mixture of different ions the signal at the output of the amplifier is a composite transient signal with a frequency spectrum that is related to the mass spectrum by the cyclotron equation, Eq. (2).

When ions are formed by electron impact or ion–molecule reactions they have low translational energies and random phases in their cyclotron motion. The first step in FT-ICR is the coherent excitation of ions to large orbits to enhance the amplitude of the image currents. Accordingly, ions in the analyzer cell are excited by a RF pulse whose frequency is swept linearly over the range of cyclotron frequencies of interest in a few milliseconds (12). The image current induced by the accelerated ions is then amplified, digitized, and stored in a computer. Often several cycles are accumulated to improve the signal-to-noise ratio. The Fourier transform of the digitized transient signal is then computed to give the mass spectrum of the mixture.

Mass resolution in FT-ICR is linearly related to the duration of signal acquisition. Using a 4-MHz analog-to-digital converter a mass resolution of 10,000 at m/z 500 can be obtained with a 7-T magnet when the transient signal is digitized for about 60 msec. However, this requires storage of about 250,000 words of computer memory. In practice, therefore, the size of the available computer memory limits the maximum mass resolution to a lower value.

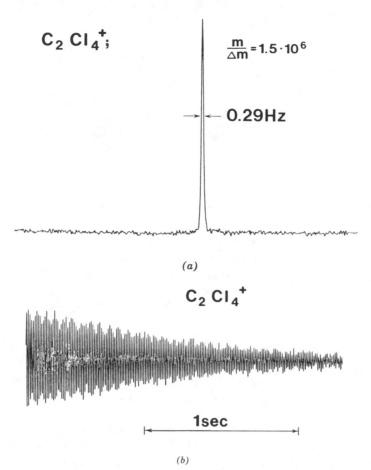

Figure 7. (a) Signal for m/z 166, from the spectrum of $C_2Cl_4^+$, taken at a resolving power of 1.5×10^6. (b) First part of the transient $C_2Cl_4^+$ signal shown in a. Both figures reprinted from Ref. 13a with the kind permission of Springer-Verlag.

This limitation is overcome (over a limited mass range) by narrow-band data acquisition. A suitable reference frequency, provided by a frequency synthesizer, is mixed electronically with the image current signals to produce a low-frequency signal that can be digitized at a much slower rate. Only a narrow range of the mass spectrum (about 10 amu) can be acquired at a time, but mass resolution greater than 1,500,000 has been reported using this technique, as shown in Fig. 7 (13). Typically, broad-band FT-ICR is used to obtain a medium-resolution spectrum over a wide mass range, and specific regions of interest are examined at high resolution using the narrow-band mode.

An alternative Fourier method developed by McIver (14,15) and co-workers, called Rapid Scan ICR, is illustrated in Fig. 8. Ions stored in the analyzer cell are subjected to an excitation RF signal that is scanned across the spectrum. When the frequency of the excitation signal matches the cyclotron frequency of an ion, a

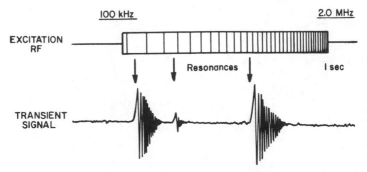

Figure 8. Schematic illustration of the technique of Rapid Scan ICR. The excitation signal (upper trace) is scanned slowly and detection (lower trace) is simultaneous with excitation.

transient signal is induced in the amplifier. The mass spectrum is recovered by cross-correlating the detected signal with a reference signal generated under identical conditions. Since the transient signals are separated in time, they can be digitized at a much slower rate than in conventional FT-ICR.

Demonstrated unique capabilities of FT-ICR spectrometers (14,15) include the following:

1. *High Mass Resolution.* To achieve high mass resolution in Fourier transform mass spectrometry (FT-MS) coherent cyclotron motion of the ions must be sustained for times of the order of a second. A resolution of 1.5×10^6 is demonstrated in Fig. 7. Such sustained coherent motion requires both efficient ion storage and high vacuum (in the 10^{-8}-torr range). Signals were accumulated for 6.8 sec to obtain the resolution shown. Since mass resolution is proportional to the magnetic-field strength and inversely proportional to mass, superconducting electromagnets are needed for high-performance FT-ICR.

2. *Accurate Mass Measurement.* Mass measurement accuracy of a few parts per million (ppm) is needed to determine the elemental composition of ions. Frequency is the primary parameter measured in FT-MS, and currently available electronics allow frequency—and hence mass—measurement accuracy of 0.8 ppm over a 2-amu mass range (16).

3. *Low-Pressure Ion–Molecule Reactions.* Long storage times (and the corollary high-vacuum requirements) imply that ion–molecule reactions can be investigated with sample partial pressures as low as 10^{-9} torr. Initial studies in this pressure regime addressed reaction mechanisms and optimum experimental parameters (17). Attempts to investigate the chemical ionization mass spectra of low-volatility samples have revealed that memory effects are a problem with this method; several hours are often required to bake out low-volatility compounds from the vacuum system, while ultraclean chambers serve as "pumps" for these classes of compounds.

4. *MS/MS Experiments.* Collision-induced dissociation (CID) is an attractive, widely employed mass spectrometric technique for ion structure determination and complex-mixture analysis (18–20). The technique consists of mass selecting a spe-

cific ion, colliding it with a target gas, and mass analyzing the daughter-ion fragments. Tandem mass spectrometers are customarily used, but similar experiments can be performed using FT-ICR (21). Parent ions are accelerated by a high-power RF pulse in the presence of a target gas such as helium. Daughter ions formed by CID are stored in the analyzer cell until they are detected by a time-delayed FT pulse. Consecutive CID processes may be investigated by accelerating a daughter ion and examining the spectrum of granddaughters, and so on.

5. *Multiphoton Ionization.* As noted in Chapter 5, scanning spectrometers are not well suited to experiments involving pulsed lasers. Since FT-ICR necessarily stores ions in the analyzer cell, it is highly suitable for many laser experiments. Laser multiphoton ionization (MPI) of polycyclic aromatic hydrocarbons has recently been reported (22). A complete mass spectrum is obtained for each laser pulse. Moreover, FT-ICR provides far higher mass resolution than do time of flight mass spectrometers, which have been used for most MPI studies. Because the open geometry of the analyzer cell samples ions over a volume of several cubic centimeters, laser alignment and focusing are not critical.

Despite these attractive features, FT-ICR is used much less widely than other mass spectrometric methods. This is at least in part because mass resolution and detection sensitivity are both degraded if the pressure in the analyzer cell reaches a few times 10^{-7} torr. This makes it difficult to interface a FT-ICR instrument with gas and liquid chromatographs, solids probes, and similar inlets involving substantial gas loads. Another problem is that the analyzer cell has a limited dynamic range, making it difficult to detect minor components. The highest ion concentration that will avoid space-change perturbation of detected signals is a few thousand ions, and storage times of the order of seconds seriously limit sensitivities for general applications.

An attractive new approach to these problems separates the ion-production and ion-detection regions as in drift-cell ICR but retains the unique features of FT-ICR. The FT-ICR is used as the product analyzer–detector stage of a tandem mass spectrometer (23). Positive and negative ions are produced in the ion source of a standard quadrupole mass filter; mass-selected ions are injected into a superconducting solenoid magnet, trapped in a modified ICR analyzer cell, and detected by FT methods.

It appears that tandem quadruple–FT-ICR sidesteps many of the above-mentioned problems. High mass resolution is possible, because the FT-MS analyzer cell is maintained at low pressure by differential pumping. Ion space-charge effects are avoided and the full dynamic range of the image current detector can be used for the ions of importance—the selected reactant ion and its reaction products.

REFERENCES

1. E. O. Lawrence and N. E. Edlefsen, *Science* **72**, 376 (1930).
2. J. A. Hipple, H. Sommer, and H. A. Thomas, *Phys. Rev.* **76**, 1877 (1949).
3. D. Wobschall, J. R. Graham, Jr., and D. Malone, *Phys. Rev.* **131**, 1565 (1963).

4. D. Wobschall, R. A. Flugge, and J. R. Graham, Jr., *J. Chem. Phys.* **47,** 4091 (1967).

5. J. L. Beauchamp, L. R. Anders, and J. D. Baldeschwieler, *J. Am. Chem. Soc.* **89,** 4569 (1967).

6. M. T. Bowers, D. D. Elleman, and J. L. Beauchamp, *J. Phys. Chem.* **72,** 3599 (1968).

7. R. P. Clow and J. H. Futrell, *Int. J. Mass Spectrom. Ion Phys.* **4,** 165 (1970).

8. J. H. Futrell, in *Dynamic Mass Spectrometry,* Vol. II, D. Price, Ed., Heydon and Son, London (1971), p. 97.

9. R. T. McIver, Jr., *Rev. Sci. Instrum.* **41,** 555 (1970).

10. T. E. Sharp, J. R. Eyler, and E. Li, *Int. J. Mass Spectrom. Ion Phys.* **9,** 421 (1972).

11. M. B. Comisarow and A. G. Marshall, *Chem. Phys. Lett.* **25,** 282 (1974).

12. M. B. Comisarow and A. G. Marshall, *Chem. Phys. Lett.* **26,** 489 (1974).

13(a). M. Allemann, Hp. Kellerhals, and K.-P. Wanczek, in *Ion Cyclotron Resonance Spectrometry II* (Lecture Notes in Chemistry Series, Vol. 31) H. Hartmann and K.-P. Wanczek, Eds., Springer-Verlag, Berlin (1982), p. 380; (b) M. Alleman, Hp. Kellerhals, and K.-P. Wanczek, *Chem. Phys. Lett.* **84,** 547 (1981).

14. R. T. McIver, Jr., *Am. Lab.* **12**(11), 19 (1980).

15. R. L. Hunter and R. T. McIver, Jr., in *Ion Cyclotron Resonance Spectrometry II* (Lecture Notes in Chemistry Series, Vol. 31), H. Hartmann and K.-P. Wanczek, Eds., Springer-Verlag, Berlin (1982), p. 464.

16. R. T. McIver, Jr., personal communication.

17. R. L. Hunter and R. T. McIver, Jr., *Anal. Chem.* **51,** 699 (1979).

18. R. W. Kondrat and R. G. Cooks, *Anal. Chem.* **50,** 81A (1978).

19. R. A. Yost and C. G. Enke, *Anal. Chem.* **51,** 1251A (1979); *J. Am. Chem. Soc.* **100,** 2274 (1978).

20. F. W. McLafferty, *Tandem Mass Spectrometry,* John Wiley & Sons, New York (1983).

21. R. B. Cody, R. C. Burnier, C. J. Cassady, and B. S. Freiser, *Anal. Chem.* **54,** 2225 (1982).

22. M. P. Irion, W. D. Bowers, R. L. Hunter, F. S. Rowland, and R. T. McIver, Jr., *Chem. Phys. Lett.* **93,** 375 (1982).

23. Donald F. Hunt, J. Shabanowitz, R. T. McIver, Jr., R. L. Hunter, and J. E. Syka, *Anal. Chem.* **57,** 765 (1985) and references cited therein.

PART 3

TECHNIQUES FOR STUDYING COLLISION PROCESSES

CHAPTER 7

Swarm Methods

WERNER LINDINGER

Institut für Experimentalphysik
Leopold-Franzens-Universität
Innsbruck, Austria

1 INTRODUCTION

Interactions between ions and neutrals at energies higher than a few electron volts are usually investigated by means of beam experiments. As indicated in Fig. 1, a beam of monoenergetic ions, A^+, with a total initial current $i(A^+)_0$ passes through a collision chamber containing the neutral target gas B at a number density [B]. Here the ion beam suffers losses due to collisions between A^+ and B along the way from $X = 0$ to $X = L$, leading to an attenuation of the ion current to a current

$$i(A^+)_L = i(A^+)_0 \, e^{-[B]\sigma L}, \tag{1}$$

where σ is the total cross section for the interaction process

$$A^+ + B \xrightarrow{\sigma} \text{Products.} \tag{2}$$

   Below a few electron volts of ion kinetic energy, and especially at thermal energies, it becomes extremely difficult to produce monoenergetic ion beams (contact potentials already can cause serious distortions of the ion energies) and thus only a few beam experiments can be done in this low-energy regime (1–5).

   However, this low-energy regime is the most important one for ion–neutral reactions occurring in natural and technical plasmas, which range in temperature from as low as a few tens of kelvins (K), as in interstellar dense and diffusive clouds, up to several thousand kelvins, as in the ionospheres of planets or gaseous discharges.

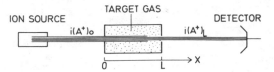

Figure 1. Beam experiment.

Thus a wide variety of so-called swarm experiments have been developed over the past three decades, such as the stationary (6,7) and flowing (8,9) afterglows and the drift tubes (10–12). They are especially suited for the investigation of ion–neutral reactions in the range from subthermal energies up to a few electron volts of center-of-mass kinetic energy (KE_{cm}) between the reactants.

In beam experiments the energy of the ions is adjusted with lenses or other devices before entry into the collision chamber, which contains only the reactant gas B. In contrast, in a swarm experiment, the reactant ions A^+ are introduced into a chamber or tube filled with a nonreactive buffer gas, such as He or Ar, at a typical pressure of a few tenths of a torr (a few tens of pascals). After a few collisions with the buffer gas the ions reach some kind of equilibrium velocity distribution, which they maintain as long as they stay in the chamber. A certain spatial distribution of the ions is also reached, the shape of which is not of interest at this point. If we now add traces of reactant gas B to the buffer gas, after many collisions with the buffer gas particles, an ion A^+ will undergo a potentially reactive collision with an atom or molecule B:

$$A^+ + B \overset{k}{\rightarrow} \text{Products.} \tag{3}$$

2 PRINCIPLES FOR OBTAINING RATE COEFFICIENTS IN SWARM EXPERIMENTS

Let us now pick out a small volume V_0 within the tube. The ion concentration in this volume is balanced by diffusion alone as long as no reactive losses occur. In the presence of reactant gas B (in amounts small enough that the ion diffusion is not changed significantly), however, an *additional* loss of A^+ ions occurs, because of a chemical reaction between A^+ and B. The rate of this reaction is

$$-\frac{d[A^+]}{dt} = k[A^+][B], \tag{4}$$

where k is the reaction rate coefficient, and brackets denote densities. Because $[B] \gg [A^+]$ and thus is essentially constant, integration gives

$$[A^+]_t = [A_0^+]e^{-[B]kt}. \tag{5}$$

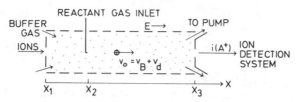

REACTANT GAS INLET

Figure 2.   Swarm experiment.

The time dependences of the ion concentration in a given volume observed without and with reactant gas B in the system yield different decay curves, from which the rate coefficient k is deduced. This was done successfully in the early investigations of the decay of stationary afterglow plasmas (6,7).

With respect to the measuring procedure, it is much easier if we can convert the above time dependence of the ion concentration $[A^+]$ into a function of a spatial coordinate. This is possible if we make the volume V within which the reaction is occurring move with a known velocity v_0 along an X-axis, so that dt is converted into dX/v_0.

This has been achieved in flow tubes [flowing afterglows (8,9) and selected-ion flow tubes (13)] and in various drift experiments (10–12). The general method is the following: Ions A^+ are introduced into a buffer gas or produced therein at a constant rate at point X_1 as indicated in Fig. 2. These ions are carried downstream by the flow of the buffer gas at a velocity v_B and/or by the influence of an electric field E, which contributes an ionic drift velocity v_d, so that the total bulk velocity of the ions is $v_0 = v_d + v_B$. On the way from the ion source toward X_3 the ions will spread out due to diffusion, but a fraction of them will reach the detection system through the orifice at X_3, yielding a current $i(A^+)_0$. If reactant gas B is added at the inlet port at X_2, the ion density $[A^+]$ and thus the ion current $i(A^+)_0$ leaving through the orifice at X_3 will decrease due to additional reactive losses on the way from X_2 to X_3 according to

$$i(A^+)_B = i(A^+)_0 \exp(-k[B]L/v_0). \qquad (6)$$

The current $i(A^+)_B$ will usually decrease exponentially with the addition of B, yielding straight lines in semilogarithmic plots (Fig. 3), the slopes of which are

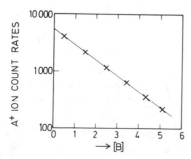

Figure 3.   Ion count rate of A^+ as a function of the density $[B]$ of the reactant gas added to the buffer.

proportional to k. It should be pointed out that these declines reflect *only* the reactive losses and are *not* influenced by diffusion. It is thus very simple to extract rate coefficients from such data, in contrast to the time-dependent decay curves of the ion densities obtained in stationary afterglow experiments, where the time constants for the decays contain both reactive *and* diffusive losses, making the extraction of rate coefficient data more complicated. This is the main reason why the investigation of ion–molecule reactions is now usually done using flow systems rather than stationary afterglows.

3   CLASSIFICATION OF SWARM EXPERIMENTS

A distinction among the different types of swarm experiments can best be made on the basis of the bulk velocity v_0 of the ions.

3.1   Stationary Afterglow

The value of v_0 is 0 for stationary afterglow experiments; thus the time dependences of the ion number densities have to be determined directly, either by continuous extraction and ion analysis or by *in situ* monitoring of the decaying plasma by microwave techniques (6,7).

   The time constant τ, which is related to the loss processes of the ions, is given by $1/\tau = D_a/\Lambda^2 + v$, where D_a is the ambipolar diffusion coefficient of the ions, Λ is the fundamental characteristic diffusion length of the plasma chamber, and v is the volume-destruction frequency of the ions and contains the rate coefficients for reactive loss processes of the ions under investigation. In the simplest case, where a single ion species is lost in a binary reaction as indicated in Eq. (3), v is proportional to $k[B]$.

3.2   Flow Tubes: The Flowing Afterglow (FA) and Selected-Ion Flow Tube (SIFT)

In the SIFT (13) and its precursor, the FA (8,9), the ions are carried through the reaction region solely by the flow of the buffer gas; thus $v_0 = v_B$ in this case, and no electric field is applied. In the early stages of development of flow tubes, in the FA (8,9), the ion source was placed upstream of the reaction region within the flow tube (Fig. 4). Discharges served as ion sources and produced fairly high densities of $He(2^3S)$ in the He buffer gas. These metastable atoms were carried downstream together with He^+ ions by the buffer gas. Interaction with impurities and recombination processes led to the production of excited particles that emitted photons and created a bright region beyond the discharge, reaching further downstream as v_B increased. This in fact appeared as a "flowing afterglow"; hence the name of the technique. Various ions could be created by addition of traces of particular gases to the buffer. N^+ and N_2^+ ions were most easily produced by addition of N_2 to the

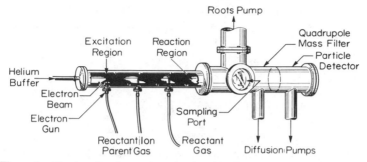

Figure 4. Flowing afterglow system, developed by Ferguson and co-workers (8,9).

He buffer. In this case all the He^+ produced in the discharge was neutralized within a small distance downstream of the source by the fast reaction

$$He^+ + N_2 \rightarrow N^+ + N + He \tag{7a}$$

$$\rightarrow N_2^+ + He, \tag{7b}$$

and all the $He(2^3S)$ was converted to N_2^+ by the Penning process

$$He(2^3S) + N_2 \rightarrow N_2^+ + He + e. \tag{8}$$

In this way only N^+ and N_2^+ ions were present at the entrance to the reaction region and addition of a reactant gas B at a reactant-gas inlet (see Fig. 4) caused a decline of the N^+ and N_2^+ ion signals detected by the ion-detection system, from which the rate coefficients k_1 and k_2 for the reactions

$$N^+ + B \xrightarrow{k_1} Products \tag{9}$$

and

$$N_2^+ + B \xrightarrow{k_2} Products \tag{10}$$

could be obtained by using Eq. (6) with $v_0 = v_B$. Besides the declines of the primary-ion signals, growth of the product-ion signals was also observed. However, the product distribution of a reaction could be determined with certainty only when only one primary ion was present. This usually was not the case, but various experimental changes and tricks allowed some progress in this respect. Discharges soon were replaced by electron impact ion sources. With the use of these, for example, the electron energy could be kept low enough that He metastables, but no He^+, were produced in the source, so that addition of N_2 beyond the electron gun resulted only in the production of N_2^+ ions by Penning ionization, as mentioned above. Thus in the reaction region only N_2^+ was present as a primary ion, so the product distribution of N_2^+ reactions with various neutrals could be obtained with great accuracy. But in many cases the situation was less fortunate, and often it was impossible to isolate a single type of ion. Furthermore, in nearly every case the parent gas of an ion, for example N_2 in the case of N_2^+, was present in the buffer gas, so that excited species such as $N_2^+(v > 0)$ were converted to the ground state by the invariably fast reaction (14)

$$M^+(v > 0) + M \rightarrow M + M^+(v = 0), \tag{11}$$

where M represents any molecule. Despite these and other drawbacks, the FA technique of Ferguson and his colleagues (8,9) has produced an enormous body of rate coefficient data (15), and allowed the calculation of variable-temperature, FA-produced ion thermodynamic data (9,16) such as heats of formation, proton affinities, and entropies (17).

Substantial progress in flow-tube techniques was achieved by Adams and Smith (13,18), who built the first selected-ion flow tube (SIFT). The main features of a SIFT are identical to those of the selected-ion flow drift tube (SIFDT), described below and shown in Fig. 5, with the exception that a simple metal tube replaces the drift rings indicated in the figure. Its main advantage over the conventional FA is its ion-production section, located in a vacuum chamber separated from the flow tube, which allows injection of a single, preselected type of ions into the flow region. Ions produced in an electron impact ion source or in a discharge pass through a quadrupole mass filter for selection of reactant ions. These ions are then introduced into the flow tube by means of a Venturi inlet, which reduces the backdiffusion of buffer gas into the ion-source region and at the same time causes the ions to be carried efficiently into the flow tube. Only one type of ion is present in the flow tube and thus it is possible in most cases to investigate the product distributions of the reactions. In addition, the ions are injected into a pure buffer gas, and because at room temperature both vibrational and electronic excitation of ions are quenched very inefficiently when He is the buffer gas, such excitation is retained as the ions travel down the flow tube. Therefore the influence of the internal excitation of ions on their reactivities and/or the quenching of this excitation by various neutrals can be investigated, thus opening up a wide new field of research.

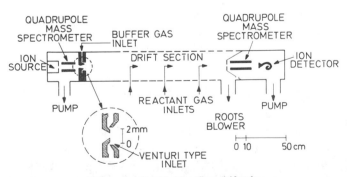

Figure 5. Selected-ion flow drift tube.

Moreover, the construction of a variable-temperature SIFT by Smith and Adams has led to an enormous increase in knowledge of the mechanisms of various types of reactions, including isotope-exchange (19) and three-body association (20) reactions.

3.3   Drift Tubes

3.3.1   Principles of Operation

The basic principles of operation of drift tubes are simple and similar to those for the FA or SIFT, and are best explained by considering the schematic drawing shown in Fig. 6. Ions are either created within the drift chamber or created in a separate ion source and introduced into the drift tube. Keeping the metal rings shown in Fig. 6 at appropriate potentials provides a homogeneous electric field E along the axis of the drift tube. The effect of E is to give the ions an average bulk velocity v_d in the direction of E.

3.3.1.1   Drift Velocities, Mobilities.   The drift velocity v_d is related to the ion mobility μ by

$$\mu = \frac{v_d}{E}. \tag{12}$$

Drift velocities are usually measured by pulsing a grid located upstream and monitoring the spectrum of the arrival times of the ions downstream. The average time t for the ions to travel the distance L from the grid to the end of the drift section gives the drift velocity: $v_d = L/t$. To facilitate comparison of ion mobilities determined at various pressures and over a small range of temperatures, the measured mobilities are usually converted to the reduced mobility μ_0, given by

$$\mu_0 = \frac{p}{760} \frac{273.16}{T} \mu, \tag{13}$$

the pressure p in torrs and temperature T in kelvins (21). The reduced mobility μ_0 is usually presented as a function of E/N, where N is the number density of neutrals.

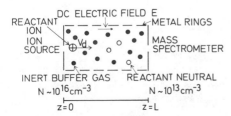

Figure 6.   Schematic representation of a drift experiment.

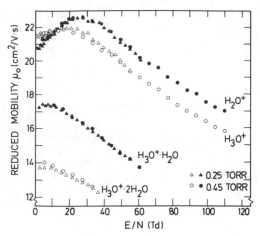

Figure 7. Reduced mobilities for various ions in a helium buffer.

The usual unit of E/N is the Townsend, $(1\ \mathrm{Td}\ =\ 10^{-17}\ \mathrm{volt} \cdot \mathrm{cm}^2)$. Typical results (22) appear in Fig. 7.

3.3.1.2 The Mean Kinetic Energy Between the Reactants, KE$_\mathrm{cm}$.

While v_d represents the bulk or swarm velocity of the ions relative to the buffer gas, the actual velocity distribution of the ions is a set of random velocities superimposed on a velocity in the direction of the electric field. Both parts contribute to the mean kinetic energy of the ions drifting in the buffer gas, KE$_\mathrm{ion}$. Wannier (23) calculated KE$_\mathrm{ion}$ from basic principles, taking into account the ion-induced dipole interaction between the drifting ions and the buffer-gas particles. His result was

$$\mathrm{KE}_\mathrm{ion}\ =\ \frac{3}{2}\,k_b T\ +\ \frac{mv_d^2}{2}\ +\ \frac{M_\mathrm{ion}}{2}v_d^2, \tag{14}$$

where m is the mass of the buffer-gas particle, M_ion the mass of the ion, and k_b is the Boltzmann constant. For reaction kinetics, it is not KE$_\mathrm{ion}$, but rather the mean relative center-of-mass kinetic energy between the ions and the neutral reactants (which are contained in small quantities in the buffer gas), KE$_\mathrm{cm}$, that is of importance. It is given (12) by

$$\mathrm{KE}_\mathrm{cm}\ =\ \frac{1}{2}\,\frac{M\cdot M_\mathrm{ion}}{M\ +\ M_\mathrm{ion}}\ (v_i^2\ +\ v_n^2), \tag{15}$$

where M is the mass of a reactant neutral particle and v_n^2 is the mean-square thermal velocity of the reactant neutrals (obtained from the relation $\frac{3}{2}k_b T\ =\ \frac{1}{2}Mv_n^2$), and v_i^2

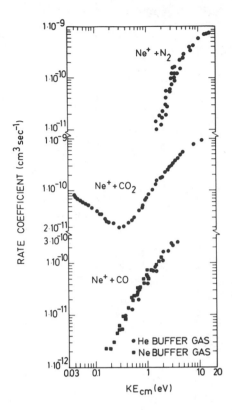

Figure 8. Dependences of rate coefficients of several ion–neutral reactions on KE_{cm}.

is the mean-square ion velocity (obtained from $\frac{1}{2}M_{ion}v_i^2 = KE_{ion}$). Combining the above equations leads to (12)

$$KE_{cm} = \frac{M}{M_{ion} + M}\left(KE_{ion} - \frac{3}{2}k_bT\right) + \frac{3}{2}k_bT. \qquad (16)$$

The value of KE_{cm} in turn is changed by varying E/N (which results in a change of v_d). The validity of Eq. (16) has been proved experimentally by Albritton and others (24–26) and it is now common for rate coefficients obtained in drift-tube experiments to be expressed as a function of KE_{cm}, derived using Eq. (16). Examples are shown in Fig. 8.

3.3.1.3 The Internal Energy of Drifting Ions. A fraction of the ions produced in an ion source may be in long-lived excited vibrational or electronic states. In many cases, this excitation is retained, surviving many collisions, even as the ions pass through the drift chamber, especially when He is used as the buffer gas and when E/N is *kept low* (27–32).

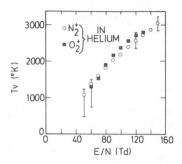

Figure 9. Dependences of vibrational temperatures (Tv) of O_2^+ and N_2^+ drifting in He on E/N.

On the other hand, heavier buffer gases, such as Ne, Ar, or Kr, are more efficient in quenching ionic excitation. However, ions entering the drift chamber in their ground state may become vibrationally excited (33–37), reaching an equilibrium distribution (when the drift tube is long) that depends on E/N. According to calculations of Viehland et al. (38–41), for molecular ions drifting in an inert buffer gas, the effective internal temperature T_{eff} attained by the ions is given by

$$\frac{3}{2} k_b T_{eff} = \left(1 + \frac{m_B}{M_{ion}} \xi\right)\left(\frac{3}{2} k_b T + \frac{1}{2} m v_d^2\right)(1 + \beta), \qquad (17)$$

where β is a small ($\ll 1$) correction term and ξ is a dimensionless ratio of collision integrals that characterizes the fractional energy loss due to inelastic collision. Relation (17) is *not* restricted to heavy buffer gases.

That in fact *at elevated E/N* even He becomes quite efficient in quenching and exciting vibrational excitation of molecular ions has been demonstrated experimentally (42). This consideration becomes important where the ion–buffer collision energy becomes comparable to or exceeds vibrational spacings in molecular ions. Quenching efficiency and vibrational-excitation efficiency are related by the principle of microscopic reversibility. Figure 9 shows measured vibrational temperatures of O_2^+ and N_2^+ in He as a function of E/N (42).

3.3.2 Quasistatic Selected-Ion Drift Tubes (SIDTs)

In contrast to the situation in a flow drift tube, in a quasistatic drift tube the buffer gas flows through the drift chamber slowly enough that $v_d \gg v_B$ even at low E/N. Thus v_B need not to be taken into account in the evaluation of the rate coefficients. The Innsbruck SIDT (43–45), shown in Fig 10, is a recently developed quasistatic drift tube having two distinct advantages:

1. It contains a highly developed ion-production section consisting of an exchangeable ion source (high- or low-pressure electron impact ion source). The ions enter an octopole storage section, where they are allowed to react with neutral gases to produce new ions in known excited states, or where they are quenched to the ground state. From the octopole the ions enter a quadrupole mass spectrometer,

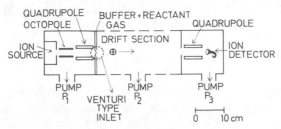

Figure 10. The Innsbruck quasistatic SIDT.

where the ion type to be investigated is selected and introduced into the drift chamber by means of a Venturi inlet. Buffer and reactant gases are also injected through the Venturi inlet. The whole drift chamber is filled with a homogeneous mixture of buffer and reactant gases, both of which are pumped away by a diffusion pump as indicated in Fig. 10. The density of the reactant gas B is obtained from the known flow rates F_{He} and F_B of the buffer and the reactant gas, respectively. Thus the rate coefficients are calculated from raw data of the type shown in Fig. 3 according to

$$k = \frac{F_{He}}{F_b} \frac{RT}{p} \frac{v_d}{L} \ln\left(\frac{i(A^+)_0}{i(A^+)_B}\right), \tag{18}$$

where p is the buffer-gas pressure in cgs units.

2.   Due to the small throughput of buffer gas (on the order of a few STP cm³/ sec), it is economically feasible to use the very desirable buffer gases Ne and Kr when the internal vibrational excitation of ions is the subject of investigation.

3.3.3   Flow Drift and Selected-Ion Flow Drift Tubes (SIFDTs)

The NOAA (National Oceanic and Atmospheric Administration laboratory at Boulder, CO) flow drift tube (12) is essentially a FA apparatus in which the metallic tube containing the reaction region is sliced into electrically isolated rings to which voltages can be applied to create a nearly homogeneous electric field E in the center of the tube. Thus the characteristics of a FA are combined with those of a drift tube. The SIFDT (Fig. 5) is a further step in the development of the flow drift tube: It is a SIFT containing a drift reaction section. Because the buffer-gas flow is so fast ($v_B \simeq 10^4$ cm/sec), its velocity has to be taken into account in the evaluation of the reaction rate coefficients. The modification of Eq. (18) for the calculation of rate coefficients in this case is

$$k = \frac{\pi a^2 v_0 v_B}{QL} \ln\left(\frac{i(A^+)_0}{i(A^+)_B}\right), \tag{19}$$

where Q is the rate of introduction of the reactant gas B at a reactant-inlet port, a is the flow-tube radius, and $v_0 = v_d + v_B$.

4 CURRENT AND FUTURE DEVELOPMENTS IN ION SWARM TECHNIQUES

4.1 Data Evaluation

We have not discussed in detail the evaluation of rate coefficient data in cases where appreciable numbers of excited ions are present with ground-state ions in the reaction region. These excited ions may react at rates different from those of ground-state ions, and in addition may be quenched to the ground state (or to a less excited state) in collisions with the reactant gas. Furthermore, molecular ions may be collisionally heated at high E/N and their internal-state distribution may change as a function of reactant-gas addition, with the consequence that the slope of the ion-decline curve changes smoothly with increasing reactant-gas flow. For all these and even more complicated cases, methods have been developed or are being developed to extract specific reaction rate coefficients for ground-state ions and ions in their various excited states.

4.2 Ion Sources

In the past much effort has been put into the technical development of ion-production systems, as was shown above, and such development is continuing in Innsbruck with the construction of a drift ion source in which a small drift tube acts as the ion source, allowing the creation of ions in different vibrational-state distributions. The Birmingham group of D. Smith has been developing various FA and microwave discharges. These developments will help workers to prepare a wide variety of ions, ranging from weakly bound cluster ions to excited multiply charged atomic ions. Future developments are expected to include the preparation of ions in specific excited states by means of laser light.

4.3 Product State Distribution

While the branching ratios of ionic reaction products can be obtained with high accuracy in selected ion systems, the actual state distributions of product ions have been obtained so far only in a few specific cases, by investigation of further reactions of the excited product ions. In some cases spectroscopic investigations have led to the identification of product-ion states. The use of lasers will almost certainly provide more insight into the product state distributions of ion–molecule reactions. Knowledge of the state distributions of both reactants and products will provide a better understanding of reaction mechanisms.

REFERENCES

1. G. Ochs and E. Teloy, *J. Chem. Phys.* **61,** 4930 (1974).
2. E. Teloy and D. Gerlich, *Chem. Phys.* **4,** 417 (1974).

3. B. Friedrich, W. Trafton, A. L. Rockwood, S. L. Howard, and J. H. Futrell, *J. Chem. Phys.* **80**, 2537 (1984).

4. Z. Herman and K. Birkinshaw, *Ber. Bunsen Ges. Phys. Chem.* **77**, 566 (1977).

5. P. Hierl, V. Pacák, and Z. Herman, *J. Chem. Phys.* **67**, 2678 (1977).

6. A. V. Phelps and S. C. Brown, *Phys. Rev.* **86**, 102 (1952).

7. T. D. Märk and H. J. Oskam, *Phys. Rev.* **A4**, 1445 (1971).

8. E. E. Ferguson, F. C. Fehsenfeld, and A. L. Schmeltekopf, in *Advances in Atomic and Molecular Physics*, Vol. 5, D. R. Bates and I. Estermann, Eds., Academic Press, New York (1969), pp. 1–56.

9. F. C. Fehsenfeld, E. E. Ferguson, and A. L. Schmeltekopf, *J. Chem. Phys.* **44**, 3022 (1966).

10. R. N. Varney, *Phys. Rev.* **174**, 165 (1968).

11. D. L. Albritton, T. M. Miller, D. W. Martin, and E. W. McDaniel, *Phys. Rev.* **171**, 94 (1969).

12. M. McFarland, D. L. Albritton, F. C. Fehsenfeld, E. E. Ferguson, and A. L. Schmeltekopf, *J. Chem. Phys.* **59**, 6620 (1973).

13. N. G. Adams and D. Smith, *Int. J. Mass Spectrom. Ion Phys.* **21**, 349 (1976).

14. W. Lindinger, F. Howorka, P. Lukac, S. Kuhn, H. Villinger, E. Alge, and H. Ramler, *Phys. Rev.* **A23**, 2319 (1981).

15. D. L. Albritton, *Atom. Data Nucl. Tables* **22**, 1 (1978).

16. W. Lindinger, F. C. Fehsenfeld, A. L. Schmeltekopf, and E. E. Ferguson, *J. Geophys. Res.* **79**, 4753 (1974).

17. F. C. Fehsenfeld, W. Lindinger, H. I. Schiff, R. S. Hemsworth, and D. K. Bohme, *J. Chem. Phys.* **64**, 4887 (1976).

18. N. G. Adams and D. Smith, *J. Phys.* **B9**, 1439 (1976).

19. D. Smith, N. G. Adams, and E. Alge, *Astrophys. J.* **263**, 123 (1982).

20. D. Smith and N. G. Adams, *Chem. Phys. Lett.* **54**, 535 (1978).

21. E. W. McDaniel and E. A. Mason, *The Mobility and Diffusion of Ions in Gases*, John Wiley & Sons, New York (1973).

22. I. Dotan, W. Lindinger, and D. L. Albritton, *J. Chem. Phys.* **67**, 5968 (1977).

23. G. H. Wannier, *Bell Syst. Tech. J.* **32**, 170 (1953).

24. D. L. Albritton, I. Dotan, W. Lindinger, M. McFarland, J. Tellinghuisen, and F. C. Fehsenfeld, *J. Chem. Phys.* **66**, 410 (1977).

25. L. A. Viehland and E. A. Mason, *J. Chem. Phys.* **66**, 422 (1977).

26. S. L. Lin and J. N. Bardsley, *J. Chem. Phys.* **66**, 435 (1977).

27. W. Lindinger, D. L. Albritton, M. McFarland, F. C. Fehsenfeld, A. L. Schmeltekopf, and E. E. Ferguson, *J. Chem. Phys.* **62**, 4101 (1975).

28. W. Dobler, H. Villinger, F. Howorka, and W. Lindinger, *Int. J. Mass Spectrom. Ion Phys.* **47**, 171 (1983).

29. H. Böhringer, M. Durup-Ferguson, D. W. Fahey, F. C. Fehsenfeld, and E. E. Ferguson, *J. Chem. Phys.* **79**, 4201 (1983).

30. W. Dobler, W. Federer, F. Howorka, W. Lindinger, M. Durup-Ferguson, and E. E. Ferguson, *J. Chem. Phys.* **79**, 1543 (1983).

31. J. Glosik, A. B. Rakshit, N. D. Twiddy, N. G. Adams, and D. Smith, *J. Phys.* **B11**, 3365 (1978).

32. D. Smith, N. G. Adams, E. Alge, H. Villinger, and W. Lindinger, *J. Phys.* **B13**, 2787 (1980).

33. M. Durup-Ferguson, H. Böhringer, D. W. Fahey, and E. E. Ferguson, *J. Chem. Phys.* **79**, 265 (1983).

34. E. Alge, H. Villinger, K. Peska, H. Ramler, H. Störi, and W. Lindinger, *J. Phys. (Paris)* **C7**, 83 (1979).

35. I. Dotan, F. C. Fehsenfeld, and D. L. Albritton, *J. Chem. Phys.* **68**, 5665 (1978).

36. E. Alge, H. Villinger, and W. Lindinger, *Plasma Chem. Plasma Proc.* **1**, 65 (1981).

37. E. E. Ferguson, in *Swarms of Ions and Electrons in Gases,* W. Lindinger, T. D. Märk, and F. Howorka, Eds., Springer-Verlag, Vienna, New York (1984), p. 126.

38. L. A. Viehland and E. A. Mason, *Ann. Phys. (N.Y.)* **110**, 287 (1978).

39. S. L. Liu, L. A. Viehland, and E. A. Mason, *Chem. Phys.* **37**, 411 (1979).

40. L. A. Viehland, S. L. Liu, and E. A. Mason, *Chem. Phys.* **54**, 341 (1981).

41. L. A. Viehland and D. W. Fahey, *J. Chem. Phys.* **78**, 435 (1983).

42. W. Federer, H. Ramler, H. Villinger, and W. Lindinger, *Phys. Rev. Lett.* **54**, 540 (1985).

43. H. Villinger, J. H. Futrell, A. Saxer, R. Richter, and W. Lindinger, *J. Chem. Phys.* **80**, 2543 (1984).

44. W. Lindinger and D. Smith, in *Reactions of Small Transient Species,* A. Fontijn and M. A. A. Clyne, Eds., Academic Press, London (1983), p. 387.

45. W. Lindinger, in *Swarms of Ions and Electrons in Gases,* W. Lindinger, T. D. Märk, and F. Howorka, Eds., Springer-Verlag, Vienna, New York (1984), p. 146.

CHAPTER 8

Beam Methods

JEAN H. FUTRELL

Department of Chemistry
University of Utah
Salt Lake City, Utah

1 INTRODUCTION

Since the beginning of systematic studies of ion–molecule reactions in the mid-1950s (1–3), a vast quantity of data on the rates of these reactions has become available. The general characteristics of these reactions are now well known and the field is a mature branch of reaction kinetics. Much of this information has been tabulated as rate constants for individual reactions, most of which are based on classical mass spectrometry measurements, flowing afterglow and drift-tube measurements, and ion cyclotron resonance studies. All these approaches to the study of ion–molecule reaction kinetics are reviewed elsewhere in this volume.

Microscopic measurements, such as those obtained by molecular beam methods, are very poorly suited for obtaining precise data on reaction rates. Rather, they are designed to obtain detailed information on reaction dynamics, from which in turn a fundamental understanding of the mechanisms by which reactants are converted to products may be obtained. These are the class of experiments closest to theory, in that they provide the experimental yardstick for potential-energy-surface calculations.

Figure 1 shows an "ideal" beam apparatus; consideration of its characteristics illustrates the kinds of information sought in beam experiments. In this hypothetical apparatus both reactants are selected with respect to mass, velocity, orientation, and internal state. All products are fully characterized as to identity, internal states, orientations, velocities, and angle of emergence from the scattering region.

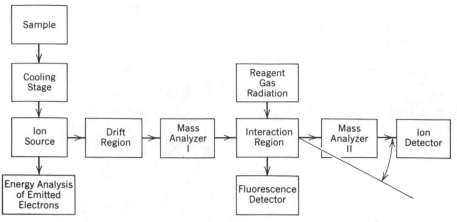

Figure 1. Block diagram showing components desirable in an "idealized" instrument for the detailed study of ion reactions. Reactants and products are fully characterized as to identity, internal states, velocities, and angle of emergence from the scattering region.

The information that can be obtained from experiments using this idealized apparatus includes the following:

1. Differential and total cross sections as a function of collision energy.
2. Differential and total cross sections as a function of the internal states and the orientation of the reactants.
3. Velocities and internal-energy distributions of the products.
4. Angular distribution of products.

Integration over velocities and angles and summation over quantum states provides the rate coefficient for the reaction in question.

Thus this microscopic approach gives exquisitely detailed information on how reactions actually occur. Unfortunately, the ultimate apparatus of Fig. 1 does not exist. If it did, the count rate of products with the complete specification invoked as a goal would certainly approach zero! Since experimentalists are obsessed with obtaining data, actual examples of single-beam (tandem mass) and crossed-beam spectrometers have fallen short of the goals embodied in Fig. 1.

2 TANDEM INSTRUMENTS

2.1 Sector Instruments

The first tandem spectrometer of in-line geometry constructed specifically for the study of ion–molecule reactions with momentum transfer is shown in Figure 2. It was constructed by C. F. Giese and W. B. Maier II at the University of Chicago (4). This apparatus essentially replaced the ion source of a sector mass spectrometer

→|1 in.|←

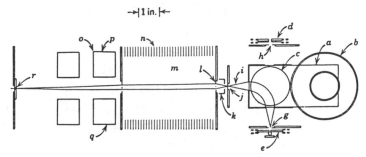

Figure 2. The single-beam apparatus of Giese and Maier (4), the first tandem spectrometer designed specifically for studying ion–molecule reaction kinetics. Reprinted from Ref. 4 with the kind permission of the American Institute of Physics. Components *a–j* are the ion gun (ion sources, electromagnet, and lenses), *k* the collision chamber, and *l–r* acceleration and focusing lenses of the ion product analyzer.

with a collision cell and an auxiliary ion gun consisting of an ion source, a 2.5-cm-radius permanent-magnet mass selector, and acceleration and deceleration lenses. The compact size resulted in quite high sensitivity, but low mass resolution precluded the investigation of any but the simplest ion–molecule reactions.

Figure 3 illustrates a more sophisticated apparatus designed and constructed by the author over the years 1962–1965 (5). This machine incorporated a moderate mass resolution ($m/\Delta m = 250$) ion gun, an energy selector that typically limited the energy spread of the ion beam to 0.3 eV, a specially designed deceleration lens that retarded the ions to near thermal energy (energy range 0.2–0.5 eV), and a high-resolution second-stage mass analyzer. Ion sources included electron impact, chemical ionization, and negative-ion sources. Still in use at Wright State University, Dayton, Ohio, this instrument's experimental capabilities have not been exceeded by other examples of this class of single-beam spectrometers.

These tandem machines have the objective of determining mechanisms (and branching ratios) for low-energy ion–molecule reactions, threshold energies for opening endothermic channels, and the kinetic-energy dependences of partial cross

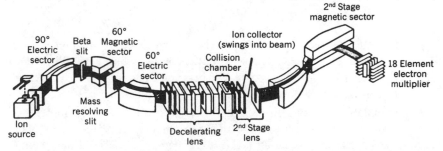

Figure 3. Schematic diagram of the ARL tandem mass spectrometer. Double-focusing mass spectrometers are used for both mass selection and product detection. The "slot" lens of Futrell and Miller (5) decelerates reactant ions to below 0.5 eV. Reprinted from Ref. 6 with the kind permission of Plenum Press.

sections. The quantity determined is a phenomenological cross section defined by the expression

$$Q = \frac{I_s}{I_p n l},\tag{1}$$

where I_s is the product-ion current, I_p the reactant-ion current, n the neutral-target number density, and l the reaction path length. Applying the usual thin-target expression for the rate constant, $k = Q v_i$, where v_i is the ion velocity, it is also possible to obtain a rate constant, via the expression

$$k = \frac{I_s}{I_p} \frac{v_i}{n l}.\tag{2}$$

(This rate coefficient is *not* a thermal value; it refers to the relative velocity of the reactants at a laboratory collision energy of the order of 0.3 eV.)

In the apparatus of Fig. 3 the number density n of the neutral target can be accurately measured and controlled (to within $\pm 10\%$) using an MKS Baratron capacitance manometer. The collision-chamber exit-slit image is refocused onto the collection slit of the analyzing mass spectrometer to improve the collection efficiency for product ions. The collision chamber is very nearly field free and well shielded to eliminate field penetration and ensure good energy definition for the collision. The reactant-ion beam current can be accurately measured with an electrometer at the collision chamber using the removable Faraday cage shown; the product-ion beam is measured with an electron multiplier operated in the pulse-counting mode. Although pulse counting minimizes multiplier discrimination, it is difficult to assess I_s/I_p absolutely. An unknown reaction rate constant is therefore referenced to a "standard" reaction having a "known" rate, both reactions being studied under identical conditions. For positive ions, the standard used is the reaction of CH_4^+ ions with CH_4 giving CH_5^+, for which the rate constant (1.2×10^{-9} cm³/molecule/sec) is relatively insensitive to collision energy below 0.5 eV (7).

All tandem mass spectrometric measurements have certain limitations intrinsic to the method. These include the following:

1.   The measured cross section is the average over the internal-energy distribution of the ion and the relative velocities of the colliding pair. When the dependence on relative velocity is appreciable between 0 and 0.6 eV, measured values of the rate constants may differ substantially from thermal rate coefficients.

2.   The internal-energy distribution is difficult to control. By operating the ion source at lower electron energies, electronic excitation can be minimized; but vibrational and rotational excitation levels are generally unspecified and unknown. High-pressure or drift-tube ion sources may be used to promote collisional relaxation of the reactant-ion internal energy, but sacrifice ion intensity.

3.   To specify the collision energy accurately, one must sometimes make large corrections for the thermal motion of the neutral target molecules. The resolution

[full width at half maximum (FWHM)] when the ion beam energy spread is negligible has been shown to be

$$\Delta E_{FWHM} = \left[11.1 \left(\frac{m}{M} + m \right) kTE_{cm} \right]^{1/2},$$ (3)

where m is the ion mass, M is the mass of the target gas, and E_{cm} is the nominal center-of-mass collision energy (8). For large m/M, a large uncertainty in collision energy results from so-called "Doppler broadening."

4. The most serious problem is that the collection efficiency of the tandem mass spectrometer (and of beam instruments in general) cannot be specified precisely. Collection efficiency depends on many factors, including the reaction mechanism and dynamics of the collision, collision-chamber and collector-slit dimensions (which together with dynamical features determine the sampling volume), product-analyzer acceptance angles, analyzer discrimination, and so forth. Obviously, it is expected that with in-line geometry and a field-free collision chamber, collection of forward-scattered products will be maximized. Crude estimates of the angular divergence of the product beam can be obtained by sweeping a lateral deflection field across two split focusing electrodes placed immediately following the chamber. Such experiments with the apparatus of Fig. 3 suggest that all product ions have essentially the same emergence-angle profiles at the low-energy limit (~0.3 eV, laboratory value). At higher collision energies this is not the case and large correction factors must be applied. Obviously, these methods do not provide complete answers to the questions regarding collection efficiency, and errors of as much as a factor of 2 could easily exist in published excitation functions (6).

2.2 Guided Beams

Many of these collection efficiency problems can be overcome by using radio frequency (RF) fields to prevent the escape of product ions from the reaction region. Such two-dimensional traps are used effectively in triple-quadrupole mass spectrometers (9) to contain product ions from collision-induced dissociation reactions. The first quadrupole mass filter is the ion selector, the second a RF-only reaction–trapping region, and the third the mass analyzer for product ions.

A more elaborate example of the same basic idea is shown in Fig. 4. This apparatus, developed at the University of Freiburg (West Germany) by Teloy, Gerlich, and Schlier (10), makes extensive use of RF fields to store and guide slow ions. By using such fields to establish either two- or three-dimensional minima of the effective potential, it is possible with this apparatus to obtain (1) ground-state primary ions with a narrow distribution of kinetic energies, (2) increased ion intensity at low collision energies, and (3) nearly 100% collection and detection efficiency for all primary and secondary ions.

In the storage ion source, ions are formed by electron impact and stored for a few milliseconds using an inhomogeneous RF field connected to alternate plates of the U-shaped, 10-plate source (4 in Fig. 4). The resulting effective potential is flat

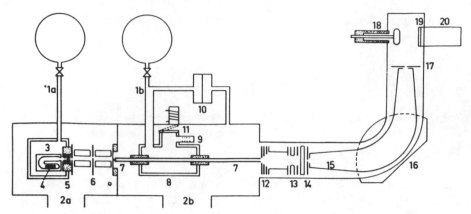

Figure 4. Simplified scheme of the guided-beam apparatus developed at the University of Freiburg and described by Teloy and Gerlich (10). 1a,b, Gas inlets for ion source and scattering chamber; 2a,b, pumping ports; 3, storage ion source; 4, cathode and electron beam; 5, electrodes for shaping and pulsing the ion beam; 6, RF mass and velocity filter (8 cm long); 7, RF octopole; 8, scattering chamber (effective length, 9.1 cm); 9, ionization gauge; 10, MKS Baratron capacitance manometer; 11, discharge valve for the scattering chamber, operated by an electromagnet; 12, 13, acceleration and shaping lenses; 14, entrance slit of the mass spectrometer (0.15 × 1.4 cm); 15, metal flight tube (−3 kV); 16, poles of the 90° magnet (radius, 12.5 cm); 17, exit slit (0.3 × 1.4 cm); 18, voltage feed-through and Daly detector dynode at −30 kV; 19, scintillator; 20, photomultiplier. Reprinted from Ref. 10 with the kind permission of North Holland Publishing Company.

near the middle of the channel and has steep repulsive walls at the borders of the enclosed volume. Ions are allowed to effuse through an outlet in the RF walls. Energy-relaxing collisions with the source gas during the long storage time generally lead to a narrow energy distribution of predominantly ground-state ions.

The combined mass and velocity filter—consisting of two RF quadrupoles arranged in series, each of which represents a lens with a mass-dependent focal length—separates out the desired ion species. Ions with the correct mass ratio *and* velocity are focused through a small tube at the exit into the adjacent collision region. The collimating entrance tube permits maximum angles of ±10° with respect to the axis.

The primary ions are accelerated or decelerated to the desired kinetic energy and injected into the reaction chamber. A RF octopole is used to guide the primary ions through the scattering chamber, and to constrain secondary ions and guide them to the detector mass spectrometer together with unreacted primary ions. It consists of eight parallel cylindrical rods equally spaced in an octagonal array. Alternate poles (or rods) are connected to opposite phases of a RF oscillator circuit. The effect is to set up a cylindrically symmetric effective potential that confines the ions to the inner ~1 cm of the guide. The effective potential of the octopole field is of the form

$$v_{\text{eff}} = \frac{e^2 V_0^{\text{RF}}}{4m\omega^2 R_0^2}\left(\frac{R}{R_0}\right)^6, \tag{4}$$

where V_0^{RF} and ω are the peak voltage and frequency of the RF, R is the distance from the octopole axis, R_0 is the radius of the octopole array, and m is the mass of the ion. This effective potential has very steep repulsive walls, close to the ideal-case analogue of a two-dimensional square well. Thus the RF trap behaves as an ion guide or pipe.

Although difficult to construct, the guided-beam spectrometer has several major advantages. The most important of these is that very few ions are lost in transit from the ion source to the reaction chamber. It is possible to achieve high trans-mission without resorting to typical lens systems, which tend to degrade energy resolution. The longitudinal velocity of the ions is not influenced by the RF field, their motion being described as a sequence of specular reflections from the cylindrical (RF) walls.

The average kinetic energy of the primary ions and its approximate distribution are determined in two independent ways: (1) by measuring the time of flight through the octopole, and (2) by retarding-potential analysis. Typical energy spreads are from 0.06 eV FWHM (Ne^+, Ar^+) to 0.2 eV FWHM (H^+, N^+). Thus, in most cases the distribution of relative collision energies is essentially determined by the thermal motion of the scattering gas, and the primary-ion energy spread can be neglected [Doppler width given by Eq. (3)].

Secondary ions formed in the scattering chamber (at 1–5×10^{-4} torr) will generally be directed forward because of center-of-mass motion. Backward-moving ions will—in most cases—be reflected at the injection end of the octopole by the potential of the collimating tube and the end plate of the mass filter. At low primary energies this potential may be too low and pulsed mode operation is useful. In this mode a burst of ions is injected to the octopole, and the potential of the mass filter is raised immediately after each pulse, so that all ions are reflected into the product-analyzer mass spectrometer. Ion transmission in this (magnetic) mass spectrometer is very high: It is estimated from indirect evidence that at least 95% of all ions of a given mass generate a counting pulse.

A mechanically less elaborate version of this spectrometer constructed in Freiburg more recently by Teloy and Gerlich (11) is shown in Fig. 5. A nominally identical apparatus has been constructed at Berkeley (9). No mass analysis is used for the first stage; the apparatus is therefore restricted to simple atomic and molecular systems for which photoionization specifies both internal quantum states and the identity of the reactant.

2.3 Merged-Beam Method

A peculiarity of ion beams is that they are relatively easy to produce at high energies, but become progressively harder to sustain in sufficient intensity as the energy decreases. Reasons for this lie in space-charge limitations and the difficulty of eliminating stray fields that can deflect and disturb a slow beam. Yet the thermal-to medium-energy range of 0.1–5 eV is of principal interest in chemical reactions.

Merging or confluent beams provide a logical means of reaching the low-collision-

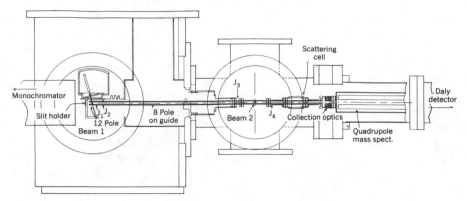

Figure 5. Schematic of the simplified guided-beam apparatus of Gerlich et al. (11). A photoionization source defines the identity and internal states of the reactant-ion beam, which is crossed by a neutral beam or reacted in a collision cell. Radio frequency-excited octopole rods confine the reactant and product ions, both of which are detected by a quadrupole mass filter. Reprinted from Ref. 11 with the kind permission of the American Institute of Physics.

energy region (11,12). Two superimposed beams of energies E_1 and E_2 having particle masses m_1 and m_2 will result in collision energies of

$$W = \left[\frac{m_1 m_2}{m_1 + m_2} \right] \left[\left(\frac{E_1}{m_2} \right)^{1/2} - \left(\frac{E_2}{m_2} \right)^{1/2} \right]^2. \tag{5}$$

As an example, consider the case where $m_1 = m_2$, $E_1 = 3000$ V, and $E_2 = 2951$ V. Then $W = 0.1$ eV. If the energy resolution of the two beams is 3 eV (FWHM), which is readily obtainable, the corresponding spread in W is only 0.017 eV. The actual energy resolution is determined more by the residual transverse velocities of the beams. In a typical case, this amounts to ~0.5 eV.

A diagram of the merged-beam apparatus constructed at the University of Minnesota, with most features drawn approximately to scale, is shown in Fig. 6 (12). Both reactant beams originate as ion beams from conventional ion sources and are extracted at the desired (high) laboratory kinetic energy. An Einzel lens focuses each beam after extraction, and quadrupole doublets are used for stigmatic correction and beam steering. The ion beam from source A is mass analyzed by a Wien filter, passed through a chicane to remove accompanying high-velocity neutrals, and neutralized by charge transfer. The ion beam from source B is mass selected and merged with the A beam in a 90° sector β-focusing magnet. After merging, the beams pass through an isolation valve and collimating apertures into a collision cell, where the ions may be accelerated or retarded relative to the fast neutrals. The collision cell is 14.6 cm in length, and corrections to the absolute total cross sections for end effects usually amount to ±3%. Because only the collision-cell voltage must be changed to vary the relative kinetic energy of collision, the energy dependence of the reaction cross section can be measured much more easily than in instruments that require retuning of the primary ion beam at each energy.

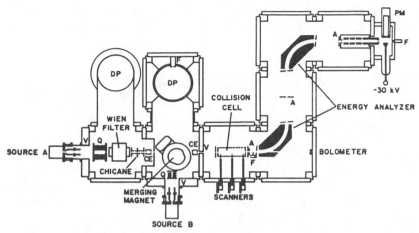

Figure 6. Schematic of merged-beam apparatus described by Gentry et al. (13). Ions from source A are mass selected by a Wien filter, steered through a chicane maze, neutralized in a charge-exchange (CE) chamber, and merged with an ion beam from source B. Adjustable apertures, A, diffusion pumps, DP, Faraday cup, F, quadrupole lenses, Q, isolation valves, V, and photomultiplier tube, PM, are designated in the figure. Final reactant relative energy is adjusted by the potential on the collision cell, and the products are identified by their velocities as measured by twin spherical energy analyzers. Reprinted with the kind permission of the American Institute of Physics.

Inside the collision cell three beam scanners are used for the measurement of the three-dimensional beam-overlap integral required for obtaining absolute reaction cross sections. The absolute ion beam intensity is measured at one of several Faraday cups, while the neutral beam intensity is monitored by secondary electron emission or by a thermistor bolometer. Relative kinetic energies of collision of ~0.005 eV have been achieved with this apparatus.

Since the reactants and products have approximately the same laboratory velocity in a merged-beam experiment, they can be separated according to mass by energy analysis. The energy analyzer consists of two 15.24-cm-radius 90° spherical electrostatic sectors in tandem, the first of which has its object point at the center of the collision cell. Energy resolutions of 0.04–1% are obtained without prior retardation. The high energy resolution in both the primary beams and the product analyzer has enabled Gentry's group to detect asymmetry in the forward–backward product angular distribution ratio at initial relative kinetic energies as low as 0.010 eV. The detector is a scintillation counter of the Daly type, with an intrinsic noise of about ~5 counts/minute and essentially unit detection efficiency.

The performance of the apparatus depends highly on the system being studied. Ion and neutral beam intensities in the collision cell are typically equivalent to 10^{-6} and 10^{-7} A, respectively, for the major constituents of the beam before mass analysis. The resolution in initial relative kinetic energy typically varies from ~0.005 eV at a nominally zero kinetic energy up to ~0.10 eV at an energy of 10 eV. Typical product signals have ranged from 1 to 100 counts/sec, while the noise from all sources has typically been less than 1 count/sec. The precision of an absolute cross section determination is about 5%, with the absolute accuracy somewhat lower. The major uncertainty in the measurement is the possibility of there being excited reactants in both the ion and neutral beams.

3 CROSSED BEAMS

Several characteristic features of ion–molecule reactions make both the angle and the velocity of the products important experimental data. Experiments are generally done at higher energy than are neutral experiments, and both forward- and backward-scattered products may be found in a small range of laboratory angles. In the absence of trapping fields, products scattered at large center-of-mass angles (which dominate the total cross section) may not be collected at all in single-beam experiments. The random velocity of the neutrals in collision-chamber experiments introduces substantial energy spreads in the center-of-mass collision frame [see Eq. (3) and related discussion]. Most importantly, crossed-beam apparatuses approach the idealized apparatus of Fig. 2 more closely than do other devices and provides the only method for measuring cross sections that are differential in both energy and angle. This information is directly related to features of the potential-energy hypersurface(s) governing the reaction dynamics of the reactive encounter.

The first apparatus to provide simultaneous energy and angular analysis, shown in Fig. 7, was constructed by Herman and Wolfgang (14). This machine, named EVA by its inventors, produced a mass-selected beam ($m/\Delta m \approx 50$) of 0.5 eV FWHM energy spread and 2° FWHM angular spread. This was crossed at 90° by a thermal-effusive-nozzle neutral beam collimated to ~15° FWHM. The critical feature of the apparatus was its ability to reach low energies (~3 eV, laboratory)

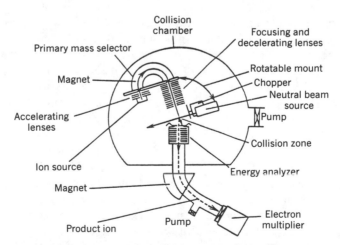

Figure 7. Schematic of the first crossed-beam apparatus with velocity and angular resolution for the study of ion–molecule reaction kinetics. Reprinted from Ref. 14 with the kind permission of the American Institute of Physics.

with sufficient resolution to allow the dynamics of several important reactions to be mapped. This apparatus permitted the first detailed studies of ion–molecule reaction dynamics.

One of the most elaborate crossed-beam apparatuses constructed for the study of ion–molecule reaction dynamics is shown schematically in Fig. 8 (15). The ion source operating pressures from 10^{-4} to 5 torr, permitting the generation of reactant ions by electron impact, charge-transfer, or ion–molecule reactions (e.g., chemical ionization conditions, generating relaxed species similar to those studied in the drift tubes described in Chapter 7). It operates near ground potential, while the flight tube of the ion-gun mass analyzer floats at accelerating voltage, typically 750 V. An exponential lens decelerates the ions to the final ion energy at the reaction region. The flux of reactant ions through the final, 1-mm-diameter defining aperture approaches the theoretical space-charge limit (15,16) at energies below 10 eV, and useful ion beams are generated at 0.5 eV and above. The neutral beam utilizes an adjustable supersonic nozzle and skimmer to generate a highly directional, nearly monoenergetic flux of neutrals at 90° to the reactant-ion beam. The neutral beam is chopped for signal-averaging purposes, and signal counts are added to a multi-channel averager when the beam flag is open, and subtracted when it is closed (50% duty cycle). A cylindrical sector, parallel-plate energy analyzer is used to measure both reactant- and product-ion velocity profiles with high resolution (typically 0.1–0.3 eV). In combination with an auxiliary electron gun that crosses the neutral beam, it measures also the neutral beam velocity distribution. The energy analyzer and a quadrupole mass filter rotate through 110° in the laboratory frame, thereby subtending all product center-of-mass scattering angles.

Data are accumulated by the multichannel analyzer at a selected angle and collision energy for a period ranging from minutes to as long as 50 hours, depending

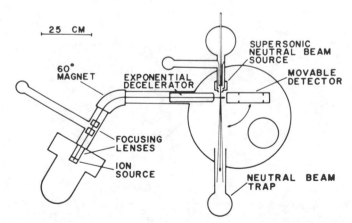

Figure 8. Schematic of the Utah crossed-beam apparatus (15). A differentially pumped ion source is used to generate reactant ions, which are mass selected by a 60° magnetic sector, decelerated to energies below 1 eV by an exponential lens, and crossed by a supersonic jet at 90°. Product ions are energy analyzed by a 90° cylindrical analyzer and mass analyzed by a quadrupole mass filter.

on the cross section for the reaction investigated, and then transferred to an on-line IBM-9000 laboratory computer for processing. Examples of data obtained with this apparatus are presented and discussed in Chapter 10.

The key features of this apparatus are summarized as follows: (1) A differentially pumped, high-pressure ion source generates reactant ions by electron impact, ion–molecule, or charge-transfer reactions; in many cases state-selected ions (e.g., fully relaxed ground-state ions) are created by suitable choice of the ion-source operating parameters. (2) The combination of a supersonic neutral beam and a well-defined ion beam accurately defines the scattering-volume and reactant-velocity vectors. (3) High-resolution energy analysis of the reactant and product ions is combined with good angular resolution. (4) Stable operation at reactant ion energies from below 1 eV to 20 eV (laboratory frame) is achieved. Collectively, these capabilities render the apparatus capable of performing "translational spectroscopy" measurements for simple systems. Provided the internal-energy states of the products are sufficiently limited and widely spaced, the measurement of translational-energy disposal will define the energy-conversion processes that occur. These capabilities are demonstrated in the results presented in Chapter 10. Similar machines incorporating most of the features of the Utah apparatus just described have been built at the University of California (Berkeley) and the University of Rochester.

The principal significance of information obtained from crossed-beam investigations is that it allows delineation of detailed dynamical mechanisms. This involves

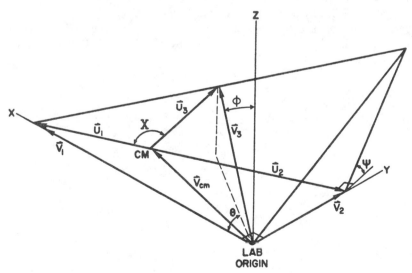

Figure 9. Three-dimensional Newton diagram. V_1 and V_2 are the laboratory-frame velocity vectors of the ion and neutral reactant, respectively; V_{cm} is the velocity of the CM; U_1 and U_2 are the velocity vectors of the ion and neutral reactant, respectively, relative to the CM; V_3 and U_3 are the velocity vectors for the product ion relative to the laboratory frame and the CM, respectively; θ and ϕ are the polar angles which define the direction of V_3 in the laboratory frame. X is the angle between U_1 and U_3; and ψ is the angle between the plane defined by U_1 and U_3 and the plane defined by V_1 and V_2.

the presentation of data as scattering (Newton) diagrams. The transformation relations involved in the presentation of data in center-of-mass velocity space are derived and the significance of data presented in this format is discussed in Chapter 2. We summarize these relationships here for the reactive scattering reaction

$$M_1^+ + M_2 \rightarrow M_3^+ + M_4. \tag{6}$$

Figure 9 shows a velocity-vector (Newton) diagram for the case in which the initial ion velocity, V_1, is perpendicular to the neutral velocity, V_2, for the general case of out-of-plane scattering. The product ion M_3^+ is detected at laboratory scattering angles θ and ϕ with laboratory velocity V_3. Center-of-mass (CM) velocity vectors U_1, U_2, and U_3 and the CM scattering angle χ are shown in the figure along with the Cartesian coordinate frame and the interplane angle ψ. Since no new information is gained by out-of-plane detection, most beam experiments are conducted in the XY-plane defined by the reactant beams. This special case is described in Fig. 10. Figure 10a shows the velocity vectors of all four particles before and after collision, Fig. 10b is the two-dimensional analogue of Fig. 9, and Fig. 10c is the Cartesian representation conventionally referred to as the CM scattering diagram.

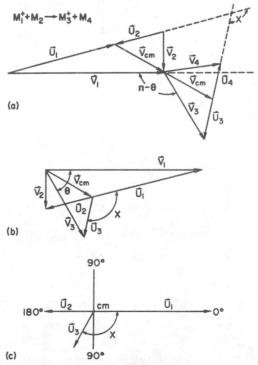

Figure 10. Two-dimensional Newton diagrams in the plane defined by the centerlines of the reactant beams. All symbols are as defined in Fig. 9. V_4 and U_4 are the velocity vectors for the neutral product relative to the laboratory frame and the CM, respectively. (a) Velocity vectors of all four particles before and after collision. (b) Two-dimensional analogue of Fig. 9. (c) Cartesian representation.

The velocity of the CM is given by

$$\mathbf{V}_{cm} = \frac{m_1}{M}\,\mathbf{V}_1 + \frac{m_2}{M}\,\mathbf{V}_2, \tag{7}$$

where

$$M = m_1 + m_2 = m_3 + m_4 \tag{8}$$

and the m_i are the masses of the indicated particles. The velocity of species M_i relative to the CM is given by

$$\mathbf{U}_i = \mathbf{V}_i - \mathbf{V}_{cm}, \qquad i = 1,2,3,4. \tag{9}$$

In Figs. 9 and 10 the X-axis is the direction of the reactant-ion beam and the Y-axis is along the reactant-neutral beam. We implicitly assume that both reactant beams can be represented by such a well-defined vector.

Coordinate transformations for the experimental data (product kinetic energy T and laboratory scattering angles) are summarized in Table 1. Cartesian coordinates

TABLE 1 Summary of Transformations

A. Coordinate Transformations

Laboratory Velocity from Laboratory Energy:[a]

$$V = \left(\frac{2eT}{m}\right)^{1/2}$$

Laboratory Cartesian Velocity from Laboratory Polar Velocity:[b]

$$V_x = V \sin\phi \cos\theta$$

$$V_y = V \sin\phi \sin\theta$$

$$V_z = V \cos\phi$$

CM Cartesian Velocity from Laboratory Cartesian Velocity:[c]

$$U_x = (V_x - W_x)\cos\gamma - (V_y - W_y)\sin\gamma$$

$$U_y = (V_x - W_x)\sin\gamma + (V_y - W_y)\cos\gamma$$

$$U_z = V_z$$

CM Polar Velocity from CM Cartesian Velocity:

$$U = (U_x^2 + U_y^2 + U_z^2)^{1/2}$$

$$X = \tan^{-1}\left[\frac{(U_y^2 + U_z^2)^{1/2}}{U_x}\right]$$

$$\psi = \tan^{-1}\left(\frac{U_z}{U_y}\right)$$

TABLE 1 *(Continued)*

Relative Energy from CM Velocity:[d]

$$T_r = \frac{m_3}{2} U_3^2 \left(1 + \frac{m_3}{m_4} \right)$$

B. Transformations of Intensities[e]

Laboratory Velocity from Laboratory Energy:

$$I(V,\theta,\phi) = mVI(T,\theta,\phi)$$

Laboratory Cartesian Velocity from Laboratory Polar Velocity:

$$I(V_x,V_y,V_z) = \frac{1}{V^2 \sin \phi} I(V,\theta,\phi) = \frac{m}{V \sin \phi} I(T,\theta,\phi)$$

CM Cartesian Velocity from Laboratory Cartesian Velocity:

$$I(U_x,U_y,U_z) = I(V_x,V_y,V_z)$$

CM Polar Velocity from CM Cartesian Velocity:

$$I(U,\chi,\psi) = U^2 \sin \chi \, I(U_x,U_y,U_z)$$

$$= \left(\frac{U}{V} \right)^2 \left(\frac{\sin \chi}{\sin \phi} \right) I(V,\theta,\phi) = \frac{U^2}{V} \frac{m \sin \chi}{\sin \phi} I(T,\theta,\phi)$$

Relative Energy From CM Velocity:

$$I(T_r,\chi,\psi) = \frac{1}{U m_3 (1 + m_3/m_4)} I(U,\chi,\psi)$$

---

[a]T is the measured laboratory translational energy (V), e the charge (coulomb), m the mass (kg), and V the velocity (m/sec) of the product. The velocity coordinates are defined in Figs. 9 and 10. The subscripts on the velocities denoting the detected product (e.g., 3 in V_3) are suppressed in this table.

[b]The laboratory Cartesian coordinate system (x,y,z) is defined in Fig. 9.

[c]The CM Cartesian coordinate system has its origin at the CM and the positive x'-axis is in the direction of U_1. W is the velocity of the CM, and $\gamma = \tan^{-1}(V_2/V_1)$.

[d]T_r is the relative translational energy of the two products, m_3 is the mass of the detected product with the relative velocity U, and m_4 is the mass of the other product.

[e]The crossed-beam experiments measure $I(T,\theta,\phi)$, the relative number of particles of a particular mass scattered with translational energy between T and $T + \Delta T$ into a solid angle $\Delta\Omega$, where $\Delta\Omega$ and ΔT are determined by the characteristics of the detector. This intensity is proportional to the differential cross section for reaction. The intensity transformations include the Jacobian elements necessary to preserve the proportionality between intensity and differential cross sections in the various coordinate systems.

are included as an intermediate stage in the transformation. Intensity contour diagrams expressed in Cartesian coordinates have the advantage that uncertainties in the CM velocity are not involved in the transformation of the data (the Jacobian is unitary). However, intensities in Cartesian coordinates are proportional to the probability that the product has velocity vectors with components between V_x and $V_x + dV_x$, V_y and $V_y + dV_y$, and V_z, and $V_z + dV_z$, rather than to the probability of a velocity of magnitude between $\mathbf{V}$ and $\mathbf{V} + d\mathbf{V}$ scattered into the element of solid angle $d\Omega(\chi,\Phi)$. Thus they minimize error propagation in the experimental data and are ideal for assessment of forward–backward asymmetry, but are misleading when used for interpretation of the relative importance of different mechanisms and energetic implications of experimental results. The transformations required for assessing translational-energy changes are given in Table 2.

These equations and relationships define the formal connections between laboratory and center-of-mass coordinates and describe the usual ways of presenting differential reaction cross-section data derived from crossed-beam experiments. To illustrate how such measurements define our concepts for kinematic models, we shall consider two hypothetial examples that are limiting cases—the long-lived-complex model and the impulsive, spectator stripping model. We anticipate that many examples may exhibit features of one or the other of these limiting cases, but that the dynamics of real systems will lie somewhere between them.

3.1   Direct Interaction Mechanisms

Direct mechanisms are generally thought to apply when the average lifetime of the collision is much less than a rotational period. Many direct mechanisms are derived from a rudimentary extension of the semiclassical (or classical) optical potential model for elastic scattering (17,18). The reactants A + BC are assumed to approach the region of chemical interaction along a two-body central force trajectory specified by an impact parameter b and kinetic energy T. A fraction $P(b,T)$ of these collisions lead to reaction, and the products AB + C depart along another two-body trajectory with impact parameter b' and kinetic energy T'. The differential scattering cross sections for elastic and reactive scattering then are given by

$$I_E(\chi) = [1 - P(b,E)] \frac{d(\pi b^2)}{d\omega},$$

$$I_R(\chi') = P(b,E) \frac{d(\pi b'^2)}{d\omega'}, \tag{10}$$

where χ and χ' are the elastic and reactive scattering angles and

$$d\omega = 2\pi \sin \chi \, d\chi,$$

$$d\omega' = 2\pi \sin \chi' \, d\chi' \tag{11}$$

TABLE 2   Summary of Equations Relating Velocities and Masses to Energetics as Determined by the Reaction Kinematics

A. Reaction:

$$M_1^+ + M_2 \rightarrow M_3^+ + M_4$$

B. Energy of CM:

$$T_{cm} = \frac{m_1}{M} T_1 + \frac{m_2}{M} T_2 = \frac{m_3}{M} T_3 + \frac{m_4}{M} T_4$$

where

$$T_{cm} = \tfrac{1}{2}(m_1 + m_2)V_{cm}^2$$

$$T_i = \tfrac{1}{2} m_i V_i^2, \qquad i = 1,2,3,4$$

$$M = m_1 + m_2 = m_3 + m_4$$

C. Initial Energy Relative to CM:

$$T_{rel}^i = \frac{m_2}{M} T_1 + \frac{m_1}{M} T_2 = \tfrac{1}{2} m_1 U_1^2 + \tfrac{1}{2} m_2 U_2^2$$

$$= \tfrac{1}{2} \frac{m_1 m_2}{M}(U_1 + U_2)^2 = \tfrac{1}{2} m_1 U_1^2 \left(1 + \frac{m_1}{m_2}\right)$$

D. Final Energy Relative to CM:

$$T_{re}^f = \tfrac{1}{2} \frac{m_3 m_4}{M}(U_3 + U_4)^2 = \tfrac{1}{2} m_3 U_3^2 + \tfrac{1}{2} m_4 U_4^2$$

$$= \tfrac{1}{2} m_3 U_3^2 \left(1 + \frac{m_3}{m_4}\right)$$

E. Translational Exoergicity

$$Q = T_3 + T_4 - T_1 - T_2 = T_{rel}^f - T_{rel}^i$$

$$= -\Delta H_R - E_{int}^f + E_{int}^i$$

where ΔH_R is the heat of reaction and E_{int}^f and E_{int}^i are the internal energies of the products and reactants, respectively.

are the corresponding solid-angle elements in the CM system. In reactive collisions the scattering angle is the sum of contributions from the reactant and product portions of the trajectory:

$$\chi' = \chi_r(b,E) + \chi_p(b',E'). \tag{12}$$

The lack of a third contribution representing the ABC complex makes apparent the *ultradirect* character of this model. Further assumptions are required to relate b' and T' and to b and T. With these relations both the elastic and reactive scattering may be evaluated from the usual semiclassical (or classical) two-body collision mechanics, once the potentials and the reaction probability function are specified.

In the spectator stripping limit the interaction occurs entirely between A and B, while C continues on its trajectory (relative to the CM) without changing speed or direction, apparently unaware that its partner has been snatched away from it (17). This description is equivalent to the assumption that the collision between the particles is sudden, so that the collision time is much shorter than the vibrational period of the target molecule. A formal basis for the spectator model is provided in quantum mechanics by the impulse approximation, which is similar to the Born approximation except that it does not assume that the interaction is small.

In the CM system the model states merely that the velocity of the spectator C is constant. If the velocity of the internal motion (i.e., vibration and rotation) of BC is neglected, then

$$\mathbf{U}'_C = \mathbf{U}_{BC}, \tag{13}$$

which by conservation of momentum implies also that

$$\mathbf{U}'_{AB} = -\frac{m_C}{m_{AB}} \mathbf{U}'_C. \tag{14}$$

The reactive scattering angle and product kinetic energies are given by

$$\chi' \simeq 0,$$
$$T' = T \frac{m_A m_C}{m_{AB} m_{BC}}, \tag{15}$$

and an explicit relation between b' and b is not required, since both are assumed to be sufficiently large that the effect of the interaction potential on the scattering can be neglected.

3.2 Long-Lived Intermediate Complex

If the constituents of the system remain within normal bonding distances of one another for periods that are long compared with the vibrational (10^{-13} sec) and rotational (10^{-12} sec) periods of the system, the system may "forget" which reactant

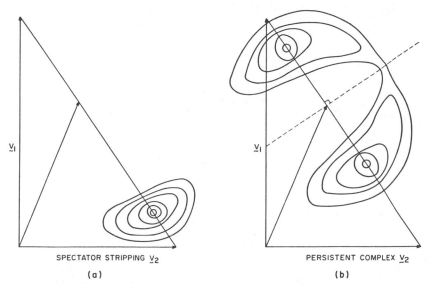

Figure 11. Representative Newton plots for two limiting-case reaction models: (*a*) pure spectator stripping model for direct reaction; (*b*) persistent-complex model with little angular momentum in the products. Both diagrams show the effect of typical velocity spreads in the ion and neutral beam at 1 eV collision energy.

came from which direction. Under these conditions the complex will execute several rotations before decomposing and the intensity distribution for the products in the barycentric system should be symmetric with respect to a plane passing through the CM normal to the initial relative velocity vector (at $\chi = \pi/2$ in Fig. 10). Furthermore, if the lifetime is sufficiently long, internal redistribution of the energy may occur.

Figures 11 and 12 illustrate the qualitative significance of these models. Figure 11*a* shows the Newton diagram for spectator stripping for the reaction D_3O^+ $(H_2O,D_2O)H_2DO^+$, while Fig. 11*b* shows the analogous distribution for a persistent complex. Distortions from the exact symmetry of these models result from the convolution of the actual spreads in ion and neutral beam kinetic energies characteristic of the Utah crossed-beam apparatus. The mean velocities of the reactants are used to define the CM reference frame, with the implicit assumption of monoenergetic particles.

Figure 12 further elaborates these models for the simpler, idealized case of monochromatic beams. In both Figs. 11 and 12 symmetry about a plane through the CM is the "signature" for the persistent-complex model. Figures 12*a–c* show examples of the different ways of conserving angular momentum in the decomposition of a persistent complex (19). Figure 12*a* illustrates the isotropic distribution that results from a low-angular-momentum (and high-temperature) intermediate complex. Figure 12*b* exhibits the symmetry characteristic of high angular momentum in the complex, and Fig. 12*c* corresponds to an oblate complex. Figures 12*d–f* depict three "direct" mechanisms. Figure 12*d* is for the spectator stripping model already

$$A^+ + BC \longrightarrow$$

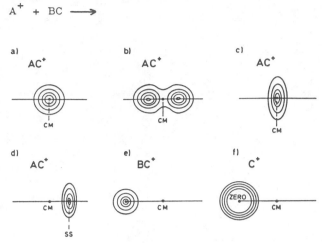

Figure 12. Simplified schematic contour diagrams resulting from different reaction mechanisms. The mass ratio is $m_A/m_B/m_C = 3:3:1$. A^+ approaches from the left and BC from the right. (a)–(c) Formation of product from the decay of a persistent complex with different couplings between orbital and internal angular momentum in the products (see text for details). (d) Diagram for spectator stripping model. (e), (f) Predicted distributions for charge transfer occurring via a long-range electron jump; f corresponds to dissociative charge transfer with kinetic-energy release giving a "hole" in the distribution centered on the CM velocity of the initial neutral BC. Reprinted from Ref. 19 with the kind permission of Plenum Press.

discussed, and Fig. 12e corresponds to the long-range electron-jump model invoked for resonance charge transfer. Figure 12f corresponds to a "hollow-sphere" model suggested for dissociative charge transfer (19).

REFERENCES

1. V. L. Tal'roze and A. K. Lyubimova, *Dokl. Akad. Nauk. SSSR* **86,** 909 (1952).

2. D. P. Stevenson and D. O. Schissler, *J. Chem. Phys.* **23,** 1353 (1955); **24,** 926 (1956).

3. F. H. Field, J. L. Franklin, and F. W Lampe, *J. Am. Chem. Soc.* **79,** 2419, 2665, 6132 (1957).

4. C. F. Giese and W. B. Maier II, *J. Chem. Phys.* **39,** 739 (1963).

5. J. H. Futrell and C. D. Miller, *Rev. Sci. Instrum.* **37,** 1521 (1966).

6. T. O. Tiernan, in *Interaction Between Ions and Molecules,* P. Ausloos, Ed., Plenum Press, New York (1975), pp. 353–386.

7. R. P. Clow and J. H. Futrell, *Int. J. Mass Spectrom. Ion Phys.* **4,** 165 (1970).

8. P. J. Chantry, *J. Chem. Phys.* **55,** 2746 (1971).

9. J. R. B. Slayback and M. S. Story, *Ind. Res. Dev.* **2,** 129 (Feb 1981).

10. E. Teloy and D. Gerlich, *Chem. Phys.* **4,** 417 (1974).

11. S. L. Anderson, F. A. Houli, D. Gerlich, and Y. T. Lee, *J. Chem. Phys.* **75,** 2153 (1981).

12. S. M. Trujillo, R. H. Neynaber, and E. W. Rothe, *Rev. Sci. Instrum.* **37,** 1655 (1966).

13. W. R. Gentry, D. J. McClure, and C. H. Douglas, *Rev. Sci. Instrum.* **46,** 367 (1975).

14. Z. Herman, J. D. Kerstetter, T. L. Rose, and R. Wolfgang, *Rev. Sci. Instrum.* **40,** 538 (1969).

15. M. L. Vestal, C. R. Blakley, P. W. Ryan, and J. H. Futrell, *Rev. Sci. Instrum.* **47,** 15 (1976).

16. E. W. McDaniel, V. Cermak, A. Dalgarno, E. E. Ferguson, and L. Friedman, *Ion Molecule Reactions,* New York, Wiley-Interscience (1970), p. 114.

17. D. R. Herschbach, *Adv. Chem. Phys.* **10,** 319 (1966).

18. K. Lacmann and A. Henglein, *Ber. Bunsenges. Phys Chem.* **69,** 286 (1965).

19. K. Birkinshaw, V. Pacak, and Z. Herman, in *Interaction Between Ions and Molecules,* P. Ausloos, Ed., Plenum Press, New York (1975), pp. 95–122.

PART 4

APPLICATIONS TO ION CHEMISTRY AND PHYSICS

CHAPTER 9

Laser Photodissociation Spectroscopy of Ions in the Gas Phase

JAMES D. MORRISON

Physical Chemistry Department
La Trobe University
Melbourne, Australia

1   INTRODUCTION

Many different kinds of ions are observed in mass spectrometry, and these are clearly the result of a variety of ionization and fragmentation processes. These processes have been studied experimentally for many years, but in many instances the mechanisms are still not fully understood. Even for the most common ions little is known of the energy states or actual structures. Some ions, once formed, undergo further, rapid rearrangement or fragmentation; others are stable and commonly appear as the end products of reaction pathways.

It is now possible, by means of theoretical calculations, to make detailed predictions of ionic structures (1). To date there is unfortunately little in the way of detailed experimental data with which to compare these results. One should not infer from this statement that a great deal cannot be deduced about ion structures from fragmentation patterns and labeling studies, or from theoretical evidence. However, to the physical chemist, a structure is not really known until it is possible to write out a list of bond lengths and bond angles for each electronic state.

Optical absorption spectroscopy is a very powerful technique for obtaining such structural information about molecules in the gas phase. Analysis of ultraviolet (UV), visible, infrared, and microwave spectra can give not only the rovibronic energy levels of a molecule, but also, in relatively simple species, the detailed molecular geometry. Even in more complex cases useful structural information can still be derived.

If it were possible to obtain such absorption spectra for ions, similarly detailed information could be obtained. Photoelectron spectroscopy (PES) goes part way to achieving this end and gives much information about vibronic states of ions, but it gives a restricted view of the molecular ion: the states are seen vertically from underneath from the neutral ground state. The energy resolution (0.01 eV) is insufficient to permit the observation of rotational structure. The technique is less well suited to the study of radical ions or other unstable ionic species.

It is not practicable, except in a very few cases of simple di- and triatomic species, to obtain a direct absorption spectrum for a pure molecular ion from a gaseous discharge (2). It is easy to produce ions in abundance by this means, but their high reactivity almost instantaneously leads to formation of numerous secondary, tertiary, and higher products. The mixtures are "hot," many excited rovibronic levels are populated, and the resulting spectra are impossible to interpret in detail.

The mass spectrometer provides a means of obtaining a beam of ions of a given m/z value. The Coulomb repulsion of like charges limits the attainable partial pressure of ions in such a beam to 10^{-8} Pa, which would be regarded by a spectroscopist as a very hard vacuum. Usually, attainable beam pressures are much lower than this. Absorption spectroscopy on such beams is not practicable. However, when the energy of irradiation is enough to cause dissociation, an indirect method of recording spectra becomes possible. Dissociation produces daughter ions of a different m/z, which therefore possess a different momentum and kinetic energy.

By monitoring the production of these daughter ions as a function of the energy of irradiation, one can detect excited states of the precursor ion. This has the double advantage that charged particles are much easier to detect than photons and it is much easier to distinguish a small ion current from a zero current than it is to detect a minute change in a large signal.

The absorption of radiation by ions can lead to several unimolecular processes:

$$AB^{\pm} \overset{h\nu}{\rightarrow} AB^{\pm *} \rightarrow AB^{\pm} + h\nu \qquad \text{(Fluorescence)}, \qquad \text{(a)}$$

$$AB^{\pm} \overset{h\nu}{\rightarrow} AB^{\pm *} \rightarrow A^{\pm} + B \qquad \text{(Direct dissociation)}, \qquad \text{(b)}$$

$$AB^{\pm} \overset{h\nu}{\rightarrow} AB^{\pm *} \rightarrow AB^{\pm **} \rightarrow A^{\pm} + B \quad \text{(Predissociation)}. \qquad \text{(c)}$$

Other processes are possible; for example, negative ions may lose an electron in photodetachment, leaving neutral products, and at higher pressures in the interaction region bimolecular processes such as charge exchange or ion–molecule reactions may also produce daughter ions of different m/z:

$$AB^{+} \overset{h\nu}{\rightarrow} AB^{+ *} \qquad \text{(Primary process)},$$

$$AB^{+ *} + C \rightarrow C^{+} + AB \quad \text{(Charge exchange)}, \qquad \text{(d)}$$

$$AB^{+ *} + C \rightarrow A^{+} + BC \quad \text{(Ion–molecule reactions of various types)}. \qquad \text{(e)}$$

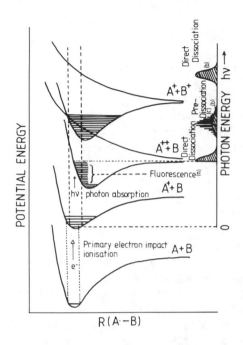

Figure 1. Potential-energy curves for a hypothetical diatomic molecule AB, showing primary ionization by electron impact and subsequent processes induced by photon impact: (*a*) fluorescence, (*b*) direct photodissociation, and (*c*) predissociation. The expected photodissociation spectrum is on the right.

All these processes may occur simultaneously and in competition with each other. Direct dissociation [reaction (b)] is the most rapid, occurring within a few vibration times ($\sim 10^{-12}$ sec); predissociation [reaction (c)] occurs on a time scale of 10^{-11}–10^{-6} sec; and fluorescence [reaction (a)] has a half-lifetime of $\sim 10^{-8}$ sec. The rates of the bimolecular processes (d) and (e) depend on the pressure. The cross sections for the two types of photodissociation, (b) and (c), as a function of photon energy differ markedly from one another.

Figure 1 illustrates the relationship between the potential-energy curves of a diatomic ion and the expected photodissociation cross sections for different types of processes. In direct dissociation, transitions take place from a bound rovibrational state of the ion to a state above its vibrational dissociation limit. The transition probability curve is given, therefore, by the vertical reflection of the expectation functions of the bound state of the precursor ion in the potential curve for the upper dissociating state. This gives a curve with a broad (usually 1–2 eV in width) maximum, which will have a sharply defined threshold only if the Franck–Condon region cuts the upper-state potential-energy curve both above and below its dissociation limit.

When photon absorption causes a transition from a bound rovibrational state of the ion to a similarly bound upper state of the ion, the absorption cross section will consist of a sharp set of peaks corresponding to the quantized levels of the upper state. If this bound excited state can then undergo a radiationless transition to a third state at a point above one of its dissociation limits, the sharp peaks of the absorption, and thus also of the photofragment production, will become broadened to a width inversely proportional to the lifetime of the bound levels prior to dissociation. For a half-lifetime of 10^{-10} sec, these absorption bands are very sharp, $\sim 10^{-3}$ cm^{-1}, permitting the resolution of rotational fine structure.

In practice transitions occur from a populated set of rovibrational levels of the lower state of the ion, and the observed photofragmentation is expected to consist of a complicated series of peaks.

Using an ion cyclotron resonance (ICR) mass spectrometer, Dunbar was able to study the photodissociation of a variety of organic ions (3). This early work was of low resolution; the light source was a xenon lamp used in conjunction with a series of interference filters, with a pass band of 10 nm, but it covered the energy range 2.0–6.0 eV. Ions can be monitored in such a cell for relatively long times, enhancing the probability of light absorption and dissociation. However, since ionization, mass analysis, light absorption, dissociation, and secondary-ion detection all take place in the same cell, interpretation of the results requires some care. In spite of these difficulties, Dunbar's work was the first to demonstrate that the plot of photodissociation probability versus photon energy obtained for molecular ions could be clearly related to the PE spectra for the corresponding neutral molecules, and the curves were therefore termed photodissociation (PD) spectra.

Some of the difficulties of ICR instruments are overcome in ion beam instruments, where the regions in which primary-ion formation, mass analysis, photon–ion interaction, and product-ion selection and detection occur can be spatially remote from each other. A separate ion source allows the use of more efficient electron impact methods and higher sample pressures, giving more intense ion beams. Further, ion–molecule reactions can be brought about in the source to create species such as CH_5^+, H_3^+, O_3^+, and so forth as primary ions. The simplest apparatus of this kind did not use a primary mass analyzer, and was limited to samples that produced only one kind of ion. Normally the first mass separation is carried out using either a magnetic sector or a quadrupole mass filter, and the product-ion selection is done either by energy analysis or by another mass analyzer. In modern practice lasers are used almost exclusively as photon sources. Continuous lasers generally give the highest average power outputs, with excellent stability and very narrow bandwidth (~ 0.1 cm^{-1}). When they are used to pump a dye laser, output ranges from the near infrared to the visible regions of the spectrum are available.

Excimer-laser-pumped pulsed dye lasers have many advantages for ion PD studies. A wider range of wavelengths, albeit with somewhat poorer resolution (0.1–0.3 cm^{-1}), is available to cover the range from 1.7 to 3.5 eV (10,000–30,000 cm^{-1}). Some 20 different dye solutions are needed, as indicated in Fig. 2. Frequency doublers can be used to extend this range to 6 eV (48,000 cm^{-1}). Etalon bandwidth narrowing can reduce the energy spread in such lasers to 0.01 cm^{-1}. The short duration of the laser pulses ($\sim 10^{-9}$ sec) limits the attainable resolution. With continuous lasers, etalons can give a line width of 10^{-5} cm^{-1}. Scanning over a range of energies is usually achieved by means of computer-controlled impulse drives, which have to change the frequency of the dye laser cavity, tune the etalon, and rotate or tilt the doubler crystal (when used) simultaneously. Only a very narrow range of energies can be covered in a given scan, and an overall PD spectrum may require many overlapping scans of as many as 10,000 data points each. The current

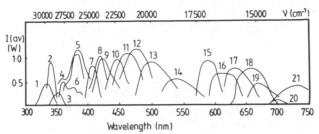

Figure 2. Output vs. wavelength for a typical dye laser pumped by an excimer laser (Lambda Physik FL 2000), using the following Lambdachrome® dye solutions (see Lambda Physik, 18 North Road, Bedford, MA 01730, Poster No. 3/82): 1, BM terphenyl; 2, PTP; 3, TMQ; 4, Bi Bu Q; 5, polyphenyl 1; 6, butyl PBD; 7, DPS; 8, stilbene 1; 9, stilbene 3; 10, coumarin 2; 11, coumarin 47; 12, coumarin 102; 13, coumarin 307; 14, coumarin 153; 15, rhodamine 6G; 16, sulforhodamine B; 17, rhodamine 101; 18, DCM; 19, cresyl violet; 20, Nile blue; 21, Oxazine 1.

of photodissociated fragment ions can range from 0.1 to 20 ions per laser pulse, making time integration of the signals essential. Pulsed lasers have the advantage that gating of the detection system can further enhance signal-to-noise statistics. Such a computer-controlled instrument provides a means of studying the energy states of ions with a resolution many times better than the best attainable by photoionization (PI) spectrometry or PES.

At the highest resolutions, calibration of the energy scale presents real problems. An absorption spectrum of I_2 may be recorded simultaneously over the wavelength region from 14,000 to 20,000 cm^{-1}, allowing calibration of energy to better than 0.01 cm^{-1} (4). In the near UV region, optogalvanic lines provide one of the few convenient calibration methods available (5). Interferometric methods can be used, but are complex to use experimentally.

The photon and ion beams can be made to interact in either crossed or coaxial mode. The interaction region is more closely defined in the former, and this method is therefore better suited to the study of fragment-ion velocities and angular distributions and to the measurement of absolute cross sections. The coaxial mode gives better overlap of the photon and ion beams, and permits the Doppler tuning technique, in which absorption frequencies of the ion can be brought into resonance with a fixed laser frequency by a change in the velocity of the ion beam. This technique, developed independently by Carrington (6) and Moseley (7), can give effective energy resolutions as low as 0.002 cm^{-1}, albeit over a very narrow energy range. A further advantage of this method is that because the ion beam is accelerated to a relatively high velocity, the percentage spread in velocities due to the thermal motions of the ions is dramatically reduced. This markedly reduces the Doppler broadening of spectral absorption lines and gives effective sample temperatures of a few degrees Kelvin.

Because of the difficulty of defining the volume of the interaction region and of determining the sample pressure of the primary ions, absolute PD cross sections usually cannot be measured to any great accuracy. In practice, a process can be detected if its cross section lies between 10^{-17} and 10^{-24} cm^2. In most ion PD experiments, great care is taken to ensure that only single-photon processes are occurring by checking the linearity of the photon-induced ion current vs. the light intensity.

In the drift-tube instrument of Beyer and Vanderhoff (8) (Fig. 3), ions are allowed after formation and separation to drift in a weak electric field through a background gas. After passing through an aperture, they are irradiated and mass analyzed. Instruments of this type have been used to examine a number of dimer and cluster ions of interest in upper-atmosphere studies.

The use of an energy analyzer, as in the ion-photofragment spectrometer of Cosby and Moseley (7) (Fig. 4), allows spectra to be recorded by determination of the kinetic energy of the photodissociated ions at a fixed photon energy. Alternatively, by setting the energy spectrometer to a low value of ion kinetic energy and varying the photon energy, one can produce a threshold photofragment spectrum.

A time-of-flight mass spectrometer has been used by Dunbar and Armentrout (9) to attempt to time resolve the PD of $C_7H_7I^+$. The dissociation was faster than the shortest time that could be observed, but the technique has possibilities for study of the fragmentation kinetics of molecular ions.

The quadrupole mass analyzer has the advantage that ion energies are low, increasing the probabilities of interaction. Also, an AC quadrupole can be used as a strong focusing lens to confine dissociation products to the interaction region. A third quadrupole mass filter is then used to detect the product ions, all three quad-

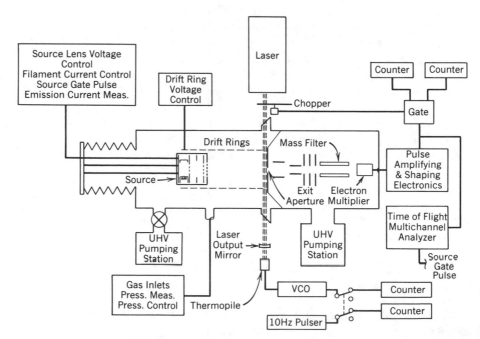

Figure 3.   Drift-tube laser PD mass spectrometer built by R. A. Beyer and J. A. Vanderhoff. UHV, ultrahigh vacuum; VCO, voltage control output. Reproduced with permission from Ref. 8.

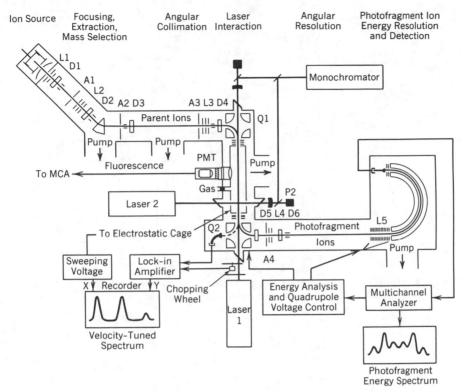

Figure 4. General-purpose ion PD spectrometer built by P. C. Cosby, J. T. Moseley, and colleagues. This device allows both coaxial and transverse photon interaction and ion energy analysis. L = lens, D = diaphragm, A = aperture, Q = quadrupole, P = photomultiplier, MCA = multichannel analyzer. Reproduced with permission from Ref. 7.

rupoles being collinear. The laser beam can be projected coaxially through the whole instrument, since photofragments formed in the first quadrupole do not reach the interaction region, and the current of product ions in the second mass filter is too low to give significant further fragmentation (Fig. 5). This instrument, designed by McGilvery and Morrison (10) gives the total PD cross section as a function of photon energy. It is not suited to study energy- or angular-dependence (10,11). A somewhat similar instrument was devised at the same time by Thomas, Dale, and Paulson (12).

2 RESULTS

The early work of Dunbar showed that most polyatomic ions photodissociate by direct processes. The PD spectra therefore correspond closely to PE spectra displaced along the energy axis by the first ionization potential. This is not, however, always the case, as can be seen from the PD curves for Cl_2^+, Br_2^+, and I_2^+ (13) (Fig. 6).

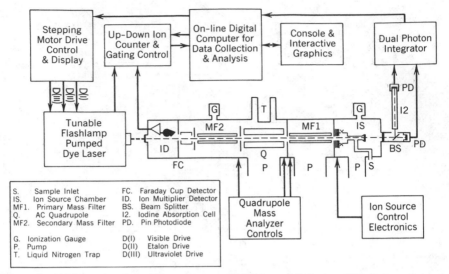

Figure 5. Triple-quadrupole ion PD spectrometer of McGilvery and Morrison. This device uses coaxial irradiation of the ion beam. Reproduced with permission from Ref. 10.

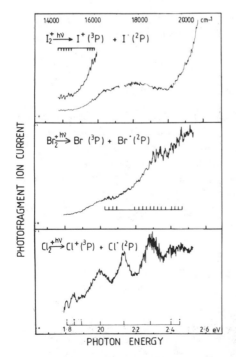

Figure 6. Photodissociation spectra for Cl_2^+, Br_2^+, and I_2^+. Weak photopredissociation is superposed on direct PD. Reproduced with permission from Ref. 13.

There is some slight indication of predissociation, but the main process appears to be direct dissociation. The potential curves for Cl_2 are shown in Fig. 7 (14). The thresholds for dissociation indicate that at least some of the primary molecular ions are in their first excited state, having survived mass analysis before the photon interaction. This can be verified by lowering the energy of the ionizing electron beam in the ion source. Below the energy of the $^2\Pi_u$ state, no fragmentation occurs (Fig. 8).

This phenomenon is even more marked in the case of O_2^+, where photofragmentation to O^+ occurs with visible light. The dissociation energy $D_0(O\text{–}O^+)$ is 6.0 eV. The lowest ionization potential of O_2 is 12.2 eV, forming the $^2\Pi_g$ state. Reducing the electron energy below 16 eV reduces the photodissociated-ion current to zero, indicating that the metastable $^4\Pi_u$ state of the molecular ion is the precursor.

The PD of O_2^+ has been studied exhaustively (15). At very low resolution, the spectrum consists of a set of narrow bands throughout the visible region superposed on a broad continuum (16) (Fig. 9). The bands indicate predissociation, while the continuum suggests a direct dissociation process. Higher-resolution studies resolve these bands into complex rotational and hyperfine structure (17) (Figs. 10 and 11) attributable to predissociation from the initially formed $(b^4\Sigma_g^-)$ state:

$$O_2^+(a^4\Pi_u) \xrightarrow{h\nu} O_2^+(b^4\Sigma_g^-) \rightarrow O^+(S^0) + O(P).$$

By the use of Doppler tuning, the rotational peaks are found to vary in linewidth, indicating differing predissociation lifetimes. Detailed measurement of these lifetimes as a function of the b-state vibration, rotation, and fine-structure levels leads

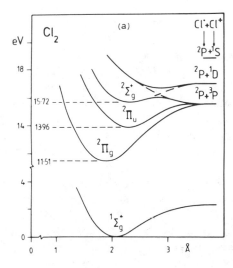

Figure 7. Potential curves for Cl_2. After Potts and Price (14).

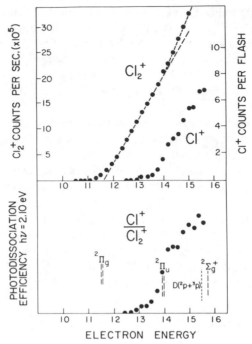

Figure 8. *Upper panel*: Ionization efficiency as a function of ionizing electron energy for Cl_2^+ and Cl^+. Data were obtained with a triple-quadrupole device. For Cl_2^+, the first and third quadrupoles were set to transmit ions with $m/z = 70$, and for Cl^+, the first quadrupole was set to transmit ions with $m/z = 70$ and the third ions with $m/z = 35$. The beam of Cl_2^+ ions was irradiated with light of energy $h\nu = 2.10$ eV. *Lower panel*: Ratio of the above curves, with the energies of the known states of Cl_2^+ and the lowest dissociation limit for Cl_2^+ superposed on the figure, showing that PD does not occur until the electron energy exceeds the energy of the $^2\Pi_u$ state of the parent ion.

to the conclusion that the predissociation mechanism for the b state involves primarily spin-orbit coupling to the $^4\Sigma_g^+$ state.

By operation in the intracavity mode (with the interaction region inside the laser cavity) weak predissociation has been detected in NO^+ (18) and in CH^+ (19). The resolution of which this experiment is capable is evident from the fact that each line occurs as a Doppler doublet, because the ion beam is traversed by the light beam traveling in the cavity in both directions.

Methyl iodide is an ideal subject for a PD study, in that the PE spectrum shows that the primary $CH_3I^{+\cdot}$ ion is formed in the ion source in very well-separated states, and the transitions occurring in PD are from the various rovibrational levels of the $\tilde{X}(^2E_{1/2})$ state to the quasibound $\tilde{A}(^2A_1)$ state, which dissociates to $CH_3^{+\cdot}$. At low resolution the spectrum consists of a well-defined series of triangular bands that are shaded to the red, indicating longer bonds in the upper state (20) (Fig. 12). Higher resolution partially resolves these bands into rotational structure, but not well enough to separate out all the detail. In such a case one can attempt to simulate one of these bands by calculation and, if a sufficiently good match can be obtained between the

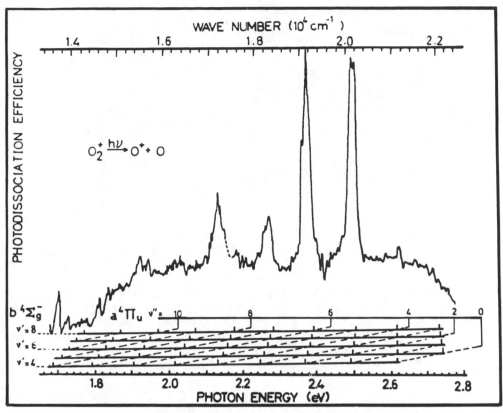

Figure 9. Photodissociation spectrum of O_2^+ at low resolution, showing predissociation superposed on a broad band due to direct dissociation. The peaks due to predissociation correspond to transitions to the $v' = 4-8$ vibrational levels of the $b(^4\Sigma_g^-)$ state from the $v'' = 0-12$ levels of the $a(^4\Pi_u)$ state. Reproduced with permission from Ref. 16.

observed and calculated spectra, then determine the change in the rotational constants on transition from the $\tilde{X}(^2E_{1/2})$ to $\tilde{A}(^2A_1)$ states (21,22) (Fig. 13).

The PE spectrum indicates that the $\tilde{X}(^2E_{1/2})$ state of the ion is very little changed in geometry from the neutral ground state. Making this assumption, the geometry of the (0,1,7) vibrational level of the $\tilde{A}(^2A_1)$ state can be calculated to be as shown in Fig. 14.

The PD of O_3^+ has been examined by several groups (23), but their interpretation of the PD spectrum differs from the one presented here. The spectrum consists of a broad maximum with some indication of diffuse vibrational structure. It is believed that this represents an instance in which several populated levels of two states of the ion undergo PD via a single upper state. The resulting PD spectrum is then equivalent to the PE spectrum displaced and reversed on the energy scale. This interpretation reconciles the apparently conflicting photoionization, thermochemical, and PD data (24).

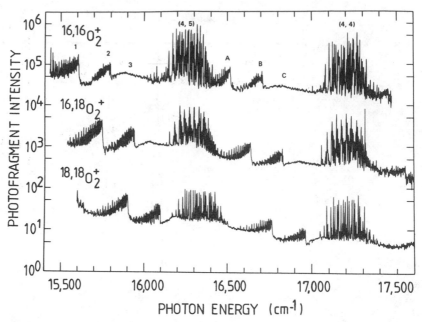

Figure 10. Wavelength dependence of the production of O^+ photofragments having center-of-mass kinetic energies in the range 0–70 meV from O_2^+, observed at intermediate (1-cm^{-1}) resolution using the instrument of Moseley et al. (7) and showing shaded bands due to predissociation. Reproduced with permission from Ref. 17.

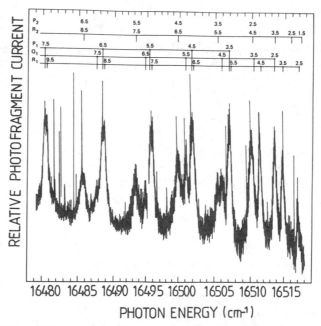

Figure 11. High-resolution spectrum of the bandhead region of the (2,5) band of the $f \rightarrow a$ system of O_2^+, with rotational assignment. For photon energies above 16,495 cm^{-1} the photofragment signal has been multiplied by a factor of 2. Reproduced with permission from Ref. 17.

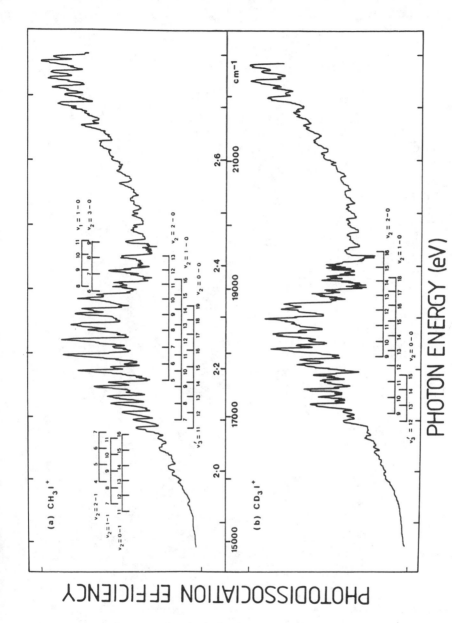

Figure 12. Laser PD spectra for CH_3I^+ and CD_3I^+ at low resolution, showing predissociation maxima. Reproduced with permission from Ref. 20.

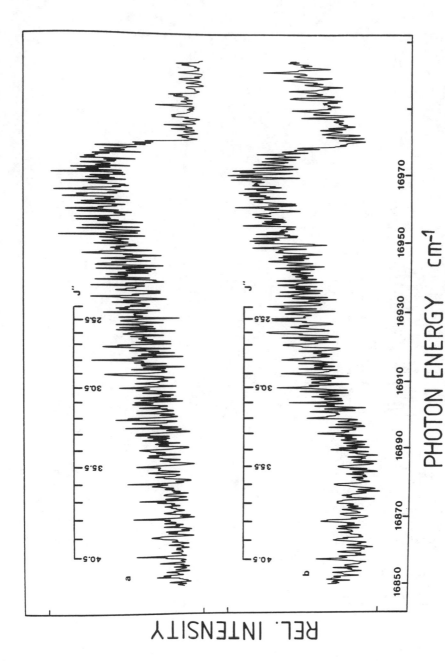

Figure 13. (a) Higher-resolution (0.3 cm⁻¹) spectrum of the $\tilde{A}(^2A_1)$$:E_{1/2}(0,1,7) \leftarrow \tilde{X}(^2E_{1/2})$$:E(0,0,0)$ vibronic band of CH_3I^+, showing partly resolved rotational structure. (b) Simulated rotational structure for this band, computed assuming $A'' = 5.119$ cm⁻¹ (634.7 μeV), $A' = 5.027$ cm⁻¹ (623.3 μeV), $B'' = 0.2500$ cm⁻¹ (31.0 μeV), $B' = 0.1854$ cm⁻¹ (22.9 μeV), $\zeta' = \zeta'' = 0.50$, and $T = 350$ K. Reproduced with permission from Ref. 22.

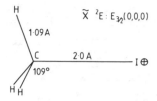

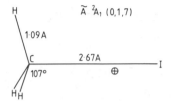

Figure 14. Calculated change in structure of the CH_3I^+ ion on transition from the $\tilde{E}$ to the $\tilde{A}$ state.

SO_2^+ photodissociates to SO^+, giving a most complex set of peaks in the visible, indicating that the fragmentation occurs by predissociation (Fig. 15). The PE spectrum (25) shows clearly that the precursor state that photodissociates cannot be the ground state of the ion, but that dissociation probably takes place from the $\tilde{A}$ and $\tilde{B}$ excited states of the ion to the $\tilde{C}$, $\tilde{D}$, and $\tilde{E}$ states. With higher resolution, all of the peaks in this PD spectrum are found to be shaded to the red (Fig. 16). Unfortunately, unlike in CH_3I^+, the transitions occur from many levels of the $\tilde{A}$ and $\tilde{B}$ states to an equally complex set of levels of the $\tilde{C}$, $\tilde{D}$, and $\tilde{E}$ states. With much effort, it has proved possible to interpret the low-photon-energy region of this spectrum and to obtain a set of vibrational frequencies for both the $\tilde{B}$ and $\tilde{C}$ ionic states that are more precise than those obtained from the PE spectra. A hitherto unobserved ionic excitation in v_2 was observed for the $\tilde{C}$ state. At higher resolution, some rotational structure becomes evident. Unfortunately, with C_{2v} symmetry, there are three rotational constants to determine, and at present this spectrum appears too complex to resolve further (26).

In such a case it is necessary to consider other ways of simplifying the spectrum. Very recently Chupka (27), using multiphoton ionization with a seeded molecular beam in the ion source, has managed to prepare CH_3I^+ ions with a much reduced vibrational and electronic distribution, and then photodissociate these ions with a second tunable dye laser. By lowering the rotational temperature of the ions in this way, he produced a much simplified spectrum, although so far the resolution in these experiments has not been sufficient for separation of the rotational fine structure (Fig. 17).

A much more elegant approach is to use two lasers in an optical–optical double-resonance experiment. Using this approach, Helm and Cosby (28) have been able to modify the populations of the various levels in the $^4\Pi_u$ state of O_2^+ before predissociation in such a way that the PD spectrum is greatly simplified, thus aiding its interpretation considerably. Optical–microwave double resonance has been used

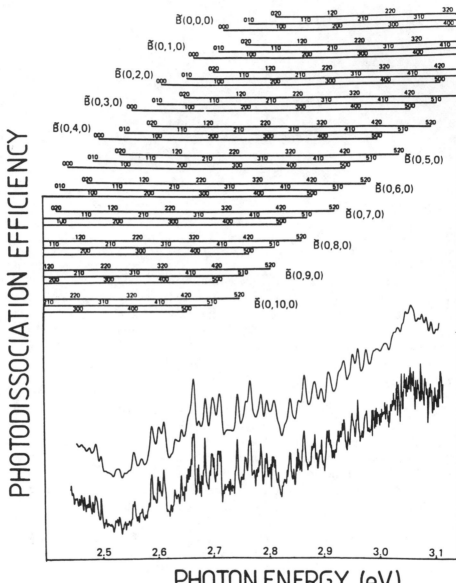

Figure 15. Photodissociation spectrum of SO_2^+, recorded from the process $SO_2^+ \rightarrow SO^+ + O$ at low resolution with and without smoothing. The spectrum consists of predissociation superposed on a broad maximum due to direct dissociation. Most of the fine structure is believed to be due to transition from $(0v_2 0)$ vibrational levels of the $\tilde{B}$ state to $(v_1 v_2 0)$ levels of the $\tilde{C}$ state of the ion. A tentative assignment of levels is shown in the upper part of the figure. Reproduced with permission from Ref. 26.

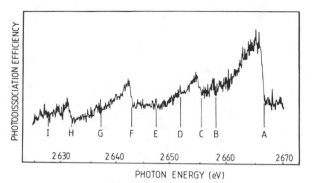

Figure 16. Photodissociation spectrum of SO_2^+ at higher resolution, showing shaded bands due to unresolved rotational fine structure. Vibrational bandheads are indicated. Reproduced with permission from Ref. 26.

(29) to saturate rotational transitions between levels of the H_2^+ ion, allowing observation of rotational hyperfine structure.

Infrared multiphoton dissociation spectra have been used to probe reaction energetics and to identify structural isomers of some molecular ions. Carrington and his colleagues have employed infrared two-photon dissociation to study vibration/rotation levels close to the dissociation limit in the ions HeH^+ (30), CH^+ (31), and H_3^+ (32). In these experiments the first photon is used to excite levels close to the dissociation limit and the second to raise the energy of the ion above the dissociation limit.

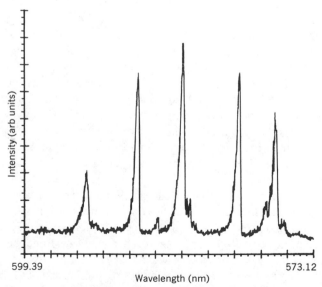

Figure 17. Photodissociation spectrum of CH_3I^+ recorded by Chupka et al., showing lower effective rotational temperature due to use of a seeded beam. Reproduced with permission from Ref. 27.

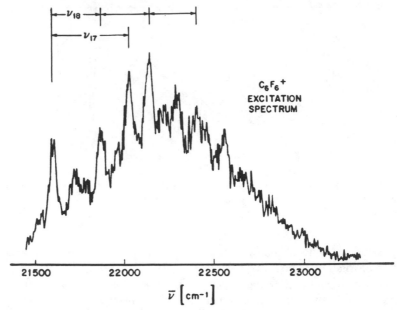

Figure 18. Laser excitation spectrum of hexafluorobenzene cation. The spectrum was obtained by reacting Ar metastables with the parent. The principal vibrational progressions are indicated. Reproduced with permission from Ref. 36.

The most serious limitation of ion PD spectroscopy, however, is that it is not a general method. Few ions predissociate, giving the very detailed PD spectra that can be used to obtain detailed structural information. It would be highly desirable to obtain a general method for detecting absorption leading to excited bound states of the ions. The bimolecular processes of charge transfer and ion–molecule reaction offer this possibility, provided a means of adding reagent gas to the absorption region at adequate pressure can be found. Both processes have been observed— charge transfer and reaction for CO^+ by Carrington (33), and charge transfer for N_2^+ by Moseley (34)—but so far this has not proved possible for more complex molecular ions. Another promising field of study is probing of the energy states of radical ions and other ionic species formed by ion–molecule processes by means of the various techniques of PD spectroscopy, since these data cannot readily be obtained by any other means.

Ions formed in bound excited electronic states can revert to the ground state by fluorescence processes; thus detection of this fluorescence as a function of the energy of the incident radiation would also give a spectrum for the ion. Experiments of this type have been carried out for some years, notably by Bondybey and Miller (35). In their work the parent cations were formed in a relatively pure state by chemiionization processes, and mass analysis was not needed. Detailed vibronic structure has been observed for a number of substituted fluorobenzenes (Fig. 18). Detection of such fluorescence from a mass-analyzed beam is much more difficult,

TABLE 1   Some Positive Ions That Have Been
Studied by Ion Photodissociation[a]

High Resolution (Species for Which Rotational/
Vibrational Detail Has Been Resolved Clearly)

O_2^+	(11,15,16,39–43)
H_2^+, HD^+, D_2^+	(29,44–51)
CH_3I^+, CD_3I^+	(4,20,27)
N_2O^+, NO^+	(12,18,52–54)
SO_2^+	(24,26)
N_2^+	(34,55)
CH^+	(19,31,56)
HeH^+	(30,57–59)
H_3^+, D_3^+	(32,60)
CO^+	(33,61)
NH^+, PH^+, SH^+	(62)

Intermediate Resolution (Species for Which Only
Vibrational Structure Is Resolved)

Cl_2^+, Br_2^+, I_2^+	(13)
O_3^+	(23)

Low Resolution (Species for Which Only Broad-Band
Absorption Is Observed)

$C_7H_7^+$ and substituted toluene cations	(9,63–69)
C_4H_8, $C_8H_{10}^+$ $C_9H_{12}^+$, $C_{10}H_{14}^+$	(70,71)
CH_4^+	(72,73)

[a]The detail of structure observable is limited by the PD mechanism, precursor-state lifetimes, and energy resolution of the photon source. Numbers in parentheses are reference numbers.

but has been achieved recently for N_2^+ by Ding et al. (37) and for CO^+ by Brown et al. (38). It will be of interest to see whether such experiments can be combined with simultaneous PD measurements to obtain branching ratios.

Table 1 lists ionic species that have been examined by PD spectroscopy.

REFERENCES

1. W. J. Bouma, R. H. Nobes, and L. Radom, *J. Am. Chem. Soc.* **105,** 1743 (1983).

2. R. J. Saykally and R. C. Woods, *Ann. Rev. Phys. Chem.* **32,** 403 (1981); J. A. Coxon and S. C. Foster, *J. Mol. Spectrosc.* **103,** 281 (1984); E. Schafer, M. H. Begemann, C. S. Gudeman, and R. J. Saykally, *J. Chem. Phys.* **79,** 3159 (1983).

3. Much of the ICR photodissociation work is reviewed by R. C. Dunbar, in *Kinetics of Ion Molecule Reactions,* P. Ausloos, Ed., Plenum Press, New York (1979), p. 463.

4. S. P. Goss, J. D. Morrison, and D. L. Smith, *J. Chem. Phys.* **75,** 757 (1981).

5. D. King, P. Schenk, I. C. Smyth, and J. Travis, *Appl. Opt.* **16**, 2617 (1977); V. Kaufman and B. Edlin, *J. Phys. Chem. Ref. Data* **3**, 825 (1974); A. Giacchetti, R. W. Stanley, and R. Zalubas, *J. Opt. Soc. Am.* **60**, 474 (1969).

6. A. R. Carrington, D. R. Milverton, and P. J. Sarre, *Mol. Phys.* **35**, 1505 (1978).

7. B. A. Huber, T. M. Miller, P. C. Cosby, H. D. Zeeman, R. L. Leon, J. T. Moseley, and J. R. Peterson, *Rev. Sci. Instrum.* **48**, 1306 (1977).

8. R. A. Beyer and J. A. Vanderhoff, *J. Chem. Phys.* **65**, 2313 (1976).

9. R. C. Dunbar and P. Armentrout, *Int. J. Mass Spectrom. Ion Phys.* **24**, 465 (1977).

10. D. C. McGilvery and J. D. Morrison, *Int. J. Mass Spectrom. Ion Phys.* **28**, 81 (1978).

11. M. L. Vestal and G. H. Mauclaire, *J. Chem. Phys.* **67**, 3758 (1977).

12. T. F. Thomas, F. Dale, and J. F. Paulson, *J. Chem. Phys.* **67**, 793 (1977).

13. R. G. McLoughlin, J. D. Morrison, and D. L. Smith, *Int. J. Mass Spectrom. Ion Proc.* **58**, 201 (1984).

14. A. W. Potts and W. C. Price, *J. Chem. Soc. Faraday Trans. II* **67**, 1242 (1971).

15. Much of the O_2^+ photodissociation literature is referenced in the reviews of J. T. Moseley and J. Durup, *J. Chimie Physique* **77**, 673 (1980); *Ann. Rev. Phys. Chem.* **32**, 53 (1981).

16. D. C. McGilvery, J. D. Morrison, and D. L. Smith, *J. Chem. Phys.* **68**, 4759 (1978).

17. H. Helm, P. Cosby, and D. L. Huestis, *J. Chem. Phys.* **73**, 2629 (1980).

18. P. C. Cosby and H. Helm, *J. Chem. Phys.* **75**, 3882 (1981).

19. P. C. Cosby, H. Helm, and J. T. Moseley, *Astrophys. J.* **235**, 52 (1980); H. Helm, P. C. Cosby, M. M. Graff, and J. T. Moseley, *Phys. Rev.* **A25**, 304 (1982).

20. D. C. McGilvery and J. D. Morrison, *J. Chem. Phys.* **67**, 368 (1977).

21. R. G. McLoughlin, J. D. Morrison, D. L. Smith, and A. L. Wahrhaftig, *J. Chem. Phys.* **82**, 1237 (1985).

22. D. L. Smith, Ph.D. thesis, La Trobe University, Melbourne (1982).

23. M. L. Vestal and G. H. Mauclaire, *J. Chem. Phys.* **67**, 3667 (1977); J. T. Moseley, J. B. Ozenne, and P. C. Cosby, *J. Chem. Phys.* **74**, 337 (1981); S. P. Goss and J. D. Morrison, *J. Chem. Phys.* **76**, 5175 (1982); J. Hiller and M. L. Vestal, *J. Chem. Phys.* **77**, 1248 (1982).

24. S. P. Goss, Ph.D. thesis, La Trobe University, Melbourne (1984).

25. D. R. Lloyd and P. J. Roberts, *Mol. Phys.* **26**, 255 (1973).

26. S. P. Goss, R. G. McLoughlin, and J. D. Morrison, *Int. J. Mass Spectrom. Ion Proc.* **64**, 213 (1985).

27. W. A. Chupka, S. D. Coulson, M. S. Searer, and A. M. Woodward, *Chem. Phys. Lett.* **95**, 171 (1983).

28. H. Helm and P. C. Cosby, *J. Chem. Phys.* **77**, 5396 (1982).

29. A. R. Carrington and J. R. Buttenshaw, *Mol. Phys.* **44**, 267 (1981).

30. A. R. Carrington, J. R. Buttenshaw, R. Kennedy, and T. P. Softley, *Mol. Phys.* **44**, 1233 (1981).

31. A. R. Carrington, J. R. Buttenshaw, R. Kennedy, and T. P. Softley, *Mol. Phys.* **45**, 747 (1982).

32. A. R. Carrington, J. R. Buttenshaw, and R. Kennedy, *Mol. Phys.* **45**, 753 (1983).

33. A. R. Carrington, D. R. Milverton, and P. J. Sarre, *Mol. Phys.* **35**, 1505 (1978).

34. F. J. Greiman, J. C. Hansen, and J. T. Moseley, *Chem. Phys. Lett.* **85**, 53 (1982).

35. T. A. Miller and V. E. Bondybey, *J. Chimie Physique* **77**, 696 (1980).

36. V. E. Bondybey and T. A. Miller, *J. Chem. Phys.* **70**, 138 (1979).

37. A. Ding, K. Richler, and M. Menzinger, *Chem. Phys. Lett.* **77**, 523 (1981).

38. R. D. Brown, R. G. Dittman, D. C. McGilvery, and P. D. Godfrey, *J. Mol. Spectrosc.* **97**, 253 (1983).

39. A. Tabche-Fouhaille, J. Durup, J. T. Moseley, J-B. Ozenne, C. Pernot, and M. Tadjeddine, *Optics Comm.* **18,** 92 (1976).

40. J-B. Ozenne, J. Durup, R. W. Odom, C. Pernot, A. Tabche-Fouhaille, and M. Tadjeddine, *Chem. Phys.* **16,** 75 (1976).

41. A. Tabche-Fouhaille, J. Durup, J. T. Moseley, J-B. Ozenne, C. Pernot, and M. Tadjeddine, *Chem. Phys.* **17,** 81 (1976).

42. J. T. Moseley, M. Tadjeddine, J. Durup, J-B. Ozenne, C. Pernot, and A. Tabche-Fouhaille, *Phys. Rev. Lett.* **37,** 891 (1976).

43. H. J. Hansen, J. T. Moseley, A. L. Roch, and P. C. Cosby, *J. Chem. Phys.* **77,** 1206 (1982).

44. F. von Busch and G. H. Dunn, *Phys. Rev. A.* **5,** 1726 (1972).

45. H. B. Dehmelt and K. B. Jefferts, *Phys. Rev.* **125,** 1318 (1962).

46. K. B. Jefferts, *Phys. Rev. Lett.* **20,** 39 (1968).

47. N. P. F. B. van Asselt, J. G. Maas, and J. Los, *Chem. Phys. Lett.* **24,** 555 (1974).

48. N. P. F. B. van Asselt, J. G. Maas, and J. Los, *Chem. Phys.* **5,** 429 (1974).

49. N. P. F. B. van Asselt, J. G. Maas, and J. Los, *Chem. Phys.* **11,** 253 (1975).

50. J-B. Ozenne, D. Pham, and J. Durup, *Chem. Phys. Lett.* **17,** 422 (1972).

51. W. H. Wing, G. A. Ruff, W. E. Lamb, and J. J. Spezoski, *Phys. Rev. Lett.* **36,** 1488 (1976).

52. A. R. Carrington, P. G. Roberts, and P. J. Sarre, *Mol. Phys.* **34,** 291 (1977).

53. M. Larzilliere, M. Carre, M. L. Gaillard, J. Rostas, M. Horani, and M. Velghe, *J. Chim. Phys. Chim. Biol.* **77,** 689 (1980).

54. J. M. Aarts and J. H. Calloman, *Chem. Phys. Lett.* **91,** 419 (1982).

55. H. Helm and P. C. Cosby, *J. Phys. Chem.* **77,** 5396 (1982).

56. A. R. Carrington, *Bull. Am. Phys. Soc.* **27,** 127 (1982).

57. D. E. Trollivier, G. A. Kyrala, and W. H. Wing, *Phys. Rev. Lett.* **43,** 1719 (1978).

58. R. Locht, J. G. Maas, N. F. B. van Asselt, and J. Los, *Chem. Phys.* **15,** 179 (1976).

59. A. Carrington, R. A. Kennedy, T. P. Softley, P. G. Fournier, and E. G. Richard, *Chem. Phys.* **81,** 251 (1983).

60. J. T. Shy, J. W. Farley, W. E. Lamb and W. H. Wing, *Phys. Rev. Lett.* **45,** 535 (1980).

61. A. R. Carrington, D. R. Milverton, P. G. Roberts, and P. J. Sarre, *J. Chem. Phys.* **68,** 5659 (1978).

62. C. P. Edward, C. S. Maclean, and P. J. Sarre, *J. Mol. Struct.* **79,** 125 (1982).

63. R. C. Dunbar, *J. Am. Chem. Soc.* **95,** 472 (1973).

64. J. M. Kramer and R. C. Dunbar, *J. Chem. Phys.* **60,** 5122 (1974).

65. P. P. Dymerski, E. Fu and R. C. Dunbar, *J. Am. Chem. Soc.* **96,** 4109 (1974).

66. R. C. Dunbar, *J. Am. Chem. Soc.* **97,** 1382 (1975).

67. E. Fu, P. P. Dymerski, and R. C. Dunbar, *J. Am. Chem. Soc.* **98,** 337 (1976).

68. J. R. Eyler, *J. Am. Chem. Soc.* **98,** 6831 (1976).

69. R. C. Dunbar and E. W. Fu, *J. Am. Chem. Soc.* **95,** 2716 (1973).

70. J. M. Kramer and R. C. Dunbar, *J. Chem. Phys.* **59,** 3092 (1973).

71. R. C. Dunbar, *J. Am. Chem. Soc.* **95,** 6191 (1973).

72. M. Riggin and R. C. Dunbar, *Chem. Phys. Lett.* **31,** 539 (1975).

73. D. C. McGilvery, J. D. Morrison and D. L. Smith, *J. Chem. Phys.* **68,** 3949 (1978).

CHAPTER 10

State-to-State Reaction Dynamics

JEAN H. FUTRELL

Department of Chemistry
University of Utah
Salt Lake City, Utah

1 INTRODUCTION

Investigation of the state-to-state reaction dynamics of ion–molecule reactions has developed steadily over the past several years, paralleling in many respects similar developments concerning the kinematics of neutral reactions. The main goal of these studies is the elucidation of the microscopic mechanism of reaction describing the precise way in which reactants evolve into products. Intimately connected with this question is the disposal of energy—internal, translation, and reaction exo- or endothermicity—in the reaction products.

Most of the experimental techniques described in previous chapters are useful for investigating various aspects of state-to-state reaction dynamics. In this chapter three techniques have been selected for discussion in the context of the complementary information that they provide on charge-transfer reaction dynamics in the nitrogen–nitrogen and argon–nitrogen systems. These techniques—crossed molecular beams, laser-induced fluorescence (LIF), and threshold-electron-photoion–secondary-ion coincidence (TEPSICO)—will be shown to provide a self-consistent experimental description of detailed mechanisms that collectively define these reactions. The crossed-molecular-beam technique is particularly useful for translational-energy and angular analysis of products, while the other techniques probe the internal state distributions of reactants (TEPSICO) and products (LIF).

The energetics of the reaction systems are summarized in Fig. 1. Charge-transfer

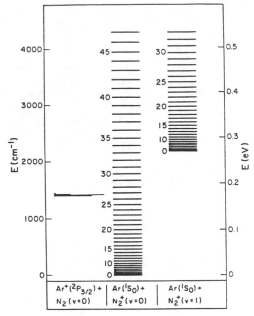

Figure 1. Energy-level diagram of reactants and products of the $Ar^+ + N_2$ charge-transfer reaction referenced to $N_2^+(^2\Sigma_g, v = 0, J = 0)$ as the zero of energy. The Boltzmann distribution of $N_2(v = 0)$ rotational levels at 7 K is indicated qualitatively by the lengths of the respective rotational lines. Reprinted from Ref. 1 with the kind permission of the American Institute of Physics.

$N_2^+(N_2,N_2)N_2^+$ obviously has an energy-resonant channel available to it. However, the charge-transfer reaction

$$Ar^+(^2P_{3/2}) + N_2(X^1\Sigma_g, v = 0) \rightarrow N_2^+(X^2\Sigma_g, v = 0) + Ar(^1S) \qquad (1)$$

is exothermic by about 0.18 eV, while generation of the product ion $N_2^+(X^2\Sigma_g, v = 1)$ is endothermic by 0.09 eV. Lindinger et al. recently resolved the enormous discrepancies in the literature values for rate coefficients for this reaction using the concept of vibrational-state specificity (2). This study inspired several of the detailed studies to be discussed here. The upper state of the argon-ion doublet, $Ar^+(^2P_{1/2})$, and vibrational levels of ground-state N are not shown in Fig. 1, which is drawn to illustrate our experimental conditions, in which a state-selected $Ar^+(^2P_{3/2})$ beam crosses a supersonic jet of N_2 at 7 K. The relative lengths of the lines correspond to the population of rotational states of N_2 from $J = 0$ to $J = 3$. The energy-level scale shown on the right is shifted by 15.58 eV, the adiabatic ionization potential of N_2.

2 EXPERIMENTAL

2.1 Crossed Beam

The apparatus used and its salient characteristics are described in Chapter 8. For the present study, the most relevant features are the high-pressure ion source, which generates relaxed $Ar^+(^2P_{3/2})$ and $N_2^+(X^2\Sigma_g^+, v = 0)$ reactant ions, and the supersonic neutral beam source, which provides a narrow beam of monoenergetic N_2 species at laboratory energies of the order of 0.5 eV; this permits the efficient detection and ready characterization of the products from resonant charge-transfer reactions.

The mass-selected ion beam is decelerated to the chosen collision energy and crossed at $90°$ with the neutral beam, which is generated by supersonic expansion through three differentially pumped vacuum chambers. The energy spread of the ion beam is approximately 0.25 eV, and its angular spread is less than $1°$ (above 2 eV laboratory energies), while the molecular beam has an angular width of about $2°$. Laboratory angular resolution is $2°$ vertically and $3°$ horizontally. The energy resolution of the differential analyzer is about 2.5% of the transmitted energy. In the present experiments, the ions are accelerated to about 3.5 eV after being resolved by the detector defining slits, resulting in an energy resolution of about 0.09 eV.

Laboratory angular and energy distributions for the product ion are measured at a series of angles from $0°$ to $90°$. These data are then transformed into Cartesian probabilities, $P_c(u_1, u_2, u_3)$—the probability density of finding the velocity of the product to be that defined by the Cartesian coordinates infinitesimally close to u_1, u_2, u_3—and scattering contour diagrams are constructed. Using the relation

$$P(T') \propto \frac{u'}{m} \int P_c(u_1, u_2, u_3) \sin \chi \, d\chi, \qquad (2)$$

where χ is the center-of-mass (CM) scattering angle, the probability densities $P(T')$ of the product relative translational energies T' are calculated.

The supersonic beam of neutral particles is critically important to the present study. First, the well-defined velocity-vector distribution is important for distinguishing among different mechanisms of charge transfer. Second, the supersonic acceleration of neutrals allows us to detect low-velocity ionic products moving with essentially the velocity of the reactant neutral. A seeded beam, consisting of about 90% He mixed with N_2, is used in the present study. This ratio was established empirically to provide the best compromise between beam flux and narrow angular distribution. The beam energy, flux, and angular distributions are measured by ionizing a small fraction of the neutral beam using an electron gun mounted orthogonally to the ion and neutral beams at the collision center.

2.2   Laser-Induced Fluorescence

One of the most sensitive and direct techniques for probing product state distributions resulting from chemical reactions is the laser-induced fluorescence (LIF) method introduced by Zare and co-workers (3). It has been reviewed in detail (4) and has been used extensively to investigate neutral reaction dynamics. A laser is used to excite transitions from the unknown population of vibrational and rotational states to electronic states, whose fluorescence is then measured. The deconvolution of the observed spectrum with the proper Franck–Condon factors and rotational degeneracies gives the desired vibration/rotation distributions of the products. The high resolution of the tunable laser source permits detailed analysis of internal-state distributions for many simple molecules. Rates of reaction and energy transfer for specific quantum states may also be investigated by this technique.

Laser-induced fluorescence was first applied to gas-phase ions by Engelking and Smith (5), who investigated the excitation of N_2^+ in a drift tube in collisions with the He buffer gas. They found that the excitation of N_2^+ rotationally proceeded very rapidly, with the rotational temperature approaching closely the mean CM energy distribution of the accelerated ions in the buffer gas. (See Chapter 7 for a discussion of drift tubes and their velocity distributions.)

A typical apparatus, used for LIF studies by the late Bruce Mahan and his co-workers (6), is shown in Figs. 2 and 3. Figure 2 shows the arrangement of the laser, fluorescence detector, computer, and related timing electronics. Figure 3 is a schematic representation of the three-dimensional quadrupole trap used for ion storage,

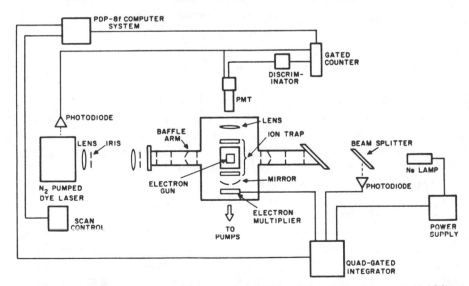

Figure 2.   Schematic representation of the trapped-ion LIF experiment of F. J. Grieman, B. H. Mahan, and A. O'Keefe, *J. Chem. Phys.* **74,** 857 (1981). Reprinted with the kind permission of the American Institute of Physics.

the electron gun ionizer, and the dye laser used for exciting $N_2^+(X)$ to the fluorescent $B^2\Sigma_u^+$ state.

The quadrupole ion trap is capable of storing up to 10^8 ions/cm^3 in a mass-selective fashion. Ions are formed by electron impact ionization of gases introduced into the vacuum housing through a variable-leak valve at a pressure ranging from 10^{-5} to 10^{-6} torr. Under these conditions, the trapped ions collide with background gas molecules on a millisecond time scale. The ions are formed by a short ($\sim$1-msec) pulse of electrons and, after a short ($\sim$0.1-msec) delay, are probed by a 10-nsec pulse from a tunable dye laser. After the fluorescence signal is recorded, the ions are ejected from the trap and a new experimental cycle is initiated with the formation of a fresh ion population. This approach ensures a reproducible initial distribution from cycle to cycle. Typically, the signal is collected for several hundred laser shots before the laser wavelength is stepped by 0.1 Å. As the dye laser frequency is varied, any resultant fluorescence from excited ions is monitored at right angles to the laser beam by a photomultiplier tube and gated photon-counting electronics. By varying the background neutral pressure over a wide range, a controlled number

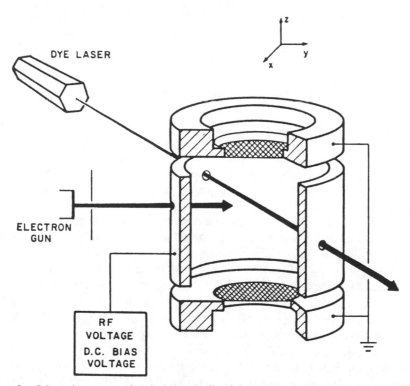

Figure 3. Schematic representation depicting details of the quadrupole ion trap shown in Fig. 2. A radio frequency (RF) voltage is applied to the center ring electrode while the top and bottom electrodes are maintained at ground potential. Reprinted from Ref. 6 with the kind permission of the American Institute of Physics.

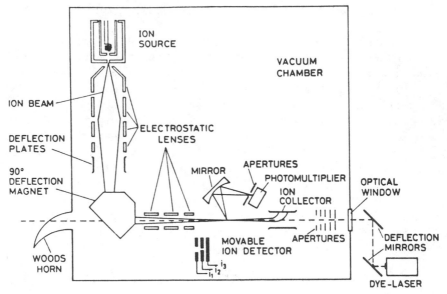

Figure 4. Experimental arrangement for the investigation of the LIF of an ion beam described in Ref. 7. Reprinted with the kind permission of Springer-Verlag.

of ion–molecule collisions are permitted to occur before the dye laser probes the ion population.

Except for very short-lived radiative states, the state distribution observed at 1.1 msec reflects the initial state distribution produced in the ionization process. The effects of ion–molecule collisions can therefore be monitored directly as changes in this distribution. A conceptually similar apparatus utilizing trapped-cell ion cyclotron resonance (ICR) for ion storage has been developed by Marx and co-workers (7).

A LIF apparatus developed by Ding and Richter (8) is shown schematically in Fig. 4. Many of the operating characteristics for the laser probing technique are similar to those for the ion-trap methods. However, the beam method has the advantage that the spectrum of primary ions is Doppler shifted from that of the product ions, making the separation of the two distributions somewhat easier. In the investigation of N_2^+/N_2 collisions, for example, the Doppler shift permitted these workers to resolve the spectra of N_2^+ charge-transfer products from elastically and inelastically scattered projectile ions (see Section 3.2).

An interesting variation of the LIF technique, developed by Leone and co-workers (9), is shown in Fig. 5. Intermediate in its characteristics between the trapped-cell and beam LIF apparatuses, this device utilizes a flowing afterglow to prepare and thermalize the reactant ions. This population is injected into a high-vacuum chamber, where the reactant gas is added by a battery of effusive nozzles. As in the previous examples, gating of the detector to the pulsed laser shot synchronizes the experiment. The principal advantages are the ability to prepare reactant ions and neutrals in

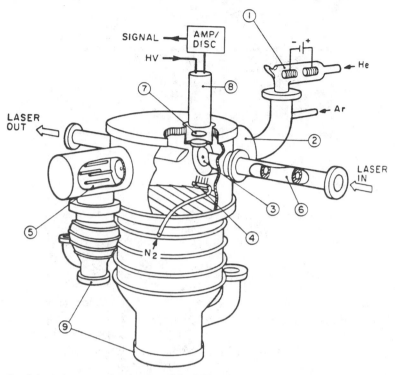

Figure 5. Schematic perspective drawing of the LIF apparatus of Ref. 9. The main components are as follows: 1, hollow cathode discharge; 2, flow tube with blower connection; 3, expansion orifice; 4, neutral reactant inlet; 5, quadrupole mass spectrometer; 6, light baffle; 7, optical filter; 8, photomultiplier; 9, diffusion pumps. Reprinted from Ref. 1 with the kind permission of the American Institute of Physics.

known populations and to carry out single-collision experiments at well-defined collision energies.

2.3   Threshold-Electron–Photoion–Secondary-Ion Coincidence (TEPSICO) Method

The TEPSICO technique was developed by Koyano and Tanaka (10) as a method for correlating ion products with specific quantum states of the reactant ion. Key features of an apparatus of this general type are shown schematically in Fig. 6. The apparatus at Orsay shown schematically in this figure utilizes Synchrotron radiation as a source of ultraviolet (UV) photons. The primary ions A^+ and corresponding electrons are produced in the ionization chamber by photoionization of the molecule by monochromatic vacuum UV radiation. The ions and electrons are repelled out of the chamber in opposite directions perpendicular to the incident photon beam. An electron-energy analyzer, together with a short-steradiancy analyzer section, selects threshold electrons and allows them to pass to the channel multiplier. Extracted ions are formed into a beam of desired velocity by a lens system and reacted

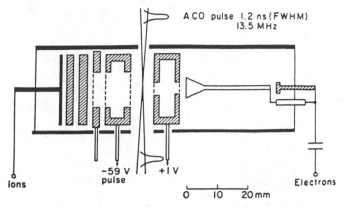

Figure 6. Experimental arrangement of the photoionization region of the Orsay synchrotron storage ring (ACO) threshold-electron coincidence spectrometer. Reprinted from I. Nenner, P. M. Guyon, T. Baer, and T. R. Govers, *J. Chem. Phys.* **72**, 6587 (1980), with the kind permission of the American Institute of Physics.

in a collision chamber. The product ions, together with unreacted primary ions, are mass analyzed by a quadrupole mass spectrometer and detected by a second channel multiplier. The ion signals are then counted in coincidence with the threshold-electron signals. The precision of the photoionization method, coupled with the limitation to the threshold-energy region, precisely specifies the state of the reactant ion for simple systems.

3   REACTION DYNAMICS OF THE CHARGE-TRANSFER REACTION OF N_2^+ WITH N_2 AT LOW AND INTERMEDIATE ENERGIES

3.1   Introduction

Symmetrically resonant charge exchange is fairly well understood for atomic systems. However, very little information exists on the same reaction between molecular ions and their parent molecules. This has several similarities to the atomic case, particularly as regards the large total charge-exchange cross sections for these reactions. However, in contrast with the situation for atomic collisions, the exact symmetry is broken by molecular orientation with respect to the collision axis. The possibility of vibrational- and rotational-energy transfer further complicates the basic dynamics of charge-exchange collisions in molecular systems.

The nitrogen system has been investigated by several techniques. Experimental measurements of the total charge-transfer cross-section have been made by Ghosh and Sheridan (11); Gustafsson and Lindholm (12); Stebbings, Turner, and Smith (13); Flannery, Cosby, and Moran (14); and Utterbach and Miller (15). Theoretical studies relating to charge transfer in the system have been carried out by Flannery et al. (14,16); by deCastro, Schaefer, and Pitzer (17); and by McAfee, Szmanda, Hozack, and Johnson (18). A dynamics investigation of the system at the relatively high impact energy of 111 eV has been carried out by McAfee et al. (18,19). Reaction energetics relevant to this investigation have been measured recently by Linn, Ono, and Ng (20) and by Stefan, Märk, Futrell, and Helm (21). State-to-state studies of energy transfer in the charge-exchange reaction of nitrogen ions with nitrogen have been carried out using the LIF method by Mahan, Martner, and O'Keefe (6) and by Ding and Richter (8).

We show below that the nitrogen system constitutes a textbook example of the case discussed by Rapp and Francis (22), who conjectured that there is a fundamental change in the mechanism of charge transfer in the low-collision-energy regime, where the possibility of the formation of transient orbiting complexes is anticipated. At high energy a two-state, parametrized treatment of charge transfer adequately describes the symmetrical resonance interaction of two identical systems that pass each other in rectilinear trajectories. As collision energies are reduced, a significant number of the charge-transfer interactions proceed via "orbiting" or "capture" trajectories; at this point charge transfer becomes a typical ion–molecule reaction channel. With decreasing energy a smooth transition to this low-energy limiting case is anticipated. The high stability of the putative N_4^+ complex [which is estimated theoretically to be 1.4 eV (17) and has been measured in recent experiments to be 0.9 eV (20,21)] is also an important parameter. Provided the system accesses the ground-state configuration of N_4^+, it is expected on statistical grounds that reducing the total energy as much as possible will increase the lifetime of the N_4^+ intermediate.

3.2   Laser-Induced Fluorescence Results for This System

Figure 7 shows the LIF spectrum of N_2^+ as formed by electron impact under collision-free conditions (6). Three vibronic bands of this sytem can be discerned in this spectrum, all of them belonging to the $\Delta v = 0$ sequence with $v'' = 0$, 1, and 2. All the bands have the same general structure, with a P-branch bandhead at the long-wavelength end of the band and an R-branch envelope extending to shorter wavelengths. Rotational components of even J'' are marked with overhead lines for the R branches of the (0,0) and (1,1) bands in the figure. The expected intensity alternation between even and odd J levels is readily discerned. Relative intensities of the observed vibronic bands reflect the relative populations of each lower vibration state. This population distribution is determined largely by the Franck-Condon overlap of the ground-state neutral molecule with various ionic and autoionizing neutral states; the observed ratio of $v = 1$ and $v = 0$ populations determines an "effective vibrational temperature" of several thousand degrees Kelvin.

In contrast to this effective vibrational temperature, the rotational distribution can be characterized by an equilibrium temperature of 300 K. This much cooler distribution results from the fact that the ionizing electron imparts very little angular momentum to the heavy molecular ion; the resultant distribution mirrors that of the neutral precursor for this room-temperature, trapped-ion experiment.

The translational temperature of the ions can be estimated from the Doppler width

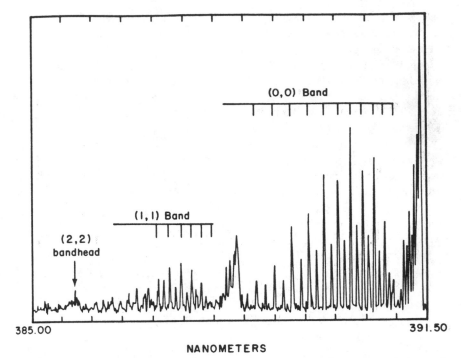

Figure 7. Laser-induced fluorescence spectrum of the N_2^+ $B^2\Sigma_u^+ \rightarrow X^2\Sigma_g^+(0,0)$ band obtained under collision-free conditions. Three vibronic bands can be discerned in the spectrum and the R-branch, even-N components are indicated by overhead ticks. Reprinted from Ref. 6 with the kind permission of the American Institute of Physics.

of well-resolved rovibronic components of each band, about 0.5 cm^{-1}. The temperature deduced is also several thousand degrees Kelvin. Such temperatures result in ion–molecule collision energies of 0.3–0.5 eV. Figure 8 is an enlarged view of part of the overlapping (0,0)- and (1,1)-band R branches of this LIF spectrum. Figure 8a shows the nascent spectrum and Fig. 8b shows the same region after an average of one N_2^+/N_2 collision has occurred. Nearly complete quenching of $N_2^+(X^2\Sigma_g,$ $v = 1)$ is observed. The responsible process is the charge-transfer reaction

$$N_2^+(X^2\Sigma_g^+, v = 1) + N_2(X^1\Sigma_g^+, v = 0) \rightarrow$$

$$N_2(X^1\Sigma_g^+, v = 1) + N_2^+(X^2\Sigma_g^+, v = 0), \tag{3}$$

Figure 8. Laser-induced fluorescence spectra of N_2^+, showing the (1,1) band in greater detail. (a) Spectrum obtained under collision-free conditions. Even-N levels of the (0,0) and (1,1) R branches are indicated by ticks. The arrows designate a splitting of N sublevels attributed to a mixing of N_2^+ $A\ ^2\Pi(v = 11)$ and $B\ ^2\Sigma^+(v = 1)$ levels (6). (b) Recording of the same spectral region after the trapped N_2^+ had experienced an average of one collision with neutral N_2. Vibrational cooling and rotational heating are evident from comparison of these figures. Reprinted from Ref. 6 with the kind permission of the American Institute of Physics.

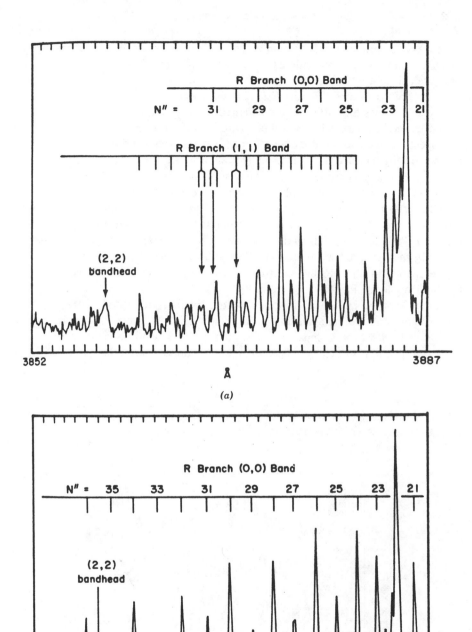

which is endothermic by only 150 cm^{-1}. This process occurs at about unit efficiency, much too rapidly for conventional vibration–translation (V-T) energy relaxation. The rate constant for this quenching reaction is known to be about 5×10^{-10} cm^3/mol/sec from drift-tube (23) and ICR (24) measurements.

Figure 9 shows the LIF spectrum obtained by Ding and Richter (8) for the same system in the beam experiment at much higher collision energy (400 eV CM). The LIF spectrum includes contributions from (a) the scattered primary N_2^+ ions, (b) the inelastically scattered N_2^+ ions resulting from the process $N_2^+(v, J)_{fast} + N_2 \rightarrow N_2^+$ $(v', J')_{fast} + N_2$, and (c) the charge-exchange process $N_2^+(v, J)_{fast} + N_2 \rightarrow N_2 + N_2^+$ $(v', J')_{fast}$. Processes (a) and (b) produce ions having roughly the same velocity, which is determined by the energy of the primary ions. The great majority of product ions from process (c) have roughly thermal velocity (no significant momentum transfer). Processes (a) and (b) display, therefore, a Doppler shift, which is determined by the ion speed and amounts to 0.94 Å in this experiment. A clean separation of the charge-transfer product spectrum generated by process (c) is made possible by this Doppler shift.

Analysis of the inelastic collisions [process (b)] is complicated by the fact that the primary ions are extracted from a plasma source and have a broad rotational distribution. However, comparison of simulated Boltzmann distributions with the observed spectra led to the conclusion that the rotational distribution of the inelastic

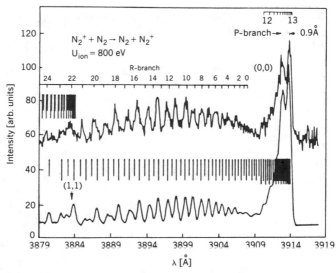

Figure 9. Experimental (top) and simulated (bottom) spectra caused by inelastic and charge-exchange collisions of N_2^+ ions with N_2. The measured spectrum is corrected for the wavelength-dependent laser intensity. The simulated spectrum was calculated assuming a laser line width of 0.5 Å. The vertical bars denote the positions of the rotational transitions of $N_2^+(^2\Sigma_u^+ \rightarrow ^2\Sigma_g^+)$. Reprinted from Ref. 8 with the kind permission of Springer-Verlag.

process (b) can be described by a temperature of 1000 K and that of the primary ions by a temperature of 650 K. This is equivalent to an average energy exchange of 30 meV or an average of about 2.5 rotational quantum jumps per collision. The lack of a permanent dipole moment and the high symmetry of the neutral and ionic molecular nitrogen are probably responsible for the small energy transfer. No vibrational excitation resulting from type-(b) collisions was detected.

For the charge-exchange reaction (c) the measured rotational distribution could be described by a room-temperature Boltzmann distribution. A significant amount (about 5%) of $N_2^+(X, v = 1)$ was also present in the charge-exchange products. It was suggested that some fractions of the two collision partners form an intermediate collision complex of N_4^+ that exchanges energy prior to dissociation. Since no exchange of linear momentum occurs, the "lifetime" of the intermediate must be extremely short. The collision time in the beam experiment is comparable to the vibrational period of N_2^+; apparently this is long enough for vibration–vibration (V-V) energy transfer to occur.

3.3   Beam Scattering Results

Figure 10 is the CM scattering diagram for the charge transfer reaction at 0.74 eV collision energy (25). At this relative kinetic energy two mechanisms are observed that are clearly distinct from each other in terms of the disposal of angular momentum in the reaction: a direct, or impulsive, mechanism, and one proceeding via a reaction complex. For the direct mechanism the locus of product vectors is nearly identical to that of the neutral beam N_2 velocity vectors. Almost no angular momentum is exchanged and the mechanism is well described as corresponding to a long-range

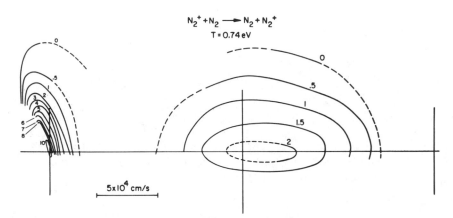

Figure 10.   Scattering diagram for the charge-transfer reaction $N_2^+(N_2, N_2)N_2^+$ at 0.74 eV collision energy. The crosses mark the velocity of the center of mass (CM) of the system and the CM velocity of N_2 prior to collision. Contours define the relative probability densities of product velocities (in Cartesian velocity space). Reprinted from Ref. 25 with the kind permission of Elsevier Science Publishers.

electron jump. Since our experiments involved only ground-state N_2^+ ions [excited species having been relaxed by collisions inside the high-pressure ion source by reaction (3)], the reaction may be written as follows:

$$\underline{N_2^+}(X^2\Sigma_g, v = 0) + N_2(X^1\Sigma_g, v = 0) \rightarrow$$
$$\underline{N_2}(X^1\Sigma_g, v = 0) + N_2^+(X^2\Sigma_g, v = 0), \quad (4)$$

where the underlined species are those moving with higher laboratory velocity before and after reaction. It therefore corresponds exactly to the symmetric resonance case.

The second, or complex, mechanism is defined by complete symmetry, within experimental error, about the location of the CM velocity vector in the hypotenuse of the Newton diagram. This high degree of forward–backward symmetry of the reaction-complex products implies a lifetime that exceeds the rotational period of the N_4^+ intermediate (5×10^{-3} sec for linear N_4^+). The clean separation of these two mechanisms is indicated in Fig. 10.

Information on the energy dependence of lifetimes of N_4^+ formed in ion–molecule collisions is available from recent studies of three-body stabilization of N_4^+ carried out in a variable-temperature drift tube by Böhringer and Arnold (26). These workers report a three-body rate coefficient for the reaction

$$N_2^+ + 2N_2 \rightarrow N_4^+ + N_2 \quad (5)$$

of $k_{3b} = 6.8 \times 10^{-29} (300/T)^{1.64}$, which describes the temperature dependence accurately over several decades. Because many ion–neutral collisions occur in drift tubes prior to reaction, this undoubtedly describes the reaction of vibrationally relaxed, ground-state N_2^+. The mechanism may be expressed as follows:

$$N_2^+ + N_2 \underset{\tau^{-1}}{\overset{k_f}{\rightleftharpoons}} (N_4^+)^*, \quad (6a)$$

$$(N_4^+)^* \overset{\tau^{-1}}{\longrightarrow} N_2^+ + N_2, \quad (6b)$$

$$(N_4^+)^* + N_2 \overset{k_s}{\longrightarrow} N_4^+ + N_2, \quad (7)$$

where k_f represents the rate coefficient for formation of the $(N_4^+)^*$ collision complex, τ its lifetime, and k_s the stabilization rate coefficient. The steady-state approximation applied to this kinetics scheme gives the result

$$k_{3b} = \frac{k_s k_f}{1/\tau + k_s(N_2)}, \quad (8)$$

which, in the low-pressure limit applicable to our crossed-beam experiment, reduces to $k_{3b} = k_s k_f \tau$. A crude estimate of the $(N_4^+)^*$ lifetime can be obtained by substituting the Langevin collision frequencies for k_s and k_f and calculating an "effective temperature" for our collision experiment from the expression

$$\tfrac{3}{2}kT_{\text{eff}} = \langle E \rangle_{\text{CM}}. \tag{9}$$

Substituting our collision energy of 0.74 eV for $\langle E \rangle_{\text{CM}}$ and the Langevin rates for k_s and k_f leads to an estimated lifetime of 5×10^{-13} sec, which coincidentally equals the rotational frequency of the linear N_4^+ complex. Although obviously imprecise, the prediction from these considerations is that some N_4^+ species formed in ion–molecule collisions at 0.74 eV should have lifetimes greater than the rotational period, and some should have shorter lifetimes. This prediction is consistent with our observations.

It is also of interest that the product ions from the second mechanism leave the CM with significantly reduced kinetic energy, implying that much of the collision energy is disposed in internal modes of the products of reaction. This is evidence for very strong coupling of the internal modes in the collision complex. Strong coupling is anticipated from the theoretical paper of deCastro et al. (17), which describes the relatively strongly bonded N_4^+ intermediate that appears to be formed in the charge-transfer reaction. These workers calculated that the linear configuration of N_4^+ was the ground-state species, bound by about 1.4 eV with respect to the separated nitrogen molecular ion and nitrogen molecule. The structure gave equal bond distances of 1.08 Å for the end nitrogen pairs and a separation distance of 1.931 Å for the central pair of nitrogens. Their calculation predicted that 0.6 of the unit charge is located in the central pair of atoms and 0.4 on the end nitrogens. Trapezoidal and T-shaped geometries were also considered and found to be slightly less bound intermediates than the linear configuration. They also noted that the linear configuration gave the proper dissociation asymptote and that there were some problems with the dissociation asymptotes for the other geometries.

Although scattering diagrams contain the most informative data regarding reaction mechanisms, it is important to note that they provide a distorted view of the relative importance of different mechanisms, as a result of the transformation Jacobians involved in constructing these diagrams from the laboratory data (see Chapter 8). The true relative importance of the direct and complex mechanisms is obtained by applying Eq. (2) to the data of Fig. 10. Figure 11 shows the results, which demonstrate that the two mechanisms are of comparable importance at 0.74 eV collision energy.

In addition to defining angular scattering—the forte of beam experiments—conservation of energy provides details on energy conversion in these reactions. The total energy E_{TOT} present in the reactants is the sum of the relative translational energy T (i.e., collision energy) and the energy in the internal degrees of freedom, U. Similarly, the total energy present in the system after reaction is given by the sum of the relative translational energy of the products, T', and the energy in the appropriate internal degrees of freedom of the products, U':

$$E_{\text{TOT}} = T + U = T' + U'. \tag{10}$$

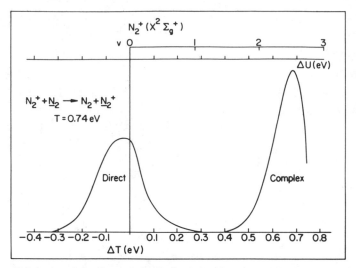

Figure 11. Relative translational-energy distributions of products expressed as translational exoergicity (see defining equations in text) at 0.74 eV collision energy for the reaction $N_2^+(N_2, N_2)N_2^+$. The correlation with ΔU is shown with a superimposed scale for $v = 0, 1, 2, 3,$ and 4 levels of $N_2^+(X^2\Sigma_g^+)$. Reprinted from Ref. 25 with the kind permission of Elsevier Science Publishers.

Consequently, conservation of energy leads to the following relationship between translational endoergicity $(\Delta T = T - T')$ and the internal energy change $(\Delta U = U' - U)$:

$$\Delta T = \Delta U. \qquad (11)$$

Hence a careful measurement of ΔT also measures the change in internal energy in the reaction.

We have superposed a ΔT scale on the bottom and a ΔU scale on the top of Fig. 11, which express the energy conversion in terms of vibrational quanta of $N_2^+(X^2\Sigma_g)$ or of $N_2(X^1\Sigma_g)$ (the vibrational spacings are nearly identical, differing by about 150 cm^{-1}). Formation of the $A^2\Pi_u$ state of N_2^+ and any other electronically excited state of N_2 can be excluded on energetic grounds (27). Clearly the direct mechanism corresponds to translationally resonant charge transfer $(\Delta T = \Delta U = 0)$, whereas the complex mechanism involves extensive internal excitation of the products. The internal energy can be accounted for if about three quanta of vibration (distributed between the N_2^+ and N_2 products in an unknown manner) are deposited in the products; more plausibly, a distribution of vibration and rotational states is formed. If all of the energy were disposed in rotation, about 150 quanta of rotation would be required (27), which is comparable to but slightly higher than is allowed by conservation of angular momentum.

As discussed in Section 3.2, the initial "vibrational temperature" observed by Mahan and co-workers (6) in their trapped-cell LIF study was several thousands of

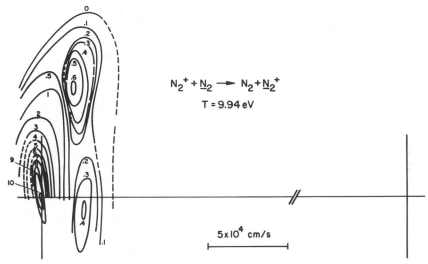

Figure 12. Scattering diagram for the charge-transfer reaction $N_2^+(N_2, N_2)N_2^+$ at 9.94 eV collision energy. The crosses mark the velocity of the CM of the system and the CM velocity of N_2 prior to collision. Contours define the relative probability densities of product velocities (in Cartesian velocity space). The broken line indicates that the vector is not to scale: The CM velocity is equal to the distance between crosses multiplied by 2.38. Reprinted from Ref. 25 with the kind permission of Elsevier Science Publishers.

degrees Kelvin, while the rotational temperature was 300 K. Rapid vibrational cooling and rotational heating of the ions was observed, removing all vibrationally excited states in a few collisions and tripling the rotational-energy distribution in one collision. Since the prediction from our beam-study analysis is that the collision-complex mechanism would have been dominant under their lower-collision-energy experimental conditions, mixing of energy rapidly redistributes the total energy into rotation and vibration of the N_2^+ and N_2 products. With the difference that our experiments involved reactant ions characterized by a room-temperature thermal distribution of internal states, the two types of experiments provide strongly complementary results.

Figure 12 is the scattering diagram at the significantly higher energy of 9.94 eV. Under these conditions the collision-complex mechanism has completely disappeared. As expected, the direct mechanism dominates high-energy collisions, while the collision-complex mechanism is dominant at collision energies below 0.7 eV CM. At a collision energy of 9.94 eV only the direct mechanism is observed. This is consistent with a dramatic decrease in the lifetime of the N_4^+ intermediate with increasing total energy of the system. Under these conditions the available energy exceeds the binding energy of the N_4^+ complex by more than an order of magnitude, and the scattering diagram provides no evidence for a contribution from the persistent-complex model observed in Fig. 10.

The direct, or impulsive, mechanism that is dominant at high energy is the same as that observed at low energy—namely, the translationally resonant reaction (4).

However, we also note the formation of satellite peaks representing scattering with significant exchange of angular momentum at CM scattering angles of approximately 5° and 8°. Both of these ridges of intensity fall inside the translationally resonant charge transfer (TRCT) circle defined by $\Delta T = \Delta U = 0$, demonstrating that they are translationally endoergic processes. The translational endoergicity noted corresponds to ΔT ($\equiv \Delta U$) values of about 0.5 and 0.8 eV, respectively. These ridges of intensity could therefore be associated with the population of $v = 2$ ($\Delta U = 0.545$ eV) and $v = 3$ ($\Delta U = 0.815$ eV) products, respectively (27). Alternatively, they could represent favored distributions of rotational energy in the products of reaction or some special distribution of rotational and vibrational energy in the two products.

It is interesting to speculate that these shoulders of intensity on the main peak may represent the same small-impact-parameter collisions as are associated with the complex mechanism at low collision energies. Although the complex mechanism has two signatures at low energy—forward–backward symmetry and energy exchange—only energy exchange is observed at higher energy. The forward–backward symmetry signature requires a lifetime longer than a rotational period. We may also conclude that the collision samples an attractive potential sufficient to produce full backscattering in the Newton diagram. The lifetime of N_4^+ is shortened dramatically with increasing energy, as we have discussed. At higher collision energy the centrifugal potential (see Chapter 2) also overwhelms the true potential and large-angle scattering can occur only from the repulsive part of the potential. This was clearly the case for the 400-eV-CM LIF study of Ding and Richter (8) and for the 111-eV beam study of McAfee et al. (18), both of which showed that vibrational excitation can accompany high-energy charge transfer in this system.

The $1/v$ dependence of the Langevin model (discussed in Chapter 2) argues that this should be the dominant collision mechanism from thermal to a few tenths of an electron volt CM collision energy. However, the centrifugal potential (also discussed in Chapter 1) overwhelms the true potential at collision energies of 1 eV and higher, and no beam experiment done with the objective of constructing a Newton diagram can be carried out using current techniques in the energy range where the Langevin mechanism is clearly expected to be dominant. (The results would be quite boring anyway.) The theoretical study of N_4^+ geometry and energetics by deCastro et al. (17) demonstrates much higher binding energy [1.4 eV calculated (17) vs. 0.9 eV experimental (20,21)] than is calculated from the Langevin model at the same separation distance (0.17 eV). This strongly attractive true potential accounts for the orbiting trajectories actually observed at 0.74 eV CM, while the existence of several geometries with similar binding energies helps to rationalize the efficient coupling of energy into internal modes. It therefore follows that the low energy Rapp–Francis–Langevin model for charge transfer is not the only possible explanation for the complex model demonstrated in Fig. 10. Finally, the separation of the charge-transfer reaction into two mechanisms at 0.74 eV is a plausible result of the interpretation discussed earlier of the three-body recombination study of Böhringer and Arnold (26), which predicts the lifetime for N_4^+ containing this total energy to be roughly equal to the inverse of the rotational frequency of

N_4^+. That our reactants are relaxed 400 K ions and 7 K neutrals may also be a relevant parameter.

The detailed mechanism of charge and energy exchange during the collision therefore remains a matter of conjecture. The coupling modes we have invoked for efficiently leaking translational energy into internal modes would trap the collision partners inside the centrifugal barrier during the first passage (at a low energy such that a well exists in the effective potential) and several "encounters" are likely prior to separation of the products. This may also be responsible in part for the clean separation of mechanisms observed in Fig. 10.

A curve-crossing mechanism between states of approximately the same g–u symmetry at small internuclear distances (1–2 Å) is invoked for N_2^+/N_2 in the theoretical treatment of charge-transfer by Flannery et al. (14). Only rectilinear trajectories are considered. This model predicts that $\Delta v = 0$ will be the dominant channel and that the product of this channel will appear predominantly at zero scattering angle, whereas significant angular deflections are predicted for vibrationally excited products. These predictions are entirely consistent with the 9.94-eV-collision-energy experiment (Fig. 12). Multiple encounters which can occur for orbiting collisions were not discussed by these researchers; however, angular information would clearly be lost in such encounters and the probability for intimate collisions responsible for vibrational excitation might be significantly enhanced by low-energy orbiting collisions. Alternatively, coupling of modes in the various geometries of N_4^+ can be invoked to explain the essentially statistical mixing of energy in all modes, including translation, suggested by Fig. 11 for the collision-complex model.

3.4 Conclusions

The conceptual picture suggested by the results discussed above is as follows: At low collision energies the Langevin cross section is quite large and separates ion–molecule trajectories into two classes—those having very large impact parameters, so that only modest deflection occurs, and those with impact parameters less than the critical value for orbiting, which causes the two scattering partners to undergo one or more intimate collisions. The Langevin cross section for charge transfer, according to the hypothesis of Rapp and Francis (22), is $(\pi/2)b_L^2$, where b_L is the impact parameter for orbiting trajectories and the factor $\frac{1}{2}$ is the probability that each of the collision partners will retreat as the charged particle. Consideration of the much stronger binding energy of N_4^+ relative to that resulting from ion-induced dipole interactions extends the energy range for orbiting trajectories to ca. 1 eV, so that it includes the range accessible to the crossed-beam technique.

The mechanism associated with the small impact parameters necessary to intermix the internal energies thoroughly is therefore manifested at low energies by both of the characteristic signatures of the orbiting-complex mechanism—scattering that is symmetric about the CM vector and energy exchange. At the high energies of Fig. 12 the cross section for orbiting collisions is smaller than the hard-core collision cross section associated with the N_4^+ intermediate itself, which is equivalent to impact parameters of ca. 2–3 Å. Hence the longer-range electron-jump mechanism is the dominant one at high energy. The small fraction of collisions that occur at internuclear distances comparable to those of the N_4^+ species may be responsible for the satellite intensities observed in addition to the main peak. Under these circumstances extensive mixing of vibrational and rotational energy is expected. A mixture of curve crossing and coupling of modes in N_4^+ rationalizes the detailed mechanisms satisfactorily.

4 THE CHARGE-TRANSFER REACTION OF Ar$^+$ WITH N$_2$ AT LOW AND INTERMEDIATE ENERGIES

4.1 Introduction

Using the TEPSICO Technique pioneered in their laboratories, Kato, Tanaka, and Koyano (28) have recently investigated the relative rate coefficients for the reactions of the $^2P_{3/2}$ states of Ar$^+$ with nitrogen. They found that the $^2P_{3/2}$ state reacts about half as fast as the $^2P_{1/2}$ state and that the ratio is independent of collision energy over the range of 2–10 eV. The selected-ion drift tube (SIDT) measurements of Lindinger et al. (2) and the 80 and 300 K flow-tube study of Smith and Adams (29) show that the overall rate coefficient for reaction of argon ions with nitrogen increases by about two orders of magnitude with increasing collisional energy over the range from thermal energies to about 3 eV. Lindinger et al. (2) and Smith and Adams (29) interpret their results in terms of the endothermicity of the reaction channel generating $N_2^+(X^2\Sigma_g^+, v = 1)$ from Ar$^+(^2P_{3/2})$. They assert that Ar$^+(^2P_{1/2})$ is rapidly converted to Ar$^+(^2P_{3/2})$ in drift tubes and that the rate coefficient for the reaction of Ar$^+(^2P_{3/2})$ with N$_2$ to generate $N_2^+(X^2\Sigma_g^+, v = 0)$ declines from 10^{-11} cm^3/mol/sec at 80 K to 7×10^{-12} cm^3/mol/sec at 140 K. The rate of the endothermic reaction of Ar$^+(^2P_{3/2})$ generating $N_2^+(X^2\Sigma_g^+, v = 1)$ rises from 10^{-12} cm^3/mol/sec at 300 K to the Langevin rate at 3 eV relative translational energy (2). The minimum in the $k(E)$ curve is obtained by extrapolating the Ar$^+(^2P_{3/2})$ and Ar$^+(^2P_{1/2})$ curves.

These experiments have resolved the order of magnitude discrepancies in the reported values for the rate coefficient of this reaction. The apparent low probability of generating ground-state ions at low collision energy, the differing reactivities for the Ar$^+(^2P_{3/2})$ and Ar$^+(^2P_{1/2})$ substates, and the large rate coefficient for the reverse reaction exhibited by vibrationally excited N$_2^+$ make rate constant measurements extremely sensitive to experimental conditions. These intriguing kinetics features also identify this reaction system as an interesting candidate for a detailed dynamics study. We show below that the complementary LIF, TEPSICO, and crossed-beam techniques reveal a very high degree of quantum-state specificity in this reaction; this specificity is responsible for the interesting kinetic features noted.

The energetic relationships for this reaction are shown in Fig. 1. In this figure the Ar$^+(^2P_{3/2})$ and $(^2P_{1/2})$ energy levels are shown along with the vibrational levels for N$_2$$^+$ and N$_2$. Only the Ar$^+(^2P_{3/2})$ lower state is present in the reactant-ion beams used in the experiments we discuss below.

4.2 Laser-Induced Fluorescence

Results from the elegant LIF study of this system by Hüwel, Guyer, Lin, and Leone (1) are shown in Fig. 13. Their flowing afterglow ion source uses helium as a carrier gas. The drift tube is terminated by a nozzle beam expansion through an orifice at the end of the flow tube that injected ions into a low-pressure chamber. The resulting beam of argon ions (and helium) crosses an effusive spray of the neutral N$_2$ reagent. Reaction occurs in the crossing zone of the two reagent beams at a mean collision energy of 0.24 eV with an energy spread of ±0.09 eV. Single-collision conditions are preserved by the efficient pumping of the reaction region, which is located in the throat of a 10,000 liter/sec diffusion pump. Excitation of the product molecules is achieved with a commercial pulsed dye laser pumped by a 10-Hz Nd:YAG laser (see also Fig. 5). The laser beam crosses perpendicular to both beams and parallel to the linear neutral-inlet array. It enters and exits the reaction chamber through a system of light baffles. Subsequent fluorescence of the N$_2$$^+$ product passes through an appropriate set of optical filters and is viewed by a fast photomultiplier. Optical filters transmit the (1,2) and (0,1) emission bands at ca. 424 and 428 nm, respectively.

Figures 13a and 13b show the LIF spectrum observed at low and high resolution, respectively. Qualitatively, the higher population of the $v = 1$ level for the N$_2$$^+$ product ion anticipated from the earlier kinetics studies [and our beam study (30)] is immediately evident in the unrelaxed, low-resolution spectrum of Fig. 13a. Figure 13b shows the unrelaxed spectrum at high spectral resolution. Deducing the quantitative ratio and the rotational distribution is relatively complex and is discussed briefly in the next section.

The "resonance rule" has often been invoked to rationalize the observed magnitude of rate coefficients for charge-transfer reactions and infer the product states that are populated, often with Franck–Condon factors added as an additional constraint. These considerations argue strongly for the formation of rotationally excited N$_2$$^+(^2\Sigma_g{}^+$, $v = 0)$. The $(2J + 1)$ multiplicity of rotational states provides a high density of states in the neighborhood of $J = 27$ that are almost exactly resonant with the Ar$^+(^2P_{3/2})$ recombination energy. [Nuclear symmetry suggests that even-J states ($J = 26, 28, \ldots$), which do not provide quite as good an energy match, would actually be favored for N$_2$$^+$ excitation; see the LIF results for N$_2$$^+$ generated by electron impact in Fig. 7.] Moreover, no angular momentum constraints preclude this kind of rotational excitation. The maximum value of the rotational quantum number can be estimated from the angular orbital momentum, given by

$$J_{max} = \frac{\mu v_{rel} (\sigma/\pi)^{1/2}}{\hbar}, \tag{12}$$

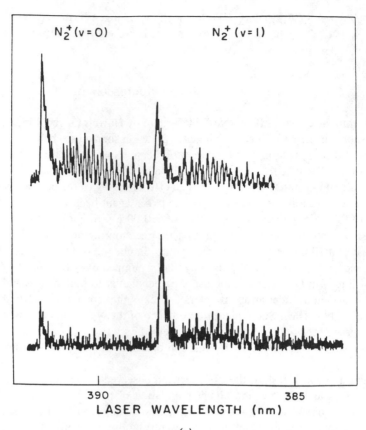

$N_2^+(v=0)$ $N_2^+(v=1)$

390   385

LASER WAVELENGTH (nm)

(a)

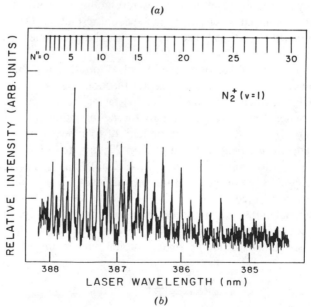

N"=0   5   10   15   20   25   30

$N_2^+(v=1)$

388   387   386   385

LASER WAVELENGTH (nm)

(b)

222

where μ is the reduced mass, σ the collision cross section, and $\hbar = h/2\pi$. This equation assumes no angular momentum is contained in the reactant N_2. For a relative velocity of 1.6×10^3 m/sec and a cross section of 5×10^{-20} m² (the *reaction* cross section at this collision energy), J_{max} is 44. The angular momentum balance needed to reach the resonant energy level is therefore easily achieved by the decreased relative velocity and smaller final impact parameters of the products. Clearly the distribution of Fig. 13 differs dramatically from that predicted by these considerations.

4.2.1 Vibrational-State Distribution

As discussed earlier, LIF is a definitive technique for determining product internal-state distributions, while the beam method provides overall information on angular and linear momentum exchange. Together they provide tight specifications for the detailed mechanism of reaction and a challenging data set for theoretical investigation of the potential-energy surface. We discuss the LIF results first, then the molecular beam study.

As stated by Leone et al. (1), the only reliable method for obtaining relative vibrational populations is to sum the intensities of all contributing rotational lines of a given transition. For $N_2^+(B-X)$, the P branch offers a convenient way to calculate this sum. All rotational lines from $J = 1$ to $J = 30$ can be excited within a spectral interval of about 30 cm^{-1} in the bandhead region. Since this includes the entire energy range actually populated, integration of the P-branch fluorescence intensities yields an accurate measure of the relative vibrational populations.

The detected fluorescence intensity following absorption of the incident laser light on any rovibronic level J'', v'' is formulated as

$$I_{LIF}(J'', v'') = n_{J'',v''}\alpha_{EXC}\beta_{EM}, \tag{13}$$

where $n_{J'v'}$ is the number density of electrons at the level of interest, α_{EXC} is the product of the laser intensity and transition strength, and β_{EM} is the product of the emission strength and overall detection efficiency. Under saturation conditions the excitation factor α_{EXC} is independent of both the vibrational overlap between the two levels and the laser intensity and reflects only the statistical weights of the excited and lower levels.

Since only P-branch excitation was used by Leone et al. (1) for the analysis of the vibrational distribution of N_2^+, the rotational excitation is governed by $J' = J'' - 1$. Subsequent emission occurs as a result of P and R transitions to all vibrational

---

Figure 13. Laser excitation spectrum of N_2^+ produced in the charge-transfer reaction $Ar^+ + N_2$ at about 0.24 eV collision energy. (*a*) Spectra taken at low resolution. The nitrogen flow is 5.2×10^{18} (result partially relaxed) and 7.1×10^{17} (unrelaxed) molecules sec^{-1} for the upper and lower traces, respectively. (*b*) Laser excitation spectrum of the $R(1,1)$ band of N_2^+ at higher resolution with rotational quantum number assignment. The nitrogen flow is 6.0×10^{17} molecules sec^{-1} (unrelaxed distribution). Reprinted from Ref. 1 with the kind permission of the American Institute of Physics.

levels of the electronic ground state, with the relative intensities governed by Franck–Condon factors. Leone et al. (1) used suitable optical filters to select the $\Delta v = -1$ progression in emission; absorption of the incident laser light on any rovibronic level was therefore followed by a *detected* emission containing only two lines. The β_{EM} factor for this situation can be written as

$$\beta_{EM} = g(v'', v')[v_R{}^4 T(v_R) + v_P{}^4 T(v_P)], \tag{14}$$

where g is the Franck–Condon factor for emission, v_R and v_P are the frequencies of the two emission lines, and the T's are the detection efficiencies at these frequencies (determined by the transmission of the optical filter and the quantum efficiency of the photomultiplier).

Since the LIF technique measures number densities, a velocity transformation is required to express relative cross sections as product fluxes in the laboratory frame. The total flux of $N_2{}^+$ products going into $v = 0$ or $v = 1$ and scattered into all directions is given by $\dot{N}_i = \sigma_i v_{rel} n_{Ar+} n_{N_2}\, dV$, where σ_i is the total of the reaction cross sections for production of $N_2{}^+(v = i)$ at velocity v_{rel}; n_{Ar+} and n_{N_2} are the projectile and target densities, respectively; and dV is the reaction volume. It follows that the branching ratio can be expressed as

$$\frac{\sigma_0}{\sigma_1} = \frac{\dot{N}_0}{\dot{N}_1} = \frac{n_0 \bar{v}_0}{n_1 \bar{v}_1} \tag{15}$$

where n_0, n_1 are the number densities of the $N_2{}^+$ $(v = 0,1)$ reaction products and $\bar{v}_0$, $\bar{v}_1$ are angle-averaged velocities in the laboratory frame for $N_2{}^+(v = 0)$ and $N_2{}^+$ $(v = 1)$. As noted earlier, the LIF-measured densities n_0, n_1 are summations of all *P*-branch excitation lines.

It turns out that the major uncertainty is in estimating the average product laboratory velocities $\bar{v}_0$ and $\bar{v}_1$. For a given pair of initial and final rovibronic states of the reagent N_2 and the product $N_2{}^+$, the CM velocity of the particular final state can be calculated from the known exothermicity and the initial translational energy. However, for each state the transformation from CM velocity u to laboratory velocity v depends critically on the CM scattering angle (discussed in Chapters 2 and 8). No experimental information exists on the differential cross section of the Ar^+–N_2 charge-transfer reaction at 0.24 eV, and it is not possible to carry out well-defined beam studies at this low energy. The most definitive results obtained to date are presented in Section 4.3.

Accordingly, Leone et al. (1) idealized the possible CM scattering patterns in $v = 0$ and $v = 1$ into forward-, sideways-, and backward-scattering Newton diagrams in order to estimate the relative cross sections using Eq. (15). Numerical values were obtained by averaging over a range of $\pm 15°$ around the principal observation direction of the LIF director, which subtends both forward- and backward-scattered products. These estimates lead to the conclusion that velocity ratios between $v = 0$ and $v = 1$ can vary by more than an order of magnitude. Assuming that the two products do not differ dramatically in their scattering properties and that forward scattering (with respect to the N_2 neutral vector) predominates for the $v = 0$ product but not for $v = 1$, they estimated a velocity-ratio range of 0.8 ± 0.5 for $v = 0/v = 1$. [The second assumption is supported by our 0.6-eV beam study (see Section 4.3). A logical extrapolation of our data to lower energy corroborates their choice of velocity ratios and suggests that the uncertainty may be reduced to ± 0.2.]

Using the appropriate Franck–Condon factors in emission, the stated range of final $v = 0/v = 1$ velocity ratios of 0.8 ± 0.5, and the values for detection efficiency and emission frequencies, a final value of 0.13 ± 0.10 is obtained from Eq. (14) for the $v = 0/v = 1$ branching ratio. Evidently about 90% of the N_2^+ molecules formed in the charge transfer at 0.24 eV acquire one quantum of vibrational excitation. The uncertainty in the branching ratio is traceable to the uncertainty in the laboratory velocities of the products. Statistical errors in the experimental data are much smaller.

4.2.2   Rotational-State Distribution

No attempt was made by Leone et al. (1) to measure the rotational-state distribution in the minor (10% or less) product channel, $N_2^+(v = 0)$. The R branch of the $B\,^2\Sigma_u^+ - X\,^2\Sigma_g^+$ transition $v = 1$ channel structure is easily discerned in Fig. 13. The qualitative shape of the spectrum approximates that expected for a Boltzmann distribution, but with a sharp cutoff at $J = 30$. From the slope of a linear-regression fit to these data, a rotational temperature of 685 ± 50 K was estimated. This corresponds to the transfer of about 5 $\hbar$ of energy into rotation, rather than the 40 h available in orbital angular momentum at this collision energy.

4.3   Beam Experiments

A CM scattering diagram that presents the Cartesian probability (the probability of locating a product with a velocity defined by the Cartesian coordinates of the product ion) for the reaction of $Ar^+(^2P_{3/2})$ at a collision energy of 1.73 eV is shown in Fig. 14. The arc of the circle drawn centered on the line CM for the collision process defines the locus of velocity vectors that would be observed if the experiment were conducted with monochromatic ion and neutral beams in their ground states and if the product ions were formed exclusively in the $X\,^2\Sigma_g^+(v = 1)$ state with no rotational energy. The inference that we reach from this diagram is that the products are formed predominantly in the $v = 1$ vibrational level with a small number of rotational quanta. Further, that the product is scattered forward (with reference to the initial neutral beam velocity vector) is evidence that very little momentum is transferred in the collision.

The product-ion state assignment is further elaborated in Fig. 15, which shows the integration of the Cartesian velocity-space diagram of Fig. 14 over angle obtained using Eq. (2) and analyzes the data in terms of the observed translational endoergicity

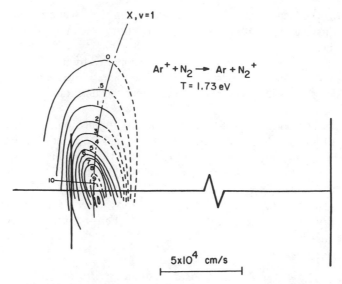

Figure 14. Scattering diagram for the charge-transfer reaction $Ar^+(N_2, Ar)N_2^+$ at 1.73 eV collision energy. The crosses mark the velocity of the CM of the system and the CM velocity of N_2 prior to collision. The break in the line indicates that the velocity-vector difference between the N_2 neutral and the CM is not to scale: The distance between the two crosses should be multiplied by about 1.7 to correspond to the indicated velocity scale. Contours define the probability densities of product velocities (in Cartesian coordinates) and demonstrate that a direct mechanism populates a slightly endothermic channel. The inscribed radius (labeled X, $v = 1$) is the locus of vectors corresponding to the translational-to internal-energy conversion required to drive the reaction $Ar^+(^2P_{3/2}) + N_2(X, v = 0) \rightarrow Ar(^1S_0) + N_2^+$ (X, $v = 1$) with no energy deposited as rotation. Reprinted from Ref. 31 with the kind permission of North Holland Publishing Company.

ΔT and the corresponding change in internal energy, ΔU. Also shown in Fig. 15 (as the cross-hatched area) is the combined dispersion of the ion and neutral beam velocities; comparison with the observed area of product-ion intensity and the superposed vibrational-energy scale for $N_2^+(X\ ^2\Sigma_g^+, v)$ at the top of the figure demonstrates that the product is generated mainly in the $v = 1$, level with a modest distribution of rotational energy. This result is completely consistent with the conclusions reached from the LIF study at 0.24 eV CM discussed in the preceding section.

Further support for the direct electron-jump mechanism in this particular system is contained in the paper by Kato et al. (28). They demonstrate that the Rapp and Francis theory (22) impact-parameter treatment (which presumes rectilinear trajectories), when combined with Franck–Condon factors for overlap of the vibrational states of the reactant neutral with the relevant vibrational states of the product ion, satisfactorily reproduces the translational-energy dependence of the cross sections for reaction of the $Ar^+(^2P_{3/2})$ substate with N_2. This calculation leads to the further conclusion that the overwhelming contribution to the cross section comes from

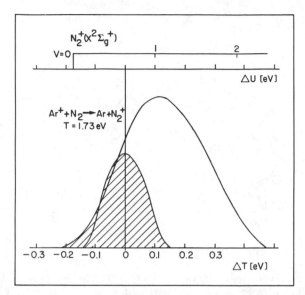

Figure 15. Relative translational-energy distribution of products observed for the reaction Ar⁺(N₂, Ar)N₂⁺ at 1.73 eV collision energy, expressed in terms of the translational exoergicity. The correlation with ΔU is shown with a superposed scale for $v = 0, 1, 2, 3$ vibrational levels for $N_2^+ (X\ ^2\Sigma_g^+)$. The crosshatched peak is the energy distribution of the neutral N_2 beam convoluted with the analyzer apparatus function and integrated over scattering angle. Reprinted from Ref. 30 with the kind permission of the American Institute of Physics.

population of the vibrational states of N_2^+ resulting in the smallest energy defects, for example in the reaction

$$Ar^+(^2P_{3/2}) + N_2(^2\Sigma_g^+, v = 0) \rightarrow N_2^+(^2\Sigma_g^+, v = 1) + Ar(^1S_0).$$

This inference from their calculation is consistent with the findings of Lindinger et al. (2) and is confirmed completely by the energy-distribution analysis in Fig. 15 and by the LIF study of Leone et al. (1)

Our discussion above of the LIF results of Leone et al. (1) pointed out that their reported uncertainty in the relative cross sections for formation of the $v = 1$ and $v = 0$ states of $N_2^+(^2\Sigma_g)$ resided almost entirely in the relative velocities of these two products defined by the angular scattering reaction dynamics for this system. We have also noted that it is impossible to carry out a crossed-beam study at a collision energy of 0.24 eV: Not only would the energy spreads in the two beams exceed the collision energy, but also angular spreads, space charge, and energy-resolution problems preclude the meaningful measurement of differential cross sections at very low ion energy. For the Ar⁺/N₂ system there is the further complication that the rate coefficient drops dramatically with decreasing energy (see Ref. 2 and Section 1).

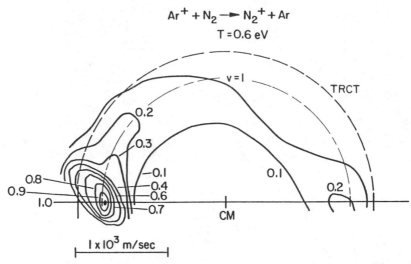

Figure 16. Scattering diagram for the charge-transfer reaction $Ar^+(N_2, Ar)N_2^+$ at 0.60 eV collision energy. Concentric circles labeled TRCT and $v = 1$ define the loci of the resonant and endothermic processes discussed in the text. The maximum intensity at $0°$ and the 20% contour at $180°$ (with respect to the initial N_2 velocity vector) both correspond to $-\Delta U = 0.1$ eV and match the endothermicity of the reaction $Ar^+(^2P_{3/2}) + N_2(X^1\Sigma_g^+, v = 0) \rightarrow Ar(^1S) + N_2^+(X^2\Sigma_g^+, v = 1)$. The symmetry beginning to develop in this diagram suggests that ca. 40% of all reactive collisions result in orbiting about the CM.

In an attempt to reach the lowest energy pragmatically attainable, we have carried out a scattering experiment at 0.60 eV CM. The results are displayed in Fig. 16. This single diagram represents the result of about 3 months of data averaging at the low count rates observed at this energy for all scattering angles. The concentric circles define the locus of vectors corresponding to elastic scattering (TRCT) and the endothermic generation of $v = 1$ products with no rotational energy. As in the higher-energy experiments already discussed, the dominant product is formed in $v = 1$. There is also some evidence in this diagram for formation of several percent of forward-scattered $v = 0$ product.

The most interesting new feature is the backscattered maximum that is connected by a ridge of large-angle-scattered intensity to the predominant forward peak. This start of the development of backward–forward symmetry is most plausibly interpreted as arising from a collision intermediate whose lifetime is about one rotational period. The observation of this partial symmetry at 0.6 eV (and its gradual disappearance with increasing collision energy) is evidence for stronger attractive forces for the ArN_2^+ species than can be accounted for by ion-induced dipole and ion–quadrupole forces. Speaking in pseudodiatomic terms, a potential-well depth of the order of 1 eV is required to rationalize the angular scattering observed, while the quantum-state specificity observed requires a direct or impulsive mechanism for the actual charge-exchange step. The well depth sampled inferred from this scattering meas-

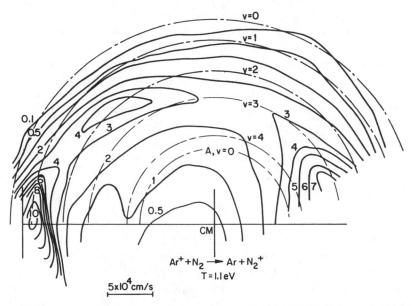

Figure 17.   Scattering diagram for the charge-transfer reaction $Ar^+(N_2, Ar)N_2^+$ at 1.1 eV collision energy. The crosses mark the velocity of the CM of the system and the CM velocity of N_2 prior to collision. Contours define the probability densities of product velocities (in Cartesian coordinates), while the dashed inscribed circles define the loci of velocity vectors for N_2^+ products leaving the collision center with conversion of translational to internal energy corresponding to the formation of $N_2^+(X\ ^2\Sigma_g)$ with 0, 1, 2, 3, and 4 quanta of vibration and of $N_2^+(A\ ^2\Pi_u, v = 0)$ with no energy deposited in rotation. Reprinted from Ref. 31 with the kind permission of North Holland Publishing Company.

urement is consistent with the 1.2-eV binding energy estimated for ArN_2^+ by Teng and Conway in a high-pressure mass spectrometry study (32).

We have also investigated this reaction at several intermediate energies. The results obtained at 1.1 eV collision energy, presented in Fig. 17 (31), are strikingly different. The efficient utilization of translational energy to populate several endothermic product channels is evident, as is the dominance of large-angle scattering under these experimental conditions. The dominance of large-angle scattering and the observation of much more extensive conversion of translational energy into internal energy in the formation of products are in striking contrast to the higher-energy results.

There are three clearly resolved centers of intensity in the scattering diagram. The probable interpretation of these maxima is indicated by the superimposed dashed circles, drawn to show the loci of CM velocity vectors that correspond to products scattered with exactly the translational to internal-energy conversion required to generate the $v = 0, 1, 2, 3,$ and 4 levels of the ground electronic state (assuming no rotational excitation). Since rotational-energy levels are closely spaced and are not resolvable in our scattering experiment, and since the centers of intensity in Fig. 17 both are well resolved and excellently match the vibrational spacing of

N_2^+, we interpret our results as evidence for vibrational rather than rotational excitation of the product ion. The shading of intensity into the regions between the superposed $v = n$ circles is evidence for some rotational excitation (particularly for $v = 3$), but we suggest that the rotational excitation is modest. Of course, beam experiments such as our are not definitive regarding the way internal energy is partitioned, and it would be highly desirable to test this interpretation of our results by the LIF method to define in detail the internal-energy states of the product ions at this collision energy.

According to this interpretation of Fig. 17, the $v = 1$ level is a prominent product with a zero scattering angle (with respect to the initial *neutral beam* velocity vector), just as is observed at both higher and lower energy. A second island of maximum probability is found at angles 30–50°, corresponding to formation of the $v = 2$ product, while $v = 3$ is found to have a maximum at approximately 180°. The locations of the maxima of the intensity contours for the apparent $v = 3$ product suggest higher rotational excitation of this level. In addition, there is some evidence of a forward-scattered maximum of lower intensity corresponding to $v = 3$ and a weak backward-scattered peak corresponding to the $v = 4$ circle. Some intensity is observed also at the $v = 0$ circle, but the resolution is such that it is not possible to make any quantitative statements about the population of the ground vibrational level.

The quantum-state-specific angular scattering observed in this three-atom system shows an intriguing analogy to the scattering reported by Lee and co-workers (33,34) for the reactive scattering of F with H_2. Although the systems are obviously quite different, and the phenomenon here is simply an electron transfer rather than an atom transfer, there are striking similarities in the results. In particular, a strong correlation of quantum-state specificity with angular scattering is observed at a particular collision energy and is not observed at higher and lower energies in both systems. Lee (33) and Neumark, Wodtke, Robinson, Hayden, and Lee (34) interpreted their results as evidence for a quantum mechanical reactive-scattering resonance. The experimental observation of such a dynamic resonance has been suggested as equivalent to obtaining a vibrational spectrum of the transition state for the reaction (35).

4.4   Conclusions

The general conclusion reached from the beam experiments at 0.6 eV and at collision energies of 1.5 eV and higher, some of which have been discussed in this chapter, and from the LIF study of Leone et al. (1) is that the predominant product from the charge-transfer reaction of $Ar^+(^2P_{3/2})$ with $N_2(X\ ^1\Sigma_g, v = 0)$ is the vibrationally excited product $N_2^+(X\ ^2\Sigma_g, v = 1)$ with a modest degree of rotational excitation. This interesting conclusion is supported by microscopic-reversibility arguments applied to the study of the reverse reaction by Kato, Tanaka, and Koyano (36) and by a detailed series of experiments by Govers, Guyon, Baer, Cole, Fröhlich, and Lavollée (37). The latter workers used a Synchrotron UV light source and time-of-flight coincidence techniques to measure the reaction cross sections for ground-state N_2^+ excited to various vibrational levels and for the first electronically excited A state. Figure 18 presents the cross sections for these vibrationally excited ions at 8 eV CM collision energy. Within experimental error the cross section for reaction of the $v' = 0$ level is 0, whereas the cross section rises rapidly to a broad maximum of about 30 Å² at $v' = 3$.

A qualitative rationalization of the results for both the forward and reverse reactions can be made in terms of crossing seams of the relevant potential surfaces. Figure 19 is a hypothetical set of pseudodiatomic potential curves for the species $N_2^+(X\,^2\Sigma_g, v = 0) + Ar$ and $N_2^+(X\,^2\Sigma_g, v = 1) + Ar$, which are similar in shape but offset (schematically) by the energy difference of O.27 eV (at $R = \infty$) (see Fig. 1). Between these curves we have sketched a hypothetical potential curve for the $N_2(X\,^1\Sigma_g, v = 0) + Ar^+(^2P_{3/2})$ that crosses the diabatic curve for $N_2^+(X\,^2\Sigma_g, v = 1) + Ar(^1S)$ at some arbitrary internuclear distance. A high probability of curve-crossing transitions of the Landau–Zener–Bauer–Fischer–Gilmore type (38,39) is indicated schematically by the dashed lines in the figure. This qualitative curve rationalizes the highly state-specific charge-transfer reactions that have been observed for the forward and reverse reactions.

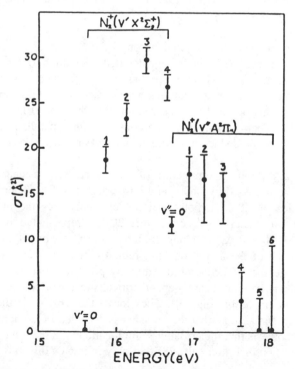

Figure 18. Cross sections for the charge-transfer reactions of state-selected $N_2^+(X\,^2\Sigma_g^+, v')$ and N_2^+ $(A\,^2\Pi_u, v'')$ with $Ar(^1S_0)$ at 8.1 eV collision energy. Reprinted from Ref. 37 with the kind permission of North Holland Publishing Company.

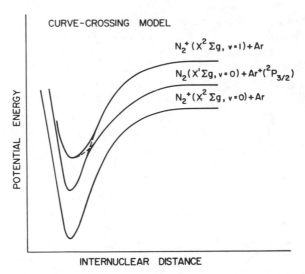

Figure 19. Qualitative pseudodiatomic potential curves suggested as a rationale for the observed charge-transfer reaction dynamics of the $(N-N-Ar)^+$ system. The dashed lines depict the high probability of an "avoided crossing" between the adiabatic and diabatic curves. Such a crossing would result in charge transfer between $Ar^+(^2P_{3/2}) + N_2$ and $N_2^+(^2\Sigma_g,\ v = 1) + Ar$. Reprinted from Ref. 30 with the kind permission of the American Institute of Physics.

Indeed, a semiquantitative rationalization of the magnitude of the cross sections is possible. Govers et al. (37) show that a detailed calculation for the reverse reaction based on the Bauer–Fischer–Gilmore model (38,39) predicts crossing distances for the relevant potential curves that rationalize satisfactorily both the magnitude of the respective cross sections and the observation that the $v = 0$ level does not react with neutral Ar. Their article should be consulted for a detailed discussion of the calculation and for the deduction of the actual crossing distances for the respective curves.

A somewhat different interpretation of a curve-crossing mechanism for this reaction has been discussed by Leone and co-workers (1). The potential surface that they use to describe their hypothetical model emphasizes the three-dimensional nature of the true potential surface. Figure 20 is a representation of the potential surfaces for Ar^+N_2 and ArN_2^+ along the internuclear axis based on the assumption that the approach of the argon ion (and retreat of the argon neutral) occurs with perpendicular geometry. Leone et al. carefully point out that the actual shape and curvature of the intersection seams depend critically on the orientation of the reactants during collision (angle θ in Fig. 20). They locate the crossing in the repulsive wall of the potential and assume that the crossing is accessed by compression of the N_2 bond. The hypothesis that the crossing-seam location is in the repulsive wall agrees with the suggestion of Kato et al. (28) that only intimate collisions lead to electron exchange.

Considerations of angular momentum exchange in the charge-transfer reaction

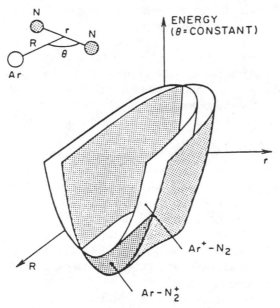

Figure 20.   Qualitative sketch of the potential-energy surfaces for reactants and products of the ArN₂⁺ charge-transfer system for a fixed angle of approach. The general features of the N₂ and N₂⁺ potentials of Ref. 27 have been used at $R = \infty$, and the attraction between the ions and neutrals as a function of R is qualitatively indicated. The charge transfer is thought to occur at the ''seam'' where the surfaces meet. Reprinted from Ref. 1 with the kind permission of the American Institute of Physics.

favor the long-range crossing mechanism of the Landau–Zener–Bauer–Fischer–Gilmore type for the higher-energy results, which exhibit little momentum exchange in the collision. The lower-energy (e.g., 0.60 eV) collisions, which are still quite specific, may very well involve crossing in the repulsive part of the curve. This might be the principal mechanism at the low collision energy c'·aracteristic of the LIF experiments. The possibility of a crossing seam extending over a range of internuclear distances, as suggested by Fig. 20, also suggests Demkov coupling as a plausible mechanism for the curve crossing between reactants and products (40). However, neither of these hypothetical models—curve crossing at larger internuclear separations or at the repulsive wall—provides a straightforward explanation of the angle-specific, quantum-state-specific scattering observed at 1.1 eV collision energy and shown in Fig. 17.

What is clearly needed is a more detailed description of the potential-energy hypersurfaces along which this nonadiabatic process evolves. The calculation of these potential surfaces is far from straightforward. The entrance channel is characterized by the ^2P configuration of Ar⁺ and the $^1\Sigma_g^+$ electronic ground state of N₂. Even though the experiments described in this chapter have generally involved only the lower-energy $^2P_{3/2}$ substate of the Ar⁺ ^2P configuration, both states lie sufficiently close in energy (0.178 eV) that both must be considered in constructing the potential-energy surfaces. The molecular orbital configuration of N_2^+ is $^2\Sigma_g^+$, which combines with the ^1S state of neutral argon to represent the exit channel. As the reactants approach each other, the quadruply degenerate $^2P_{3/2}$ state and doubly degenerate $^2P_{1/2}$ state combine to form states of definite spatial symmetry. The six states form three degenerate levels, which give rise to two surfaces of A' symmetry and one of A'' symmetry. These three entrance-channel surfaces and the exit-channel surface are all functions of three coordinates [$R(Ar-N_2)$, $R(N_2)$, and θ]. All four curves must be evaluated in all three dimensions and their nonadiabatic coupling matrices must be calculated.

The dynamics calculation is complicated by the same effects. Several potential crossing seams are possible. The topology of these seams may turn out to be somewhat simpler than our general description suggests. If so, generalizations of the Demkov (40) and Landau–Zener–Stueckelberg (38,39,41) surface-crossing models might be employed in a quasiclassical trajectory calculation. However, if there are several seams with complicated topology, a quantum mechanical close-coupling calculation might be the only way to obtain accurate dynamical information. Until we have at least preliminary potential-surface calculations, it is impossible to tell how complex the surface intersections are.

The first steps toward achieving a realistic theoretical characterization of the ArN_2^+ system have recently been taken in papers by Spalburg and co-workers (42,43). Their time-dependent quantum mechanical treatment accurately predicts experimental results for the total reactive cross section as a function of collision energy, the specific generation of $N_2^+(^2\Sigma_g, v = 1)$ as the dominant channel at most energies (their prediction is that this would be true at all energies), and the absence of significant angular scattering at energies above 1.7 eV. However, the predictions in these papers (42,43) of the magnitude and relative abundances of both reactive *and* inelastic channels at low energy (1.2 eV collision energy) are incorrect. Finally, no angular scattering predictions are possible with their theory in its present form, which assumes rectilinear trajectories. Some of the problems with the theory are obvious: It ignores long-range ion-induced dipole and ion–quadrupole forces and any valence forces responsible for the ca. 1.2-eV well depth for ArN_2^+ (32). Incorporation of these features and the carrying out of appropriate trajectory calculations to deduce angular scattering for all channels as a function of energy are logical extensions that must precede more serious calculations involving the coupling between potential hypersurfaces.

In conclusion, the first detailed dynamical investigations of charge transfer in a simple triatomic system to use state-specific measurements to determine the probability of charge transfer as a function of collision energy and angular-scattering properties has revealed a wealth of experimental details. Clearly the simplistic models invoking energy-resonance, Franck–Condon factors and simplified treatments of nonadiabatic coupling must be abandoned, along with the conjecture discussed earlier in this chapter that a statistical model is the most appropriate model for charge transfer at low energy. It is suggested that nonadiabatic transitions in localized crossing seams of the reactant and product potential-energy surfaces is responsible for the quantum-state specificity observed. Understanding the details of these processes, which exhibit pronounced quantum-state and angular-scattering specificity, will depend on the interplay between detailed theoretical studies and further detailed experiments.

REFERENCES

1. L. Hüwel, D. R. Guyer, G.-H. Lin, and S. R. Leone, *J. Chem. Phys.* **81**, 3520 (1984).

2. W. Lindinger, F. Howorka, P. Lukac, S. Kuken, H. Villinger, E. Alge, and H. Ramler, *Phys. Rev.* **A23**, 2327 (1981).

3. A. Schultz, H. W. Cruse, and R. N. Zare, *J. Chem. Phys.* **57**, 1354 (1972).

4. J. L. Kinsey, *Ann. Rev. Phys. Chem.* **28**, 349 (1977).

5. P. C. Engelking and A. L. Smith, *Chem. Phys. Lett.* **36**, 21 (1975).

6. B. H. Mahan, C. Martner, and A. O'Keefe, *J. Chem. Phys.* **76**, 4433 (1982).

7. T. R. Govers, M. Gerard, G. Mauclaire, and R. Marx, *Chem. Phys.* **23**, 411 (1977).

8. A. Ding and K. Richter, *Z. Phys.* **A307**, 31 (1982).

9. D. R. Guyer, L. Hüwel, and S. R. Leone, *J. Chem. Phys.* **79**, 1259 (1983).

10. K. Tanaka, J. Durup, T. Kato, and I. Koyano, *J. Chem. Phys.* **73**, 586 (1980).

11. N. Ghosh and W. F. Sheridan, *Ind. J. Phys.* **31**, 337 (1957).

12. E. Gustafsson and E. Lindholm, *Ark. Fys.* **18**, 219 (1960).

13. R. F. Stebbings, B. R. Turner, and A. C. Smith, *J. Chem. Phys.* **38**, 2277 (1963).

14. M. R. Flannery, P. C. Cosby, and T. F. Moran, *J. Chem. Phys.* **59**, 5494 (1973).

15. N. G. Utterbach and G. H. Miller, *Rev. Sci. Instrum.* **32**, 1101 (1961).

16. K. J. McCann, M. R. Flannery, J. V. Hornstead, and T. F. Moran, *J. Chem. Phys.* **63**, 4998 (1975).

17. S. C. deCastro, H. F. Schaefer III, and R. M. Pitzer, *J. Chem. Phys.* **74**, 550 (1981).

18. K. B. McAfee, Jr., C. R. Szmanda, R. S. Hozack, and R. E. Johnson, *J. Chem. Phys.* **77**, 2399 (1982).

19. K. B. McAfee, Jr., C. R. Szmanda, and R. S. Hozack, *J. Phys. B* **14**, L243 (1981).

20. S. H. Linn, Y. Ono, and C. Y. Ng, *J. Chem. Phys.* **74**, 3342 (1981).

21. K. Stephan, T. D. Märk, J. H. Futrell, and H. P. Helm, *J. Chem. Phys.* **80**, 3185 (1984).

22. D. Rapp and W. E. Francis, *J. Chem. Phys.* **37**, 2631 (1961).

23. W. Lindinger, F. Howorka, P. Lukas, S. Kuhn, H. Villinger, E. Alge, and H. Ramler, *Phys. Rev.* **A23**, 2319 (1981).

24. R. G. Orth, J. H. Futrell, and Y. Nishimura, *J. Chem. Phys.* **75**, 3345 (1981).

25. B. Friedrich, S. L. Howard, A. L. Rockwood, W. E. Trafton, Jr., Du Wen-Hu, and J. H. Futrell, *Int. J. Mass Spectrom. Ion Proc.* **59**, 203 (1984).

26. H. Böhringer and F. Arnold, *J. Chem. Phys.* **77**, 5534 (1982).

27. A. Loftus and P. H. Krupenie, *J. Phys. Chem. Ref. Data* **6**, 113 (1977).

28. T. Kato, K. Tanaka, and I. Koyano, *J. Chem. Phys.* **77**, 337 (1982).

29. D. Smith and N. G. Adams, *Phys. Rev.* **A23**, 2327 (1981).

30. B. Friedrich, W. Trafton, A. Rockwood, S. Howard, and J. H. Futrell, *J. Chem. Phys.* **80**, 2537 (1984).

31. A. L. Rockwood, S. L. Howard, Du Wen-Hu, P. Tosi, W. Lindinger, and J. H. Futrell, *Chem. Phys. Lett.* **114**, 486 (1985).

32. H. H. Teng and D. C. Conway, *J. Chem. Phys.* **59**, 2316 (1973).

33. Y. T. Lee, *Ber. Bunsenges. Phys. Chem.* **86**, 378 (1982).

34. D. M. Neumark, A. M. Wodtke, G. N. Robinson, C. C. Hayden, and Y. T. Lee, *Phys. Rev. Lett.* **53,** 226 (1984).

35. A. Kupperman, in *Potential Energy Surfaces and Dynamic Calculations*, D. G. Truhlar, Ed., Plenum Press, New York (1981), p. 375.

36. T. Kato, K. Tanaka, and I. Koyano, *J. Chem. Phys.* **77,** 834 (1982).

37. T. R. Govers, P. M. Guyon, T. Baer, K. Cole, H. Fröhlich, and M. Lavollée, *Chem. Phys.* **87,** 373 (1984).

38. L. Landau, *Z. Phys. Sowjet* **2,** 46 (1932); C. Zener, *Proc. Roy. Soc.* **A137,** 696 (1982).

39. E. Bauer, E. R. Fischer, and F. R. Gilmore, *J. Chem. Phys.* **51,** 4173 (1969).

40. Y. Demkov, *Sov. Phys. JETP* **18,** 328 (1964).

41. E. C. G. Stueckelberg, *Helv. Phys. Acta* **5,** 369 (1932).

42. M. R. Spalburg, J. Los, and E. A. Gislason, *Chem. Phys.* **94,** 327 (1985).

43. M. R. Spalburg and E. A. Gislason, *Chem. Phys.* **94,** 339 (1985).

CHAPTER 11

Ion–Molecule Reaction Kinetics in Plasma: Rate Coefficients and Internal-Energy and Translational-Energy Effects

WERNER LINDINGER

Institut für Experimentalphysik
Leopold-Franzens-Universität
Innsbruck, Austria

1 HOW FAST CAN ION–MOLECULE REACTIONS PROCEED?

1.1 The Langevin and Average Dipole Orientation Limiting Values, k_L or k_{ADO}, and Related Theoretical Limits for Rate Coefficents

Collisions between ions A^+ and neutrals B leading to products other than A^+ and B,

$$A^+ + B \xrightarrow{k} \text{Products,} \qquad (1)$$

are usually called reactive collisions, and proceed with a reaction rate coefficient k, defined by

$$\frac{d[A^+]}{dt} = -k[A^+][B], \qquad (2)$$

where $d[A^+]/dt$ represents the change in the ion density as a function of time in a given volume due to reactive collisions with B, which is present with a density [B].

237

The dimension of k is cubic centimeters per second (in binary collisions) and its relation to the cross section $\sigma(v)$ is

$$k = \int \sigma(v)f(v)v \, dv, \tag{3}$$

where $f(v)$ is the velocity distribution function. Thus in a first approximation we may write

$$k \simeq \langle \sigma v \rangle. \tag{4}$$

How large then can values of k be for specific reactions? Each neutral atom or molecule has a polarizability α, which means that an electric dipole is induced whenever the neutral particle is put into an electric field (Fig. 1). Whereas in a homogeneous electric field (Fig. 1a) no force is acting on the dipole, in an inhomogeneous field (Fig. 1b) such as the coulombic field of an ion, an attractive force always exists, the potential of which is

$$V(r) = -\frac{\alpha e^2}{2r^4}. \tag{5}$$

Neutrals approaching at large impact parameters are only slightly deflected, as shown in Fig. 2, but those with small impact parameters proceed along spiraling orbits[1,2]. The limit between these types of orbits is a critical impact parameter b_0, which is related to the relative velocity v_0 between A^+ and B by the equation

$$b_0 = \left(\frac{4e^2\alpha}{m_r v_0^2}\right)^{1/4}, \tag{6}$$

where m_r is the reduced mass of the reactants. This yields a cross section

$$\sigma(v_0) = \pi b_0^2 = \frac{2\pi}{v_0}\left(\frac{e^2\alpha}{m_r}\right)^{1/2}. \tag{7}$$

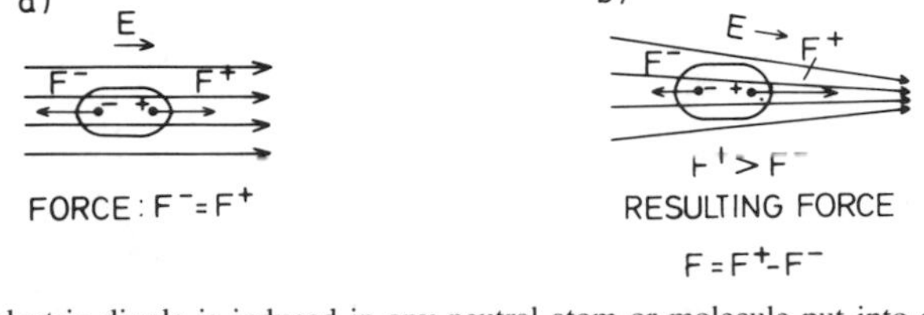

FORCE: $F^- = F^+$ RESULTING FORCE $F = F^+ - F^-$

Figure 1. An electric dipole is induced in any neutral atom or molecule put into an electric field E, due to the polarizability α of the atom or molecule. In a homogeneous field (a) no net force is acting on the neutral particle, whereas in an inhomogeneous field (b) a net force F is acting, forcing the neutral in the direction of higher field density.

After integration over the velocity according to Eq. (2), we get

$$k = 2\pi e\left(\frac{\alpha}{m_r}\right)^{1/2} = k_L, \tag{8}$$

which is called the Langevin limiting value, and can be seen as a capture rate coefficient; that is, the value indicates the rate at which the reactants are captured in spiraling orbits.

In cases where neutral reactants already possess a permanent dipole moment μ_D, the capture rate coefficient is larger than k_L. Su and Bowers (3) have derived the expression

$$k = 2\pi e\left(\frac{\alpha}{m_r}\right)^{1/2} + C\left(\frac{2\pi e\mu_D}{m_r}\right)\left(\frac{2}{\pi kT}\right)^{1/2} \tag{9}$$

which is called the average dipole orientation (ADO) limit k_{ADO}. The factor C weights the orientation of the permanent dipole and depends on the ratio $\mu_D/\alpha^{1/2}$. Values of C are listed in Ref. 3. The rotational motion of the molecule is hindered by the presence of permanent dipoles and, in general, in systems having strongly anisotropic potentials. Thus a variety of more complex theories have been developed to account not only for permanent dipole but also for quadrupole moments (3–6). Recently a new computational technique involving a combination of adiabatic capture and centrifugal sudden approximations (ACCSA) was applied by Clary (7). This theory predicts sharply increasing rate coefficients as the temperature decreases. Its predictions may have considerable importance for interstellar-cloud chemistry.

So far we have discussed theoretical capture rate coefficients, but only comparison

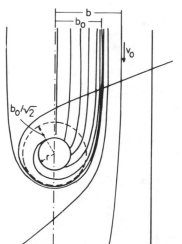

Figure 2. Trajectories for a neutral B approaching an ion A$^+$ (held at a fixed position) with constant velocity v_0 at various impact parameters b. After Gioumousis and Stevenson (2).

with experimental data can tell us whether these have any meaning. The values of k_L, k_{ADO}, and k_{ACCSA} (in the following we represent all of these by k_C) for most ion–neutral combinations lie in the vicinity of 1×10^{-9} cm³/sec (at room temperature), which corresponds to capture cross sections about 100–1000 times larger than gas kinetic cross sections for neutrals.

1.2   Comparison Between Theoretical Limits k_C and Measured Reaction Rate Coefficients

We have seen above that in collisions with large impact parameters, only slight deflections of the paths of the particles occur, whereas at small enough impact parameters spiraling orbits result. Thus we may regard the latter as a kind of complex formation, in which the reaction partners stay together long enough that *reactions may occur*. Indeed, whole classes of ion–neutral reactions proceed with rate coefficients as fast as k_C. Among these are practically all exothermic proton-transfer reactions and a large number of charge-transfer, hydrogen-abstraction, and isotopic-

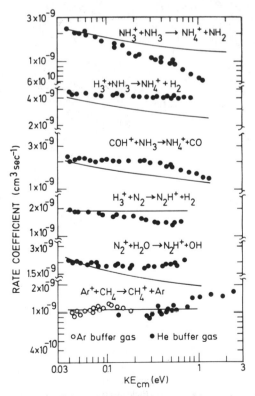

Figure 3.   Experimental rate coefficients (dots) of various reactions as dependent on center-of-mass kinetic energy KE_{cm}. Solid lines represent k_L or k_{ADO}.

exchange reactions. Figure 3 compares measured rate coefficients (8,9) with the respective Langevin or ADO limiting values k_C. Thus it is useful to compare measured values to these theoretical upper limits. In some other cases (Fig. 8 of Chapter 7), where the measured rate coefficients are considerably smaller than k_C, explanations for this low reactivity can be found on the basis of potential curves or rearrangements of bonds. This leads to an understanding of the mechanism governing the reaction, as will be shown by several examples.

1.3 Effects of Internal and Translational Energy on Ion–Molecule Reactions

Whereas reactions that are fast at room temperature ($k_{exp} \simeq k_C$) hardly ever show significant changes as a function of the relative kinetic energy of the reactants, KE_{cm} (Fig. 3), slow reactions practically always exhibit drastic dependences of k on KE_{cm} (see Fig. 8 of Chapter 7). In addition, slow reactions usually also show drastic changes both in their overall reaction rate coefficients and in the product distribution when the internal energy (vibrational excitation) of either the ionic or the neutral reaction partners varies. Strong increases in k as a function of increasing KE_{cm} usually indicate access to new (endothermic) reaction channels. This is observed especially in simple charge-transfer reactions, as is shown in Section 3.2.

Reactions of complex (molecular) reactants involving considerable rearrangement of particles tend to be slow compared with the rate given by k_C and their rates practically always show strong dependences on the internal excitation of the ions and/or neutrals participating in the reaction. This can be rationalized by a general model developed by Polanyi and Wong (10) on the basis of the shapes of the reaction surfaces. In reactions with energy minima (or maxima) in the entrance valley of the potential-energy surface, kinetic energy favors the reaction but vibrational excitation acts as an obstacle. In reactions with "late" minima (or maxima), vibrational excitation helps to overcome them and kinetic energy is mainly a hindrance. Typical examples will be discussed in Section 3.3.

2 THREE-BODY REACTIONS

Not only binary reactions, which involve only the collisions of two particles, but also three-body reactions may significantly contribute to or even dominate the balances of ion species in plasmas at low temperatures or high pressures. These three-body reactions usually result in the association of an ion A^+ with the neutral collision partner B. In the initial collision, A^+ and B form a long-lived excited complex $(AB)^{+*}$ with a mean lifetime τ until unimolecular decomposition:

$$A^+ + B \underset{\tau}{\overset{k_1}{\rightleftarrows}} (AB)^{+*}. \tag{10}$$

The complex $(AB)^{+*}$ is then stabilized or dissociated in a subsequent collision with a third body M (usually a buffer-gas atom or molecule) (11–13):

$$(AB)^{+*} + M \xrightarrow{fk_2} (AB)^+ + M, \xrightarrow{(1-f)k_2} A^+ + B + M, \tag{11}$$

where f is the fraction stabilized to form stable molecules. The overall reaction observed is described by the three-body rate coefficient for association, k_a, which has the dimension (centimeters)^6 per second:

$$A^+ + B + M \xrightarrow{k_a} AB^+ + M. \tag{12}$$

However, at a given buffer-gas density [M] we observe the usual semilogarithmic decline of the ion signal $i(A^+)$ when we add the reactant gas B to the buffer. From this we calculate an effective binary rate coefficient:

$$k_{\text{bineff}} = k_a[M]. \tag{13}$$

Using Eqs. (10)–(13) we get

$$k_a = \frac{k_1 f k_2 \tau}{1 + k_2 \tau[M]}. \tag{14}$$

Therefore k_{bineff} depends on [M], as shown in Fig. 4. From measured values of k_{bineff} one now is able to obtain both τ and f, after making the reasonable assumption that k_1 and k_2 are the Langevin or ADO rate coefficients.

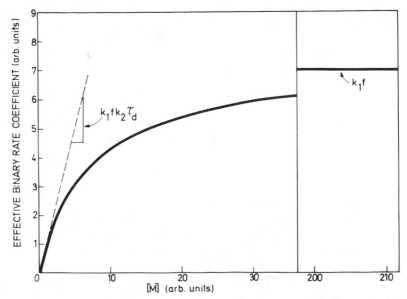

Figure 4. Dependence of the effective binary rate coefficient k_{bineff} on the buffer-gas density [M].

3 CLASSIFICATION OF POSITIVE-ION REACTIONS AND RESULTS

Prepared by Sections 1 and 2, we now establish a classification of ion–molecule reactions in order to make some generalizations concerning ion–neutral interactions. We shall proceed from reactions with high exoergicities toward interactions involving small exoergicities.

3.1 Reactions of Multiply Charged Ions

Typical potentials in the charge-transfer reaction of a multiply charged ion X^{u+} with a neutral Y,

$$X^{u+} + Y \rightarrow X^{(u-1)+} + Y, \tag{15}$$

are shown in Fig. 5 (for the case of single charge transfer between A^{2+} and B). The approach of the ion A^{2+} toward the neutral B occurs along the potential curve $(A^{2+} + B)$ governed by the ion-induced dipole interaction; the product ions A^+ and B^+ repel each other along a coulombic potential curve. According to a model developed by Landau (14), Zener (15), and Stueckelberg (16), the probability of transition from one potential curve to the other at the intersection of the potential curves (see Fig. 5) depends strongly on the internuclear distance R_x at this crossing. A maximum transition probability typically occurs when R_x has values of a few angstroms. Though this model was derived under the assumption that the relative velocity of the interacting particles is considerably higher than thermal velocity, experimental results show that the model also holds at thermal energies. Under the assumption of a constant value for the potential of the entrance channel $A^{2+} + B$, the crossing distance R_x is calculated according to

$$R_x = \frac{14.3}{\Delta E} \text{ Å}, \tag{16}$$

where ΔE is the exoergicity of the charge transfer in electron volts. In Fig. 6, the measured rate coefficients (17) of a variety of ion–neutral interactions of doubly charged rare-gas ions in both the ground state and metastable excited states with

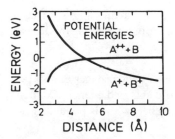

Figure 5. Dependence of the potentials between the reactants A^{2+} and B and between the products A^+ and B^+ on the internuclear distance.

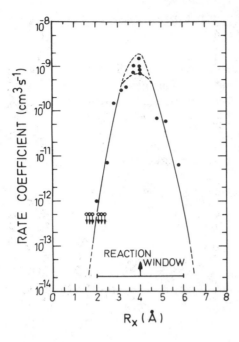

Figure 6. Dependence of rate coefficients for various charge-transfer processes involving doubly charged rare-gas ions on R_x. After Smith et al. (17).

neutral rare gases are plotted as a function of R_x. The curve shows a strong maximum in the range $3 \text{ Å} \le R_x \le 5 \text{ Å}$. At the maximum the rate coefficients reach values as high as the Langevin limits, indicating unit reaction probability in this regime.

In nearly all interactions between doubly charged ions and *molecular neutrals* many curve crossings exist in the range $3 \text{ Å} \le R_x \le 5 \text{ Å}$, due to the large number of electronic and vibrational levels present in molecular systems. Therefore one would generally expect these reactions to be fast ($k \simeq k_L$). A variety of data (18) prove this to be the case.

3.2 Charge Transfer of Singly Charged Atomic Ions

The reactions of singly charged atomic ions with molecules can be divided into two groups:

1. Reactions that are fast at room temperature and show no energy dependences (k is constant up to a few electron volts).
2. Reactions with thermal rate coefficients considerably smaller than k_L. These reactions usually show strong energy dependences, often reflecting the access to endothermic reaction channels at elevated KE_{cm} values.

For charge transfer between atomic ions and molecules Laudenslager et al. (19) have found a strong correlation between the rate coefficients and Franck–Condon factors (FCFs) from the molecular ground state into the energy levels populated by

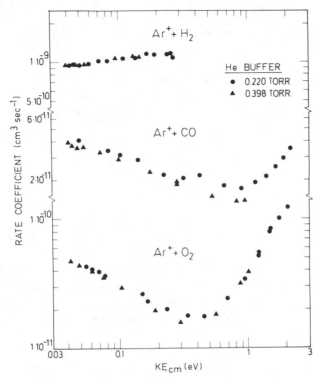

Figure 7.   Dependence of rate coefficients for the reactions of Ar$^+$ with H$_2$, CO, and O$_2$ on KE$_{cm}$.

resonant charge transfer at thermal energies. Large FCFs ($\geq 10^{-5}$) are often (but not always) connected with large rate coefficients ($k \simeq k_C$), whereas small ones seem never to allow fast charge transfer.

Typical examples of strong changes of k as a function of KE$_{cm}$ for reactions having thermal rate coefficients of $\ll k_C$ are shown in Fig. 8 of Chapter 7 and in Fig. 7 here. Reactions of Ne$^+$ with N$_2$ and CO, respectively, exhibit typical endothermic behavior despite the nominal exoergicity of the reactions. In both cases the energy dependences of the ion *product ratios* show that the increase in the overall rate coefficients is accounting for practically entirely by the opening of dissociative channels (20).

A qualitative interpretation for the reaction with N$_2$ is provided by the potential curves for N$_2^+$. The endoergicity inferred from the data in Fig. 8 of Chapter 7 approximately matches that of the $C\Sigma_u^+$ state of N$_2^+$, which tends strongly to predissociate into N$^+$(^3P) and N($^4S_0$). Accordingly, we suggest that provided KE$_{cm}$ is high enough, a fast charge transfer to form N$_2^+$($C\Sigma_u^+$) occurs, followed by a prompt dissociation into N$^+$ and N.

The reaction of Ne$^+$ with CO shows similar behavior. Further, the branching ratio to form C$^+$ + O exhibits a rapid increase in the energy region (around 1 eV) of rapid increase of the total rate coefficient. This behavior clearly reflects access to the channel forming C^+ from CO, which is endothermic for Ne^+ by ~ 1 eV (the appearance potential of C^+ from CO is 22.5 eV; the recombination energy of Ne^+ is 21.56 eV).

The energy dependence of k for the reaction of Ne^+ with CO_2, which goes through a pronounced minimum, is typical of complex formation by reactants. A faster relative velocity at elevated KE_{cm} shortens the complex lifetime and therefore diminishes the transition probability. The increase at higher KE_{cm} is caused by access to new product channels.

In the charge-transfer reaction of Ar^+ with O_2, the rate coefficients show a minimum at $KE_{cm} \simeq 0.4$ eV (Fig. 7). At higher KE_{cm} the endoergicity for the channel

$$Ar^+ + O_2 \rightarrow O_2^+(a\ ^4\Pi_u) + Ar + 0.35\ \text{eV} \tag{17}$$

is bridged, leading to the strong further increase of k. The production of $O_2^+(a\ ^4\Pi_u)$ in reaction (17) at $KE_{cm} > 0.5$ eV was proven by subsequent reaction of the $O_2^+(a\ ^4\Pi_u)$ ions in a drift experiment (21). It should be pointed out that the FCFs to the O_2^+ ground state (at vibrational levels in the vicinity of resonant thermal charge transfer) are poor, whereas those to the $a\ ^4\Pi_u$ state are very favorable.

The above examples show that adding kinetic energy to a system $A^+ + B$ has about the same consequences as increasing the internal energy (e.g., the recombination energy of A^+) of the system.

3.3   Reactions of Molecular Ions in the Ground State and in Vibrationally and Electronically Excited States

A typical reaction involving considerable rearrangement of atoms that is quite slow at room temperature ($k \simeq 6 \times 10^{-12}$ cm³/sec) is

$$
\begin{aligned}
O_2^+ + CH_4 &\rightarrow CH_3O_2^+ + H + 99\ \text{kcal/mol} &\text{(a)} \\
&\rightarrow H_3O^+ + CO + 143\ \text{kcal/mol} &\text{(b)} \\
&\rightarrow COH^+ + H_2O + 124\ \text{kcal/mol} &\text{(c)} \qquad (18) \\
&\rightarrow CH_3^+ + HO_2 - 4\ \text{kcal/mol} &\text{(d)} \\
&\rightarrow CH_4^+ + O_2 - 15\ \text{kcal/mol,} &\text{(e)}
\end{aligned}
$$

with channel (a) being the dominant one at room temperature (22,23). It is not surprising that strong changes in both the overall reaction rate and the product distributions are caused by vibrational excitation of the O_2^+ reactant ion. One can observe this by investigating the above reaction in various buffer gases, such as He, Ne, Ar, Kr, and CO_2. Some of these results are shown in Fig. 8. In a helium buffer an E/N value of about 50 Td is sufficient to enable a KE_{cm} value of about 0.4 eV between O_2^+ and CH_4 to be reached; therefore internal vibrations of O_2^+ will not be excited significantly by collisions with helium under these conditions. Argon, on the other hand, requires an E/N value of ~ 200 Td to enable a $KE_{cm}(O_2^+-CH_4)$ value of ~ 0.4 eV to be reached. This means that in the collisions between O_2^+ and

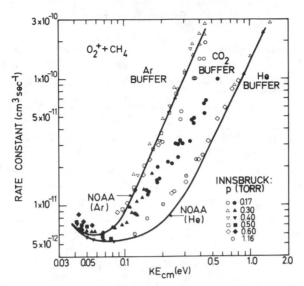

Figure 8. Dependence of rate coefficient for the reaction of O_2^+ with CH_4 on KE_{cm}, measured using various buffer gases. NOAA stands for the National Oceanic and Atmospheric Administration laboratory in Boulder, CO. After E. Alge et al., *J. de Phys.* **C7**, 83 (1979).

the argon buffer-gas atoms considerable transfer of momentum will take place, leading to internal excitation of O_2^+. Thus the strong enhancement of k in the heavier buffer gases (Fig. 8) with respect to the values obtained in helium reflects the influence of the O_2^+ vibrational excitation on the reaction. Albritton (23) proved this directly by causing the O_2^+ vibration to be quenched in the argon buffer by adding neutral O_2. This quenching proceeded according to the fast reaction

$$O_2^+(v > 0) + O_2 \rightarrow O_2^+(v = 0) + O_2. \tag{19}$$

As a consequence of this "cooling" of the vibration of O_2^+, a remarkable decline of k toward the values characteristic of helium-buffer data was observed. When O_2^+ is excited to the $a^4\Pi_u$ state, it not only reacts fast with CH_4 ($k_{exp} = 1.2 \times 10^{-9}$ cm^3/sec $\simeq k_L$), but also gives CH_3^+ and CH_4^+ as the dominant products, in a ratio of 2:1.

While the reaction of O_2^+ with CH_4 is favored by O_2^+ excitation, the hydrogen abstraction

$$CO_2^+ + H_2 \rightarrow CO_2H^+ + H \tag{20}$$

is hindered by CO_2^+ vibrational excitation. Here the decline of k with increasing KE_{cm} is more rapid in N_2 and Ar buffers than in helium (23). Again, reducing the vibrational excitation of CO_2^+ by adding CO_2 to the Ar buffer shifts k toward the He-buffer data, leading in this case to an increase in k (23).

3.4 Proton Transfer

Proton-transfer reactions play a major role in practically any system involving hydrocarbons (24). Any neutral possesses a proton affinity (PA), and therefore if
PA(Y) is significantly larger than PA(X), the reaction

$$XH^+ + Y \rightarrow YH^+ + X \tag{21}$$

is exothermic. All exothermic proton-transfer reactions are fast and proceed with
rate coefficients close to k_C. To know in which direction a proton transfer will occur,
it is often sufficient to know the PAs of the respective atoms or molecules (radicals)
involved. Proton affinities have been measured for hundreds of neutral species (25);
here we mention briefly how these PAs are obtained in swarm experiments, preferably in variable-temperature afterglows. For each proton-transfer reaction, the
common relation

$$-RT \ln K = \Delta H^0 - T \Delta S^0 \tag{22}$$

holds, when the process in forward and reverse direction

$$XH^+ + Y \underset{k_r}{\overset{k_f}{\rightleftarrows}} YH^+ + X \tag{23}$$

has the equilibrium constant $K = k_f/k_r$. It follows that

$$-\frac{d[XH^+]}{dt} = k_f[XH^+][Y] = -\frac{d[YH^+]}{dt} = k_r[YH^+][X], \tag{24}$$

and therefore

$$\frac{[YH^+][X]}{[XH^+][Y]} = \frac{k_f}{k_r} = K. \tag{25}$$

Thus K can be deduced directly in the experiment by measuring the ratio $[YH^+]/$
$[XH^+]$ as a function of $[Y]$ at a given density $[X]$. One thus obtains linear increases,
the slopes of which are a direct measure of K. Values of K obtained at different
temperatures are then plotted in the form of van't Hoff diagrams, from which both
ΔS and ΔH are obtained. ΔH is by definition the difference in the PAs. Another
approach is the direct measurement of the forward and reverse rate coefficients, the
ratio of which yields K and ΔPA ($= \Delta H$) as described above.

3.5 Isotopic-Exchange Reactions

Isotopic-exchange reactions are frequently used to determine structures of ions (22);
they are of special interest also in low-temperature plasmas. The isotopic enrichment

of heavier isotopes in interstellar clouds results from such processes (13,26–28). Reactions of the type

$$XH_n{}^+ + HD \underset{k_r}{\overset{k_f}{\rightleftarrows}} XH_{(n-1)}D^+ + H_2 \tag{26}$$

are often fast in the forward direction at low temperatures because the zero-point-energy difference between H_2 and HD is usually larger than that between $XH_n{}^+$ and $XH_{n-1}D^+$. Detailed investigations by Smith and his group (13,26–28) indicate that many of the isotopic abundances observed in interstellar molecules, which differ from the respective abundances in terrestrial molecules, can be understood on the basis of isotopic enrichment resulting from exchange reactions at low temperatures.

3.6 Quenching and Creation of Ion Vibrational Excitation in Collisions with Neutrals

Although the quenching of vibrational excitation of neutral molecules in collisions with neutrals has been the subject of intensive investigations for many years, the quenching of ionic vibrational excitation could be investigated thoroughly only after the development of selected-ion swarm experiments. In the early swarm experiments, such as the flowing afterglow (FA) and the flow drift tube (FDT), the parent gas M of an ion M^+ usually was present in small quantities within the buffer gas, thus causing any excited ions $M^+(v > 0)$ to become relaxed to the ground state by the invariably fast interaction

$$M^{+*}(v > 0) + M \rightarrow M^+(v = 0) + M^*. \tag{27}$$

In modern selected-ion swarm experiments preselected ions are introduced into a *pure* helium buffer, where the ions in most cases retain their vibrational excitation. The use of monitor ion probes for vibrational states now allows the investigation of quenching processes in selected-ion flow drift tube (SIFDT) experiments at collision energies from thermal values to ~ 1 eV. In this way the relaxation of $N_2{}^+(v = 1)$ by Ne, N_2, O_2, and CO_2 (29) and by Kr and NO (30) was measured in Innsbruck. The vibrational relaxation of $O_2{}^+(v = 1$ and $v = 2)$ by various neutrals has been investigated in Boulder (31), and recently results were obtained in Innsbruck (32,23) on the quenching of $NO^+(v = 3, 2,$ and $1)$. Typical results for the energy dependences of quenching rate coefficients $k_q(v)$ are shown in Fig. 9 for the case of $NO^+(v) + CH_4$. The decline in $k_q(v = 1)$ and $k_q(v = 3)$ with increasing KE_{cm} shown in Fig. 9 is typical for transfer of vibrational into translational energy (V-T transfer), which requires the formation of a complex. Coupling between the NO^+ vibration and the NO^+–neutral vibrational modes enables NO^+ vibrational energy to be transferred to the neutral collision partner. This in turn leaves the complex with enhanced translational energy. This coupling should be stronger the higher the dissociation energy of the bound ion–neutral complex is. The latter is proportional (in a first-order approximation) to the polarizability α of the neutral and thus an

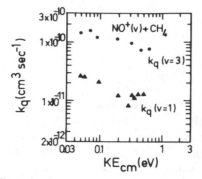

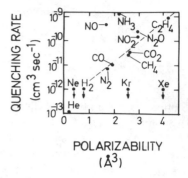

Figure 9. Dependence of rate coefficients for the quenching of NO$^+$($v = 3$ and 1) in collisions with CH$_4$ on KE$_{cm}$.

Figure 10. Thermal rate coefficients for the quenching of NO$^+$($v = 1$) in collisions with various neutrals, plotted vs. the respective polarizabilities of these neutrals.

increase of k_q with increasing α of the neutral collision partner is observed. Figure 10, which shows the thermal values of k_q for the quenching of NO$^+$($v = 1$) by a variety of neutrals, demonstrates a strong correlation between k_q and the respective values of α. However, the rare gases do not seem to fit into this picture. A detailed model of the quenching of ionic vibration based on these principles has been developed recently by Ferguson (34).

Vibrational–vibrational (V–V) transfer may also occur for cases of near resonance between ionic and neutral vibrational modes, resulting in enhanced values of k_q. In addition to the quenching of ionic vibration in nonreactive collisions, as in the examples shown involving NO$^+$, quenching also may occur in parallel to reactive collisions. This was observed in collisions of N$_2^+$ with various neutrals (29,30). However, the data available are too scarce to allow generalizations to be made about the mechanisms involved.

3.7 Three-Body Reactions

The complex lifetimes τ in collisions between ions and neutrals at room temperature typically lie below 10^{-10} sec. Therefore association reactions involving stabilizing collisions with a third partner do not play a significant role in hot, low-density plasmas. Specific ions, such as CH$_3^+$, form much longer-lived complexes ($\tau > 10^{-6}$ sec) even at room temperature, thus their associations can compete with binary reactions even at low pressures. Moreover, when τ becomes very large (e.g., at low temperature), stabilization of the complex may occur by the emission of an infrared photon and collisions with a third body are no longer required for association. Thus under interstellar conditions association reactions are important even though densities are extremely low. Three-body rate coefficients obtained by Adams and Smith (12) for reactions of CH$_3^+$ are shown along with their temperature dependences in Fig. 11. These rate coefficients show typical strong temperature (T) dependences

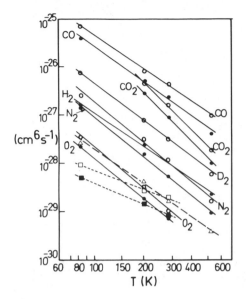

Figure 11. Dependence of rate coefficients for the association of CH_3^+ with various neutrals on temperature. After Adams and Smith (12).

of T^{-u}, with u in the range 2.7–4.4. The absolute values of k vary greatly from one reaction partner to another, but correlate strongly with the complex binding energies D [e.g., $D(CH_3^+ \cdot CO) > D(CH_3^+ \cdot O_2)$].

4 ION–MOLECULE REACTIONS IN PLASMAS

The vast amount of data available on ion–molecule reactions inspired the development of a variety of models of and theories on ion chemistry within specific plasmas, the details of which lie far beyond the scope of this paper. However, we shall discuss, in this final section, a few practical aspects of ion–molecule reactions, especially their importance for the understanding of technical and natural plasmas.

4.1 Plasma Sampling

To understand the processes balancing ion densities in plasmas, one needs information on the composition of the plasma, especially on the densities of the different types of ions present. These can be obtained by mass spectrometric plasma sampling, in which ions are extracted by various means from the plasma and then analyzed. By use of a hole probe system (35) developed for the investigation of the radial density distributions of ions in hollow-cathode discharges, the results shown in Fig. 12 were obtained from an argon discharge with traces of H_2O present. Although the density of neutral Ar dominated that of H_2O by three orders of magnitude, H_3O^+

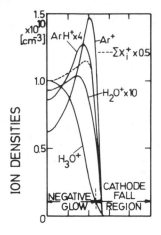

Figure 12. Radial density profiles of the ions Ar^+, ArH^+, H_2O^+, and H_3O^+ and of the sum of all ions in a hollow-cathode discharge in argon containing traces of H_2O. After Lindinger (35).

was the dominant ion at the center of the discharge. This ion resulted from the fast reactions

$$Ar^+ + H_2O \rightarrow ArH^+ + OH \tag{28}$$

and

$$ArH^+ + H_2O \rightarrow H_3O^+ + Ar, \tag{29}$$

both of which proceed with $k \simeq 10^{-9}$ sec^{-1}.

The region within the negative glow of the hollow-cathode discharge was practically field free; hence the steady-state condition for each ion species X^+ could be applied in the following form:

$$\frac{d[X^+]}{dt} = 0 = D_a \, \Delta[X^+] + v_p - v_l - \alpha[X^+][e_s], \tag{30}$$

where $D_a \, \Delta[X^+]$ is the diffusion term. Here D_a is the ambipolar diffusion coefficient, and in calculating $\Delta[X^+]$ only the radial component in cylindrical coordinates needs to be taken into account; that is,

$$\Delta[X^+] = \frac{\partial^2[X^+]}{\partial r^2} + \frac{1}{r} \frac{\partial[X^+]}{\partial r}. \tag{31}$$

The production of X^+ is described by the term v_p, which is the product of a rate coefficient for the formation process, k_a, and the densities of the reaction partners leading to the production of X^+ (or the sum of such products in the case of parallel reactions). Losses due to ion–molecule reactions are given by

$$v_l = k_b[X^+][Y], \tag{32}$$

and those due to electron recombination by

$$\alpha[X^+][e_s], \tag{33}$$

where $[e_s]$ denotes the density of slow plasma electrons. Extracting $\Delta[X^+]$ from measured radial ion-density profiles and introducing this information into the respective steady-state equation allowed the calculation of the radial density profile of the precursor ion of X^+. Comparison of this calculated profile with the measured profile showed the internal consistency of the model developed.

In the same fashion it was possible to calculate the radial distribution of fast electrons ($E_e \simeq 100$ eV, originating in the cathode fall) in the negative glow that produced the primary ions (Ar^+ in the situation represented by the data shown in Fig. 12) in the discharge by electron impact. In this way an internally consistent model was developed for the radial profiles of the dominant ions and their production and loss processes. Applications of this model have been made to other systems, such as N_2, He, and Ne discharges with admixtures of impurities (36,37), showing that plasma sampling is a versatile tool for obtaining information on the macroscopic structure of plasmas.

4.2 Ionospheric Plasma

Ultraviolet (UV) radiation causes photoionization, leading to the creation of plasmas in the higher regions of the atmospheres of the Earth and other planets. The densities of these ionospheres are balanced by the rates of ion production and loss by ion–molecule reactions and recombination. The charge-carrier density of the Earth's ionosphere is considerably higher than those of other planets (in the F region at a height of 200–300 km), because of the peculiarity that O^+, the dominant ion in this region (38), reacts very slowly with both N_2 and O_2 ($k = 10^{-12}$ and 2×10^{-11} cm^3/sec, respectively), the most abundant reactants in this region. Thus high densities of O^+ ($\sim 10^6/cm^3$; see Fig. 13) are built up (39). Since O^+ is an atomic ion,

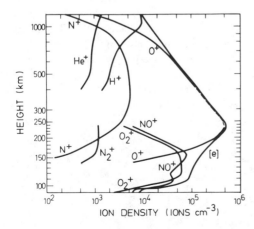

Figure 13. Density profiles of ions in the Earth's ionosphere. After C. Y. Johnson, *J. Geophys. Res.* **71**, 330 (1966).

its recombination with electrons is very slow compared with that of molecular ions; consequently the reactive losses, though proceeding slowly, are the dominant loss processes.

The dominant ions in the E region (~ 100 km) are O_2^+ and NO^+, both of which are destroyed by dissociative recombination. Because of meteorite impact, considerable concentrations of Mg and Fe can build up locally; charge transfer between O_2^+ and NO^+ and these metals converts the latter into Mg^+ and Fe^+. These atomic ions have recombination coefficients three to four orders of magnitude lower than those of O_2^+ and NO^+; thus much higher ion densities (by about two orders of magnitude) are built up temporarily. This effect is known as the appearance of "sporadic dense E layers." At lower regions (below 80 km), three-body reactions can compete with ion–electron recombination, converting O_2^+ and NO^+ into a variety of cluster ions, mainly water adducts. Very detailed models have been developed by Ferguson (38) and his group to explain the ion chemistry of the several layers of Earth's ionosphere.

4.3   The Role of Ion–Molecule Reactions in Interstellar Molecular Synthesis

In the past few years research astronomers have detected a wide variety of neutral molecules in interstellar clouds (see Table 1). Besides H, H_2, and He, considerable amounts of H_2O, NH_3, and CH_4 are present in nearly all types of dark and dense clouds; various acids, alcohols, aldehydes, and ethers have lower densities. It is clear that under the unfavorable conditions in interstellar media (low temperature and low density) neutral chemistry cannot be responsible for the build-up of these neutrals.

Whereas the ion chemistry of the ionosphere is now nearly fully understood, an understanding of molecular synthesis in interstellar clouds is just now being gained on the basis of laboratory experiments and model calculations. Smith and his group at Birmingham (41) have shown in a number of laboratory experiments how many of the detected neutrals are produced by ion–molecule reactions followed by dissociative recombinations.

Ion–molecule reaction sequences leading to the observed interstellar molecules must contain a few steps as possible preceding the final recombination step, all of which have to be fast. For example, when O^+ is formed by UV ionization or by the associative charge transfer

$$He^+ + O_2 \rightarrow O^+ + O + He, \tag{34}$$

it reacts in consecutive steps with the most dominant neutral reactant H_2:

$$O^+ + H_2 \rightarrow OH^+ + H, \tag{35}$$

$$OH^+ + H_2 \rightarrow OH_2^+ + H, \tag{36}$$

$$OH_2^+ + H_2 \rightarrow H_3O^+ + H. \tag{37}$$

TABLE 1   Molecules Observed in Interstellar Clouds [After Thaddeus (40)]

Inorganic Molecules

Diatomic	Triatomic	Tetraatomic
H_2	H_2O	NH_3
CO	H_2S	
CS	SO_2	
NO	HNO	
NS	OCS	
SiO		
SiS		

Organic Molecules

Alcohols		Aldehydes and Ketones		Acids	
CH_2OH	methanol	H_2CO	formaldehyde	HCN	hydrocyanic
CH_3CH_2OH	ethanol	CH_3CHO	acetaldehyde	HCOOH	formic
		H_2CCO	ketene	HNCO	isocyanic

Amides		Esters and Ethers		Sulfur Compounds	
NH_2CHO	formamide	CH_3OCHO	methyl formate	H_2CS	thioformaldehyde
NH_2CN	cyanamide	$(CH_3)_2O$	dimethyl ether	HNCS	isothiocyanic acid
NH_2CH_3	methylamine			CH_3SH	methyl mercaptan

Paraffin Derivatives		Acetylene Derivatives		Others	
CH_3CN	methyl cyanide	HCCCN	cyanoacetylene	CH_2NH	methylenimine
CH_3CH_2CN	ethyl cyanide	$HCCCH_3$	methylacetylene	CH_2CHCN	vinyl cyanide

Finally H_3O^+, which cannot react further with H_2, recombines dissociatively to form a water molecule:

$$H_3O^+ + e \rightarrow H_2O + H. \qquad (38)$$

In a similar way high abundances of NH_3 are produced. Besides H and He (the sum of which accounts for 99% of the matter), C, N, and O are most abundant elements in the universe, and in interstellar clouds these predominantly appear bound in the molecules CH_4, CO, NH_3, and H_2O. These molecules can act as reactants for the build-up of more complex molecules, such as those listed in Table 1.

The presence of formic acid implies the existence of its most probable ionic precursor, protonated formic acid [$HC(OH)_2^+$] in interstellar clouds. The further fast association of $HC(OH)_2^+$ with water (42) may lead to the formation of methanol (CH_3OH), an important interstellar species:

$$HC(OH)_2^+ \xrightarrow{H_2O} (CH_5O_3)^+ \xrightarrow{e} HO_2 + CH_3OH. \qquad (39)$$

(Table 1) The pathways of production of many of the observed interstellar molecules are not yet known. However, recent efforts, especially those of Smith's group at the University of Birmingham, lead us to expect a consistent picture of the overall ion chemistry to interstellar clouds to be developed in the near future.

Although all the different molecules listed in Table 1 make up only a tiny fraction

(<1%) of the matter in interstellar space, their distributions and relative abundances provide all the information we have on interstellar clouds. The shapes, densities, and temperature profiles are now known for different stages of the history of many interstellar clouds. From these we can learn about the very early stages of their gravitational contraction, which leads to the birth of new solar systems.

ACKNOWLEDGMENT

All the work performed in Innsbruck on this subject was sponsored by the Öster-reichischer Fonds zur Förderung der Wissenschaftlichen Forschung (e.g., Project P 5149).

REFERENCES

1. P. M. Langevin, *Ann. Chim. Phys.* **5**, 245 (1905).

2. G. Gioumousis and D. P. Stevenson, *J. Chem. Phys.* **29**, 294 (1958).

3. T. Su and M. T. Bowers, *Int. J. Mass Spectrom. Ion Phys.* **12**, 347 (1973).

4. T. Su and M. T. Bowers, *Int. J. Mass Spectrom. Ion Phys.* **17**, 211 (1975).

5. T. Su, E. C. F. Su, and M. T. Bowers, *J. Chem. Phys.* **69**, 2243 (1978).

6. L. Bass, T. Su, and M. T. Bowers, *Int. J. Mass Spectrom. Ion Phys.* **28**, 389 (1978).

7. D. C. Clary, *Mol. Phys.* **53**, 3 (1984).

8. W. Lindinger, D. L. Albritton, F. C. Fehsenfeld, A. L. Schmeltekopf, and E. E. Ferguson, *J. Chem. Phys.* **62**, 3549 (1975).

9. W. Lindinger and D. Smith, in *Reactions of Small Transient Species*, A. Fontijn, Ed., Academic Press, London (1983), p. 387.

10. C. P. Polanyi and W. H. Wong, *J. Chem. Phys.* **51**, 1439 (1969).

11. D. Smith and N. G. Adams, *Chem. Phys. Lett.* **53**, 535 (1978).

12. N. G. Adams and D. Smith, *Chem. Phys. Lett.* **79**, 563 (1981).

13. D. Smith, N. G. Adams, and E. Alge, *J. Chem. Phys.* **77**, 126 (1982).

14. L. Landau, *Z. Phys. Sowjet* **2**, 46 (1932).

15. C. Zener, *Proc. Roy. Soc. A* **137**, 696 (1932).

16. E. C. G. Stueckelberg, *Helv. Phys. Acta* **5**, 370 (1932).

17. D. Smith, N. G. Adams, E. Alge, H. Villinger, and W. Lindinger, *J. Phys. B* **13**, 2787 (1980).

18. H. Störi, E. Alge, H. Villinger, F. Egger, and W. Lindinger, *Int. J. Mass Spectrom. Ion Phys.* **30**, 263 (1979).

19. J. B. Laudenslager, W. T. Huntress, and M. T. Bowers, *J. Chem. Phys.* **61**, 4600 (1974).

20. H. Villinger, J. H. Futrell, R. Richter, A. Saxer, S. Niccolini, and W. Lindinger, *Int. J. Mass Spectrom. Ion Phys.* **47**, 175 (1983).

21. W. Lindinger, H. Villinger, and F. Howorka, *Int. J. Mass Spectrom. Ion Phys.* **41**, 89 (1981).

22. H. Villinger, R. Richter, and W. Lindinger, *Int. J. Mass Spectrom. Ion Phys.* **51**, 25 (1983).

23. D. L. Albritton, in *Kinetics of Ion Molecule Reactions*, P. Ausloos, Ed., Plenum Press, New York (1979), p. 119.

24. J. M. Goodings, C.-W. Ng, and D. K. Bohme, *Int. J. Mass Spectrom. Ion Phys.* **29**, 57 (1979).

25. R. Walder and J. L. Franklin, *Int. J. Mass Spectrom. Ion Phys.* **36**, 85 (1980).

26. D. Smith and N. G. Adams, *Ap. J.* **248,** 373 (1981).

27. D. Smith, *Phil. Trans. Roy. Soc. London A* **303,** 535 (1981).

28. D. Smith and N. G. Adams, *Ap. J.* **242,** 424 (1980).

29. W. Dobler, F. Howorka, and W. Lindinger, *Plasma Chem. Plasma Proc.* **2,** 353 (1982).

30. W. Dobler, H. Ramler, H. Villinger, F. Howorka, and W. Lindinger, *Chem. Phys. Lett.* **97,** 553 (1983).

31. H. Böhringer, M. Durup-Ferguson, D. W. Fahey, F. C. Fehsenfeld, and E. E. Ferguson, *J. Chem. Phys.* **79,** 4201 (1983).

32. H. Dobler, W. Federer, F. Howorka, W. Lindinger, M. Durup-Ferguson, and E. E. Ferguson, *J. Chem. Phys.* **79,** 1543 (1983).

33. W. Federer, W. Dobler, F. Howorka, W. Lindinger, M. Durup-Ferguson, and E. E. Ferguson, *J. Chem. Phys.* **82,** 1032 (1985).

34. E. E. Ferguson, "Vibrational Excitation and Deexcitation and Charge Transfer of Molecular Ions in Drift Tubes," in *Swarms of Ions and Electrons in Gases,* W. Lindinger, T. D. Märk, and F. Howorka, Eds., Springer-Verlag, Vienna (1984), p. 126.

35. W. Lindinger, *Phys. Rev. A* **7,** 328 (1973).

36. F. Howorka, W. Lindinger, and R. N. Varney, *J. Chem. Phys.* **61,** 1180 (1974).

37. F. Howorka, *J. Chem. Phys.* **64,** 5314 (1976).

38. E. E. Ferguson, in *Kinetics of Ion Molecule Reactions,* P. Ausloos, Ed., Plenum Press, New York (1979), p. 377.

39. M. McEwan and L. F. Phillips, *Chemistry of the Atmosphere,* Edward Arnold, London (1975).

40. P. Thaddeus, *Phil. Trans. Roy. Soc. London A* **303,** 469 (1981).

41. D. Smith and N. G. Adams, *Int. Rev. Phys. Chem.* **1,** 271 (1981).

42. H. Villinger, A. Saxer, E. E. Ferguson, H. C. Bryant, and W. Lindinger, in *Proceedings of the 3rd International Swarm Seminar Innsbruck,* W. Lindinger, Ed. (1983), p. 127.

CHAPTER 12

Cluster Ions: Their Formation, Properties, and Role in Elucidating the Properties of Matter in the Condensed State

A. WELFORD CASTLEMAN, JR.
Department of Chemistry
The Pennsylvania State University
University Park, Pennsylvania

TILMANN D. MÄRK
Institut für Experimentalphysik
Leopold-Franzens-Universität
Innsbruck, Austria

1 INTRODUCTION: THE IMPORTANCE OF CLUSTER IONS

Cluster ions are ubiquitous, existing in such diverse circumstances as in the immediate environs of an ion in the liquid phase and in interstellar regions. The study of ion clustering, in terms of both the ionic properties and the ion–molecule association reactions (i.e., the successive addition of neutrals to an ion) responsible for formation of cluster ions, constitutes a very active area of research in the field of

This chapter was drafted while T. D. Märk was Visiting Professor of Chemistry at the Pennsylvania State University, University Park, Pennsylvania, and completed while A. W. Castleman, Jr., was Guest Professor of Physics at the Leopold-Franzens-Universität, Innsbruck, Austria.

gaseous ion chemistry. Interest in this topic stems in part from the realization that the results contribute to a better understanding of the forces between ions and neutrals. Furthermore, studies of the properties of ion clusters are particularly helpful in elucidating the molecular details of the collective effects responsible for phase transitions (nucleation phenomena), the development of surfaces, solvation phenomena, and formation of the condensed state. Potential fields of application include catalysis, crystal growth, photography, combustion processes, high-temperature plasma injection, nuclear reactor safety, and atmospheric and astrophysics processes.

The first question to be addressed is, what constitutes a cluster ion? The most general definition is that clusters (whether neutral or ionized) are entities comprising a nonrigid assembly of components and having properties between those of large gas-phase molecules and the bulk condensed state (1). In principle, studies of clusters allow one to bridge the gap between atomic and molecular physics on the one hand and condensed-matter physics on the other. In this regard, work on systems of increasingly higher degrees of aggregation has had direct bearing on the field of interphase physics, and some researchers have called small aggregates composed of atoms and molecules the *fifth state of matter* (2), in recognition of their potential role in phase transformations.

Another definition, given by Hoare (3), is that a *microcluster* may be defined as an aggregate, whether of atoms, ions, or molecules, that is so small that an appreciable proportion of its constituent units must be present on its surface at any given time. According to Hoare this sets an upper size limit of the order of $n = 10^4$ to 10^5 units. It is interesting to note in this conjunction that important thresholds for the beginning of effectively macroscopic or bulk properties already lie below this size limit, somewhat against preconceived notions. For example, melting transitions occur already at the $n = 100$ level; other examples will be discussed in detail in Section 4. Of course, these classification schemes leave open the question of what constitutes the lower size limit. Obviously a dimer has no surface tension associated with it (4,5) and is not strictly a cluster. Nevertheless, since higher degrees of aggregation develop from the first, dimerization stage, we include this in our treatment.

To acquire insight into the linear, surface, and volume dimensions of microclusters as a function of the number n of units making up the cluster, we present a crude estimate of these values. Let r and R be the radii of a single unit and a whole cluster (both assumed to be hard spheres), respectively; v and V the volumes of the unit and the whole cluster, respectively; and n_s the surface number. Assuming that n units make up a sphere with the minimum possible surface/volume ratio and neglecting the packing fraction and geometrical details of the structure, it follows [after Hoare (3)] that

$$n \sim \left(\frac{R}{r}\right)^3 \quad \text{with} \quad R \sim N^{1/3}r, \tag{1}$$

$$n_s \sim 4\left(\frac{R}{r}\right)^2. \tag{2}$$

TABLE 1 **Comparison of Linear, Surface, and Volume Relationships for Clusters Composed of Hard Spheres of 1 Å Radius**

n	R (Å)	$\dfrac{n_s}{n}$ (Surface Occupancy Ratio in %)	V (Å³)
125	4	80	5.2×10^2
10^3	10	40	4.2×10^3
10^4	21.5	10	4.2×10^4
10^6	100	4	4.2×10^6

Taking a standard radius $r = 10^{-8}$ cm (1 Å), the values given in Table 1 result. With only a few exceptions, discussion in this chapter is confined to clusters composed of less than 100 primary units, in which the vast majority of units are indeed on the surface.

A further restriction is that we will treat only a selected class of clusters, namely *ionic clusters*. They can be categorized into two groups: cluster ions with a central ion about which neutral ligands are clustered (with an upper limit to the bonding energy of ~1 eV), and cluster ions consisting of ionized van der Waals molecules (see Table 2). As the neutrals are largely surface entities, in the case of nondelocalized charge and moderate cluster size, the ion often occupies the "bulk" rather than the surface sites (e.g., see "Ionic Clusters" in Table 2).

Information on the transition to the condensed state is frequently obtained through investigations of the properties of cluster ions. A particularly interesting situation arises in the case of protonated clusters of systems exhibiting hydrogen bonding; in

TABLE 2 **Ionic Microclusters ($2 < n < 10^6$)[a]**

Single-Component Systems

1. Atomic clusters:	$A_n{}^+$	with A = rare gases, alkalis, metals
2. Molecular clusters:	$(A_2)_n{}^+$	with $A_2 = O_2$, N_2, etc.
	$(AB)_n{}^+$	with AB = NO, CO, etc.
	$(ABC)_n{}^+$, etc.	

Multicomponent Systems

1. Mixed clusters:	$(X_n \cdot Y)^+$	with X and Y-atoms or molecules
	$Z^+ \cdot X_n \cdot Y_m$	with X, Y, and Z-atoms or molecules
2. Protonated clusters:	$X_{n-1} \cdot XH^+$	with X = H_2O, NH_3, HI, etc.
3. Ionic clusters:	$I^+ \cdot N_n$	with I^+-central ion (e.g., alkali metals) and N-neutral atom or molecule

[a]Some of these ionic microclusters exist also as negatively charged ions (1).

this case the proton may become mobile, as discussed in Section 4.3. Finally, it is necessary to point out that in this chapter the single bonding of a proton to a neutral is excluded from the definition of a cluster ion.

2 PRODUCTION OF CLUSTER IONS

Depending on the degree of physical isolation, four general types of clusters may be produced and studied:

1. Gas-phase clusters.
2. Clusters supported on a substrate surface.
3. Clusters existing in liquids and solids.
4. Clusters introduced into the gas phase through vaporization, or via bombardment of the condensed phase.

In this chapter, we concentrate largely on those produced and studied in the gas phase.

Cluster ions in the gas phase have been studied in molecular beams, in swarm experiments, or by methods in which the production process involves the transfer of energy to the surface of a sample, leading to the emission of clusters (e.g., Knudsen-cell evaporation, laser evaporation, field evaporation, spark evaporation, and sputtering). An extensive review of the literature pertaining to these techniques is given in Ref. 1 (see also Refs. 6 and 7).

Swarm experiments include a variety of (selected-ion) drift- and flow-tube experiments described in detail in the literature (8,9). The cluster ions are usually produced in an ion or discharge source. As an example, we give the principal steps leading to formation of the hydronium-series cluster ions in a He–H_2O plasma afterglow (10): The first step is direct electron impact ionization of H_2O and/or metastable-atom formation by electron impact in the helium buffer gas,

$$He + e \rightarrow He^m + e_s, \tag{3}$$

followed by Penning ionization of water vapor:

$$He^m + H_2O \rightarrow He + H_2O^+ + e. \tag{4}$$

The H_2O^+ is rapidly converted to H_3O^+ by the fast two-body ion–molecule reaction

$$H_2O^+ + H_2O \rightarrow H_3O^+ + OH. \tag{5}$$

Thereafter hydrated cluster ions are formed by a sequence of three-body association reactions of the type

$$H_3O^+ \cdot (H_2O)_n + H_2O + He \rightleftarrows H_3O^+(H_2O)_{n+1} + He. \tag{6}$$

The gas temperature, the partial pressure of H_2O, the total gas pressure, and the reaction time are all of importance in determining which cluster ions dominate. Under appropriate conditions, an equilibrium distribution can be attained, from which thermodynamic properties of the ion clusters can be derived (11). At lower temperatures and higher partial pressures, the equilibrium of cluster-ion association reactions will generally be shifted to the right (e.g., see the distribution of metal-ion hydrates in Fig. 1).

Figures 2 and 3 show typical examples of a high-pressure drift cell and a variable-temperature selected-ion flow tube (SIFT) apparatus, respectively, used for the study of production and loss processes of cluster ions. The first of these is representative of the most useful and widely used technique for measuring thermochemical properties (11,14,15), while the second finds application in the study of the kinetics of cluster formation (8). The vast majority of data have come from the more common

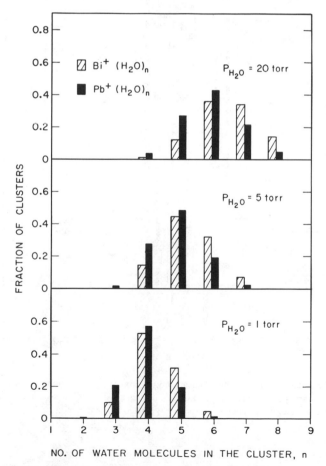

Figure 1.   Distribution of bismuth- and lead-ion hydrates as a function of water-vapor partial pressure at 300 K. After Tang and Castleman (12).

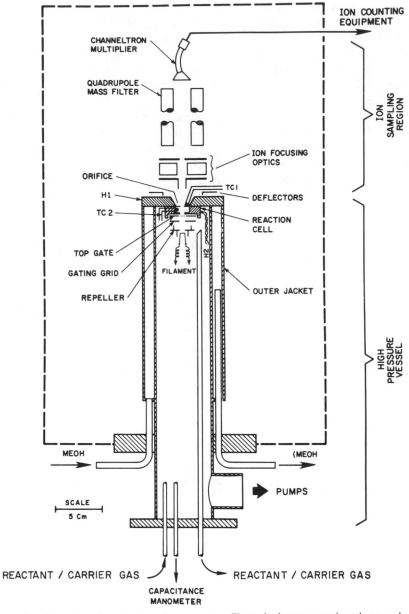

Figure 2. Schematic of an ion-clustering apparatus. Figure is drawn approximately to scale, with internal supports omitted for clarity. Broken line delineates a region of high vacuum. H and T designate locations of heaters and thermocouples, respectively. After Castleman et al. (13).

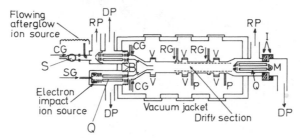

Figure 3. Schematic diagram of the SIFT–DRIFT apparatus. S, microwave discharge; CG, carrier gas; SG, ion-source gas; RP, roots pump; DP, diffusion pump; B, stainless-steel bellows; Q, quadrupole mass filter; V, optical viewport; RG, reactant gas; P, pressure-measurement port; I, electrical insulator; M, channel multiplier. After Smith and Adams (8).

flowing afterglow method (16–19). Other techniques for kinetic studies include the stationary afterglow method (20).

Other cluster-production techniques include supersonic expansion, in which a molecular beam is produced when a gas expands from a stagnation chamber at a pressure p_0 and a temperature T_0 through a small orifice (diameter D) into an evacuated region ($p \ll 10^{-3}$ torr). Ions are then produced by various techniques and subjected to mass spectrometric analysis. The interested reader is referred to Ref. 1 and references therein as a suitable introduction to the extensive literature. A notable advance in this area has been reported by Echt et al. (21), who combined the advantages of the supersonic pulsed-valve molecule beam technique with laser-induced multiphoton ionization and a reflection time-of-flight (TOF) mass spectrometer. The apparatus (Fig. 4) generates cluster ions, in favorable cases with control of the internal energy in the cluster, and uses a TOF method (22) to investigate the unimolecular dissociation processes that occur on a microsecond time scale. An example of the cluster ions produced by this method are the protonated ammonia

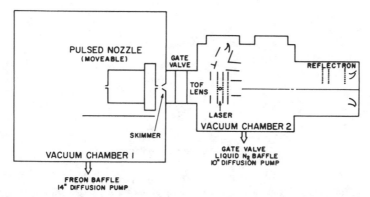

Figure 4. Schematic of the experimental setup of Echt et al. (21). The ionizing laser intercepts the neutral-cluster beam in the TOF lens. Various methods of monitoring the ion current are possible with the particle detectors shown.

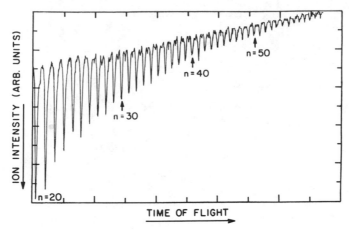

Figure 5. "Normal" TOF mass spectrum of ions with m/ze ratios corresponding to cluster sizes of $20 \leq n \leq 63$. The signal is accumulated for 256 laser shots. After Echt et al. (21).

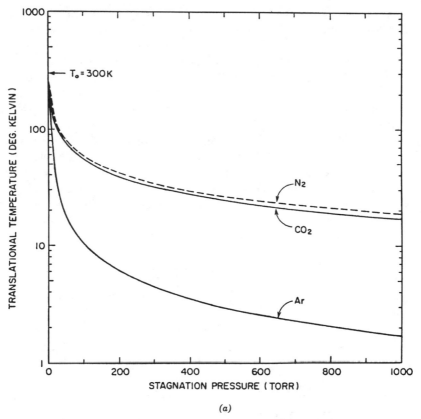

(a)

Figure 6. (*a*) Translational temperature vs. stagnation pressure for $D = 100$ μm and a stagnation temperature $T_0 = 300$ K. (*b*) Translational temperature vs. axial distance x/D for $D = 100$ μm, $T_0 = 300$ K, and a stagnation pressure p_0 of 1000 torr. Locations of the "freeze-in surfaces" are indicated by vertical bars.

series with more than 50 coclustered ammonia molecules (see Fig. 5). One other method for cluster production and study that is worthy of mention but has not been widely used is the crossed-beam charge-exchange technique developed by Ding and co-workers (23).

Molecular beam sources can be characterized by the Knudsen number, $K = \lambda_0/D$, where λ_0 is the mean free path of the gas in the reservoir. If $K > 10$ (*effusive flow*), the particles pass through the orifice without undergoing collisions; if $K \ll 1$ (high stagnation-source pressure), *supersonic flow* results and the particles undergo many collisions while flowing through the orifice. These collisions during the hydrodynamic expansion give rise to the unique properties of a supersonic beam, namely cooling of the rotational, vibrational, and translational (usually $T_{\mathrm{vib}} > T_{\mathrm{rot}} > T_{\mathrm{trans}}$) degrees of freedom and condensation via supersaturation. Translational temperatures of a few degrees kelvin are routinely achievable. Some typical values calculated for the case of a free jet expansion from a 100-μm-diameter flat plate housing a source gas maintained at 300 K are shown in Fig. 6. The freeze-in

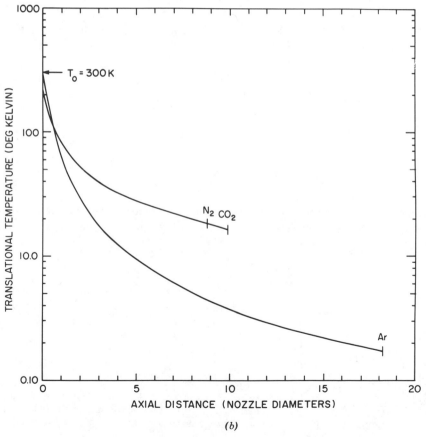

(b)

Figure 6 *(Continued)*

surface concept used for these calculations is an approximation that arises in considering the transition from continuous to collisionless, one-dimensional, effusive flow after an expansion distance of only 10–20 nozzle diameters into a perfect vacuum (see, e.g., Ref. 24–27).

The primary effect of this isentropic expansion is the conversion of random thermal energy into directed mass flow, thereby causing a decrease in translational temperature with a corresponding increase in mass-flow (stream) velocity u. Consider, for the sake of simplicity, the adiabatic expansion of an ideal gas of mass m from a source with heat capacity C_{p_0} and temperature T_0. By conservation of energy

$$C_{p_0}T_0 = \tfrac{1}{2}mu^2 + C_{p_b}T_b, \tag{7}$$

where the subscript b designates the beam. Since all the numbers are positive, it follows that $T_b < T_0$. The sonic velocity is given by

$$v = \left(\frac{\gamma RT}{m}\right)^{1/2}, \tag{8}$$

where $\gamma = C_p/C_v$ of the respective particles. For an ideal gas, the temperature T_b of the supersonic beam is (28)

$$T_b = T_0\left(1 + \frac{\gamma - 1}{2}M^2\right)^{-1}. \tag{9}$$

The Mach number M, defined by u/v, is given by (29)

$$M = c\left(\frac{x}{D}\right)^{\gamma - 1}, \tag{10}$$

where x is the distance from the nozzle and c is a constant characteristic of the beam gas. As the expansion proceeds, M increases until it approaches a terminal

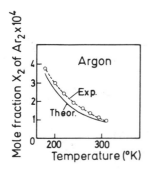

Figure 7. The equilibrium mole fraction X_2 of the argon dimer Ar_2 as a function of stagnation gas temperature for a stagnation gas pressure of 100 torr. After Andres (33).

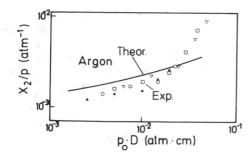

Figure 8. The equilibrium mole fraction X_2 of the argon dimer Ar_2 as a function of nozzle stagnation gas pressure p_0 times nozzle diameter D for a stagnation gas temperature of 300 K. After Andres (33).

value M_t (when collisions between molecules become negligible). Anderson and Fenn (30) found that

$$M_t \approx 1.17 \text{ K}^{-0.4} \approx (p_0 D)^{0.4}. \tag{11}$$

The amount of condensation depends on the time of the expansion [which can be lengthened by adopting a proper nozzle design (31)] and on the degree of cooling during the expansion; it scales roughly by $p_0^2 D$ (32). Figures 7 and 8 show the equilibrium mole fraction X_2 of the argon dimer as a function of T_0 and $p_0 D$. X_2 has a greater power dependence on T^{-n} than on pressure (33,34); higher intensities of dimers (and larger clusters) are acquired by lowering T_0 than by increasing p_0. Neutral clusters extending into the $n = 10^8$ range are observed on decreasing the stagnation temperature. For conditions of constant cluster size, the scaling law

$$p_0 D_0^{1.5} T_0^{-2.4} = \text{Const.} \tag{12}$$

applies (see Fig. 9), where D_0 is a reduced diameter (35).

An interesting question has arisen concerning the role in the formation and growth of larger (neutral) clusters of the preexisting dimer present in the source, compared with new dimers formed in the expansion. Using a new method of studying cluster similarity profiles, Breen et al. (36) have shown that the dominant contribution to

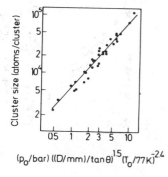

Figure 9. Average cluster size of hydrogen clusters for different nozzles (of diameter D and cone angle θ) and different nozzle gas temperatures T_0 as a function of $p_0(D/\tan \theta)^{1.5}(T_0/77)^{-2.4}$. After Obert (35).

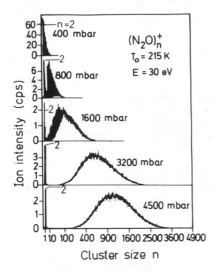

Figure 10. Size distributions of $(N_2O)_n^+$ clusters for different stagnation gas pressures. T_0, nozzle temperature; E_i, electron ionization energy. After Sattler (6).

cluster formation involves the preexisting dimer, which finding is supported by other work (e.g., see Ref. 37).

Since condensation involves production via statistical processes, it is difficult to obtain narrow distributions of cluster size unless some specifically devised source conditions are employed. Figure 10 shows as an example cluster-size distributions obtained by electron impact ionization of N_2O at different p_0, and Fig. 11 gives the dependence of the signal for Ar_n^+ clusters (produced by photoionization) on p_0. The dramatic change in cluster size with changing stagnation conditions is clearly evident

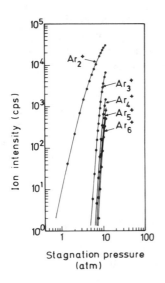

Figure 11. Dependence of the Ar_n^+ cluster-ion signal on stagnation gas pressure at a photoionization wavelength of 772.5 Å with room-temperature expansion gas. $D = 10\mu m$. After Dehmer and Pratt (38).

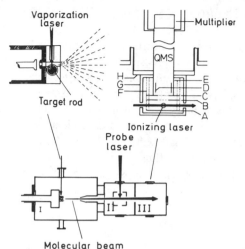

Vaporization
laser

Multiplier

QMS

Target rod

Ionizing laser
Probe
laser

I

III

Molecular beam

Figure 12. Schematic view of supersonic molecular two-photon ionization mass spectrometer. I, main chamber, containing pulsed nozzle with laser vaporization; II, probe chamber; III, detection chamber, containing TOF mass spectrometer; A, repeller plate; B, draw-out grid; C, flight-tube grid; D, deflection plates; E, flight tube; F, support rods; G, copper cryoshield; H, stainless-steel liquid-nitrogen dewar; QMS, quadrupole mass spectrometer. After Hopkins et al. (39).

from the figures. Beyond a certain set of conditions where the dimer is favored, a large set of clusters suddenly develops, as is clearly shown for Ar_n^+ in Fig. 11.

One final general method of cluster-ion formation that warrants mention is that involving laser interaction with a surface and subsequent multiphoton ionization. Figure 12 shows an overview of a rather novel technique involving pulsed laser vaporization within a pulsed supersonic nozzle with subsequent mass-selective resonant two-photon laser ionization (39). In this method, a (metal) rod of the desired cluster species is slowly rotated to expose a fresh surface to a tightly focused pulsed laser beam. A carrier gas that serves as a third body for effecting further cluster cooling and growth is pulsed over the vaporizing surface. Clusters of a number of different metals, including Cu, Mo, and Fe, have been generated with this technique.

In principle, these neutral-cluster beams can be ionized by any of several means and the cluster ions produced may be analyzed directly by a mass spectrometer. Although different ionization (electron, single- and multiphoton, Penning ionization, etc.) and detection techniques have been used, the basic setups of so-called molecular beam ionization mass spectrometers are quite similar. Figure 22 in Chapter 3 shows as an example a detailed cross section of the molecular beam-production, electron impact ionization, and ion optics elements of the double-focusing mass spectrometer used in the Innsbruck laboratory of Märk and co-workers (40). A major advantage of this instrument is its ability to investigate dissociation processes occurring several microseconds after the ionization event. In fact, these workers were among the first to observe metastable clusters (see below), even though hints of the importance of dissociation had been observed earlier (41). Other than magnetic sector mass spectrometers, the only other variety that permits ready detection of unimolecular dissociation are the TOF variations (22) equipped with provisions for reflecting (21,42) or accelerating the cluster ions (21,43).

3   STABILITY, REACTIVITY, ENERGETICS, AND STRUCTURE

A very important question in cluster-ion physics concerns the stability of cluster ions: Is the intensity of a given species in a cluster-mass spectrum representative of the neutral precursor (or the initial ionic distribution probed, if cluster ions are detected directly)? This question does not have a simple answer. An ionization mass spectrum is influenced by many factors: the stability of the neutral clusters entering the ionization region, the partial ionization cross sections, mass-dependent discrimination in the ion source and/or analyzer, and, finally, the stability of the ionized products. In the following, prompt and metastable fragmentations and how they relate to the structure are discussed in detail. Ionization cross sections for clusters were recently reviewed by Märk (44–47).

3.1   Fragmentation

Although the fragmentation of molecules following their ionization is a well-known and accepted process (40), the subject of fragmentation has been a controversial one in cluster research (48). In fact, it has been commonly assumed that there is a direct correspondence between detected cluster-ion intensities and the distributions of neutral (cluster) precursors (49). In studies of the pressure dependence of observed metastable peaks in argon, ammonia, and carbon dioxide clusters, Märk and co-workers (50–52) reported the first quantitative studies of cluster ion dissociation. For the case of NH_3, the protonated cluster ions are much more abundant than the unprotonated cluster ions (52). In this regard, it is interesting that certain clusters (e.g., H_2 and HF) cannot even be observed at their parent mass: Internal ion–molecule reactions lead to rapid atom-rearrangement reactions. In a photoionization study of $(H_2)_n$ clusters, Anderson et al. (53) found no evidence for the existence of a stable $(H_2)_2^+$ species. The reaction $H_2^+ + H_2 \rightarrow H_3^+$ is exoergic by 1.7 eV, and the ground-state surface has no barrier going down from $H_2^+–H_2$ to $H_3^+–H$ and only a shallow well at a geometry corresponding to $H_3^+–H$. For the production of H_4^+ via collisional dissociation of a H_5^+ beam, see Ref. 54.

Employing techniques similar to those of Märk, Stace and his colleagues (55–58) also found evidence of metastables in cluster ions formed by the electron impact ionization of neutral aggregates. In interpreting the dependence of the photoelectron spectra of argon clusters and of the species Xe_2^+ and Xe_3^+ as a function of the conditions in their expansion source, Dehmer and collaborators (59,60) obtained evidence that cluster fragmentation contributes significantly to features of the smaller species. Finally, Ding and co-workers (23), in charge-transfer experiments, and Buck and Meyer (61,62), employing molecular beam-scattering techniques with Ar and NH_3 clusters, also demonstrated the importance of fragmentation processes for rare-gas clusters (see also Ref. 41). More recently, Echt et al. (21,63) found that up to five NH_3 molecules were lost from a protonated cluster ion containing eight monomer units, and that at least one water molecule was lost from $H^+(H_2O)_{21}$ (43), in agreement with the predictions of Holland and Castleman (64) and with other experimental observations (58,65). Very recently, Castleman and Kay (66) employed

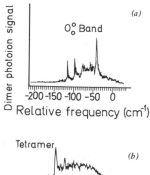

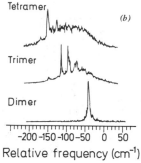

Relative frequency (cm⁻¹)

Figure 13. (*a*) One-color resonant two-photon ionization spectrum of a benzene cluster beam, obtained by monitoring the photoion signal in the benzene-dimer channel. The frequency scale is relative to the monomer origin at 38086.1 cm^{-1}. Note by comparison with *b* that most spectral features in this scan result from fragmentation of the trimer into the dimer signal channel. (*b*) Two-color resonant two-photon ionization scan of the absorption bands of the benzene dimer, trimer, and tetramer in the region of the $^1B_{2u}(\pi\pi^*)$ ^1A_g origin of the benzene monomer. The intensity of each spectrum has been normalized to a constant level. Actually, the observed two-color peak photoion signals were in the ratio $1:0.05:0.03$ for dimer/trimer/tetramer, respectively. The zero of the relative frequency scale is set to the position of the forbidden origin of the benzene monomer at 38086.1 cm^{-1}. The red shifts of the most prominent features with respect to the monomer origin are -40, -115, and -149 cm^{-1} for the dimer, trimer, and tetramer, respectively. The strong features in the trimer spectrum centered 22 and 40 cm^{-1} to the blue of the origin are due to progression activity in the van der Waals modes of the cluster. Both parts after Hopkins et al. (67).

an electric technique for the deflection of neutral clusters to demonstrate fragmentation of cluster ions in the $(SO_2)m\cdot(H_2O)_n$ system without the need for confining the observations to a specific time frame. They suggest that this method may prove useful for isolating features of polar dimers in the presence of higher-order nonpolar species.

If the cluster ion is produced in an excited state above the dissociation limit and this energy cannot be accommodated by prompt relaxation processes, fragmentation is inevitable. It therefore follows that certain vertical ionizing transitions and ion–atom association (without collisional stabilization by a third body) reactions will lead to prompt dissociation (41). Considerable direct fragmentation of small clusters occurs even at relatively small excess energies. [For more details see Märk and Castleman (1).]

Multiphoton ionization techniques are becoming especially useful in elucidating details of dissociation processes, particularly the role of excess energy. Figure 13

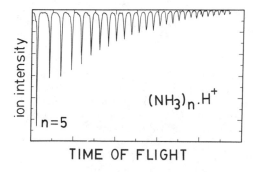

Figure 14. Mass spectrum with $p_0 = 1300$ torr of neat NH_3. After Echt et al. (63).

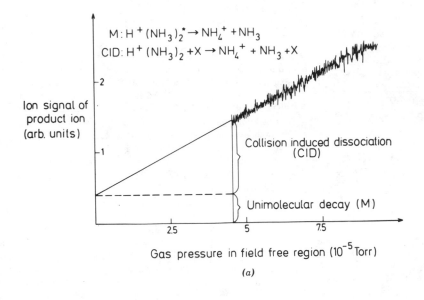

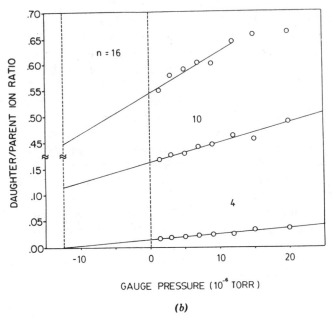

Figure 15. (a) Schematic representation of pressure dependence of metastable and collision-induced dissociations in the field-free region of a sector field mass spectrometer. After Stephan and Märk (70). (b) Pressure dependence of daughter/parent-ion intensity ratio. The plotted ratios correspond to the signal for the daughter ions accumulated via loss of one NH_3 unit due to dissociation in the field-free region before reflection. The effective region of zero pressure lies between the dashed lines. After Echt et al. (21).

274

is a particularly dramatic example of the power of the new methods; it demonstrates that the benzene trimer obscures the true benzene-dimer signal in one-color experiments. In Fig. 13*a* the second laser photon reaches 4500 cm^{-1} above the ionization potential. In Fig. 13*b* the energies of the two-color photons were chosen to just ionize the dimer, thereby avoiding fragmentation of higher clusters. Dao et al. (68,69) have demonstrated similar advantages of multiphoton ionization (MPI) for a variety of other cluster systems.

Another instructive example of the importance of fragmentation of ion clusters is the pattern of "magic numbers" observed for particularly stable species. In MPI studies of ammonia clusters using a TOF mass spectrometer detector (63), an unusually prominent species corresponding to $(NH_3)_4 \cdot NH_4^+$ was found under all expansion conditions (see Fig. 14). A systematic investigation of the conditions leading to the rather stable protonated pentamer showed that, in general, this cluster ion is most prominent under conditions of low laser fluence and under MPI conditions close to those that just ionize the clusters. The results pertain directly to the problem of cluster energetics. They are discussed and references are given in Section 3.3.

In the investigation of metastable cluster ions it is important to distinguish between true metastables (whose properties are defined by clustering and ionization) and collision-induced decompositions. Metastable ions are most commonly detected and studied in the field-free region of sector field mass spectrometers. Figure 15 shows as an example the detected metastable-ion signal divided by the precursor-ion signal as a function of gas pressure in the field-free region. The intercept of such a plot gives a direct measure of the fraction attributable to unimolecular (metastable) dissociation (21,50,51,70,71). The collisional dissociations of other cluster ions have been studied, and their general behavior and dependences have been characterized (72,73).

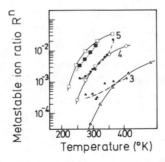

Figure 16. Comparison of metastable-ion ratio R^n predicted from RRKM theory calculations (open symbols) and measured experimentally (solid symbols) by Sunner and Kebarle (75). R^n is the ratio of metastable ions $[K^+(H_2O)_n \rightarrow K^+(H_2O)_{n-1}]$ to ions detected as $K^+(H_2O)_n$. The different shapes of R^n (exp.) and R^n(RRKM) are believed to be due to contributions from collision-induced dissociations not accounted for in the experimental procedure. The abscissa gives the temperature of the high-pressure ion source.

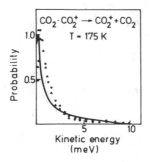

Figure 17. Metastable kinetic-energy-release distributions for the reaction $CO_2 \cdot CO_2^+ \rightarrow CO_2^+ + CO_2$. The points are the experimental data (high-pressure ion source) and the lines are the results of statistical phase-space-theory calculations. After Illies et al. (76).

Three different unimolecular dissociation mechanisms can be distinguished:

(i) *Statistical (Vibrational) Predissociation.* The abundances and properties of certain metastable cluster ions are in agreement with the hypothesis that their metastability results from statistical (vibrational) predissociation (i.e., if a molecular ion is complex enough, extensive Lissajous motion on the potential-energy hyper-surface precedes dissociation of the activated complex) (50,56,74–76). Figure 16 shows the metastable-ion ratios for $K^+(H_2O)_n$ clusters as a function of temperature, and Fig. 17 the kinetic-energy distribution of the metastable dissociation products for $(CO_2)_2^+$. These examples show the monotonic variation of the metastable ion ratios, with temperature and equipartitioning of energy characteristic of long-lived, statistical complexes. In addition, neutral and ionized metastable giant clusters of He have been reported (77). As shown in Fig. 18, liquid-phase fluctuation phenomena explain the ejection of miniclusters.

(ii) *Electronic Predissociation and* (iii) *Barrier-Penetration Predissociation.* On the other hand, we have also observed recently the existence of very small atomic (or pseudoatomic) cluster ions, such as He_2^+, Ar_2^+, Ar_3^+, ArN_2^+, and N_4^+ (7,50,51,70,74,78–80). Their existence has been rationalized as being due to predissociation through an electronic mechanism and/or (rotational) barrier penetration. Figure 19 shows the measured metastable/precursor-ion ratios of Ar_2^+ produced by direct ionization of neutral dimers (Fig. 19a) and associative ionization (Fig. 19b) (80). Finally, Fig. 20, which shows the temperature dependence of the metastable fragmentation of Ar_3^+ (41), demonstrates that this ratio can change from $<10^{-2}$ to >1.

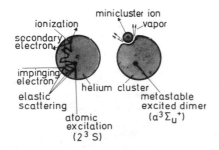

Figure 18. Schematic outline of the primary events during an electron's passage through a large (n up to 10^7 atoms) He_n cluster (left side), and during a later stage, showing the ejection of a charged minicluster. After Gspann (77).

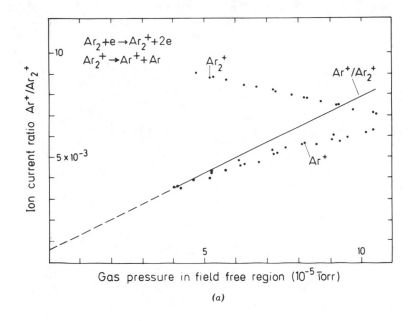

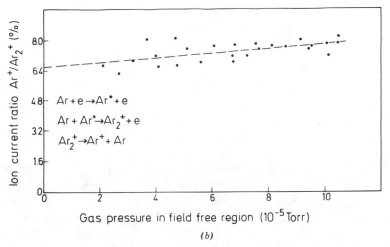

Figure 19. (a) Pressure dependence of ion currents of Ar_2^+ (parent ion produced by electron impact) and Ar^+ (daughter ion produced via the dissociative reactions $Ar_2^{+*} \rightarrow Ar^+ + Ar$ and $Ar_2^+ + Ar \rightarrow Ar^+ + 2Ar$). Also plotted is the ratio of daughter- to parent-ion intensity (scale given at left-hand side). The electron energy for the ionization process $Ar_2 + e \rightarrow Ar_2^+ + 2e$ is 70 eV. Repeated measurements prove consistently the existence of a metastable dissociation reaction (finite intercept at zero pressure). After Stephan et al. (80a). (b) Same as a, but with the parent ion produced by associative ionization. The electron energy in the associative-ionization sequence is 70 eV. It can be seen that the metastable transition is ~100 times larger than in a (6.5% compared to 6×10^{-4}). After Stephan and Märk (80b).

277

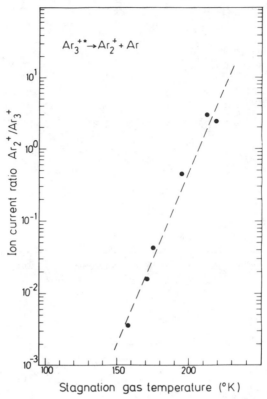

Figure 20. Intensity of the daughter ion Ar_2^+ (including only metastable dissocations) normalized by the precursor ion Ar_3^+ as a function of the stagnation gas temperature. After Stephan and Märk (70).

3.2 Multiply Charged Clusters

The recent observation of several multiply charged cluster ions raises an interesting point about their production and stability (7). According to observations and calculations (81), doubly ionized heteroatomic rare-gas dimers can be produced by direct ionization from the neutral ground state to either (i) weakly bound attractive-potential curves lying below the Coulomb dissociation limits of the respective atomic ions (e.g., $NeXe^{2+}$; see Fig. 20 in Chapter 3) or (ii) quasibound molecular levels above the dissociation limits (e.g., $ArXe^{2+}$, $NeKr^{2+}$; see Figs. 21 and 22) (81,82). The existence of stable mononuclear metal trimer ions (e.g., Ni_3^{2+}, Au_3^{2+}, W_3^{2+}) has been explained (83) by a simple empirical rule stating that the ratio of Coulomb energy to neutral bond energy has to be <1.3 to allow formation of stable doubly charged clusters. Figure 23 shows some recently obtained mass spectra of $(CO_2)_n$ clusters (84), demonstrating the appearance of strong $(CO_2)_n^{2+}$ peaks at around $n = 45$. In addition, the existence of very large multiply charged clusters has been reported and their stability explained by the fact that multiply charged clusters are

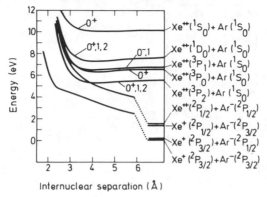

Figure 21. Potential-energy curves for the lowest states of $ArXe^{2+}$. After Helm et al. (81).

stable only if their size exceeds a certain critical value at which the repulsive Coulomb energy (minimized by distributing the charges uniformly on the surface) is smaller than the binding energy (85). Recently we have reported, for the first time, the existence of dinegative cluster anions (7).

3.3 Energetics and Structure

Total binding energy, the binding energy per unit, and the individual successive binding energies of added units are of great importance in determining the stability and structure of a given cluster ion. A wealth of data on these properties, as

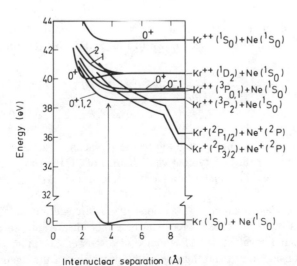

Figure 22. Potential-energy curves for the lowest states of $NeKr^{2+}$ and NeKr. After Stephan et al. (82).

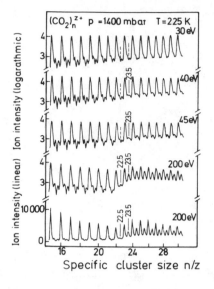

Figure 23. CO_2-cluster mass spectra for different electron impact ionization energies. After Echt et al. (84).

determined from van't Hoff plots in high-pressure mass spectrometry experiments, are available (11,14,15). These precise thermodynamic data are in strong contrast with our knowledge of the structure of cluster ions, most of which is indirect and depends on comparison of experimental results with calculations.

Equilibrium thermodynamic properties are measured by establishing a balance between a forward and reverse reaction,

$$A^+-B_{n-1} + B \rightleftharpoons A^+-B_n, \tag{13}$$

for the species of interest (A and B) and measuring the ratio of cluster-ion intensities in an apparatus similar to the one shown in Fig. 2. For this reaction,

$$K = \frac{[A^+-B_n]}{[A^+-B_{n-1}]p_B} = \exp\left(-\Delta G^0/RT\right)$$
$$= \exp\left(-\Delta H^0/RT\right)\exp\left(\Delta S^0/R\right), \tag{14}$$

where the usual assumptions pertaining to the replacement of activities by partial pressure p_B are made (86). A typical van't Hoff plot (87) is shown in Fig. 24, the slope of which gives the standard enthalpy of clustering, $\Delta H_{n-1,n}^0$; the intercept gives the entropy, $\Delta S_{n-1,n}^0$.

According to Castleman et al. (88), ion–ligand forces dominate the structure at the smallest cluster size. These forces can be either electrostatic or a combination of electrostatic and covalent ones. Reference 11 discusses the applicability of simple electrostatic considerations and the role of the properties of the ligand ion (see also Ref. 89). As a first approximation, clusters involving species with large dipole moments display the strongest bonding, although among similar species the polar-

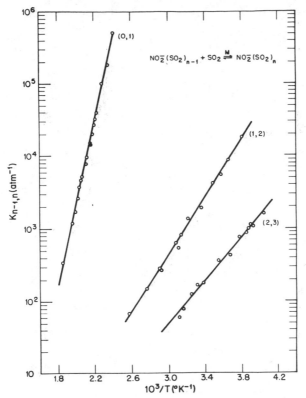

Figure 24. Van't Hoff plots of the gas-phase equilibria for the reaction $NO_2^-(SO_2)_{n-1}$ + $SO_2 \overset{M}{\rightleftharpoons} NO_2^-(SO_2)_n$. Numbers in parentheses are: $(n - 1, n)$. After Keesee et al. (87).

izability is also important. The role of the quadrupole moment is seen by comparing the bonding of SO_2 with that of NH_3; both have approximately comparable dipole moments and polarizabilities, but the quadrupole moment of SO_2 leads to a repulsive interaction, compared to an attractive one for NH_3. This accounts for the much lower bonding strength of SO_2.

At larger sizes the ligand molecules, although still oriented with respect to the central ion, can begin to interact sterically with each other and eventually form "solvation" shells. As clusters become larger and the enthalpy of reaction begins to approach the heat of condensation for the bulk phase, ligand–ligand interactions and the nature of the ligand itself become more important. Calculations suggest that in the case of positive ions, symmetrical ordering of polar molecules occurs about a central charge, whereas in the case of negative ions chainlike bond linkages resulting from hydrogen bonding between ligands are predicted (90).

Figure 25 plots the enthalpy of reaction (which is directly related to bond energy) for successive clustering as a function of cluster size. The enthalpy decreases monotonically with cluster size; furthermore, the enthalpy for a particular n value varies inversely with ionic radius (91).

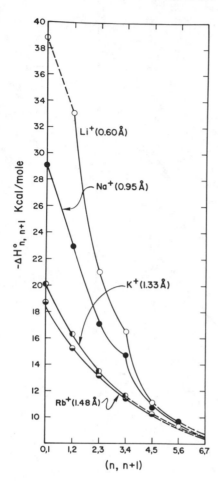

Figure 25. Plot of $\Delta H_{n, n+1}^{0}$ vs. cluster size for the reaction $M^+(NH_3)_n + NH_3 \rightleftharpoons M^+(NH_3)_{n+1}$. After Castleman et al. (91).

Most of the evidence for cluster structures comes from theoretical calculations (90,92–94). The experimental data on cluster bonding have been especially useful for comparison with theory, and provide encouraging evidence that the predicted structures are reasonable. In this regard there is need for more effort in the area of cluster-ion spectroscopy.

Finally, information about stability, structure, and energetics may be used to explain observed anomalies in mass spectra of cluster distributions. We discuss below three well-documented examples of particular interest.

3.3.1 Protonated Cluster Ions of Ammonia

Recently, we observed a particularly stable cluster ion in clusters of ammonia following multiphoton ionization (21,63). An unusually prominent and stable species is expected whenever a closed solvation shell forms. In the case of NH_4^+, it is expected that four ammonia molecules bonded to each of the hydrogen atoms of the central moiety would represent a closed solvation shell. Varying the distance of the pulsed nozzle (employed as the source) to the ionization region and the gas-expansion parameters did not affect the prominence of the protonated pentamer, suggesting that the ion structure, rather than a preferred neutral cluster distribution, is responsible for the observed "magic number." When the laser ionizing energy and fluence are relatively low, it is expected that excess energy deposited in the cluster will lead simply to the "boiling away" of a few ammonia molecules. This loss process leads to a "piling up" of the intensities for cluster ions with thermodynamically more stable structures. A discontinuity in the intensity for the locally more stable cluster ion results.

3.3.2 (H₂O)ₙ-Cluster Distributions

Mass spectra for cluster ions of water have been recorded by a number of authors under quite different experimental conditions. Several experiments have indicated

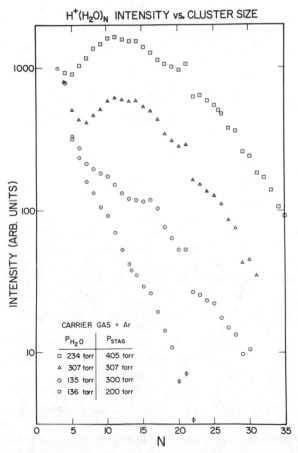

Figure 26. Semilogarithmic plots of $H^+(H_2O)_n$ cluster intensity vs. cluster size for various stagnation conditions. Uncertainties (determined based on experimental reproducibility) are smaller than symbols except where indicated. After Hermann et al. (65).

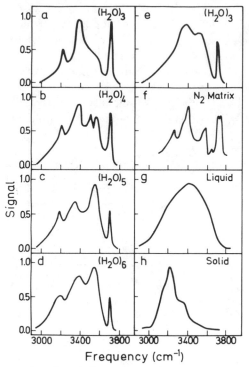

Figure 27. Water-cluster and condensed-phase spectra. After Vernon et al. (99). For details see text and Ref. 99.

the existence of particularly stable structures of cluster ions, for instance the Eigen structure $H_3O^+(H_2O)_3$ (95), which is based on both experimental findings (65) and theoretical computations (93). Yet thermodynamic measurements on the bonding of water to H_3O^+ (96) fail to reveal a particularly dramatic change in bond energies for the addition of additional water molecules to the Eigen structure.

Indirect evidence has also been reported (see Ref. 43, 58, and 64 and references contained therein) for the unusual stability of $H_3O^+(H_2O)_{20}$. However, bond-energy measurements on clusters of this size do not support the inferences from the various experimental studies. For example, Fig. 26 shows $(H_2O)_nH^+$ intensities vs. cluster size as a function of stagnation conditions (65). The most prominent irregularity occurs at $n = 21$. Since this irregularity is observed at the same cluster size irrespective of how the clusters are produced [expansion plus ionization (43,58,65), ion expansion (97), or secondary ion mass spectrometry (SIMS) of ice (98)], it was concluded that this discontinuity is due to processes occurring during and after the ionization process. These conclusions are supported by recent findings (99) on direct fragmentation of water clusters following ionization (seen in Fig. 27, which presents spectra as a function of cluster size (Fig. 27a–d), stagnation pressure (Fig. 27a,e) and form of aggregation (Fig. 27f–h)), and in the metastable decay of water clusters

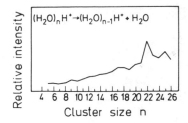

Figure 28.   Relative peak intensities resulting from the uni-molecular decomposition of $(H_2O)_nH^+$ ion clusters at an electron energy of 70 eV. After Stace and Moore (58).

(Fig. 28). The stabilization of $(H_2O)_{21}H^+$ apparently occurs via a dodecahedral clathrate structure in which an "excess" proton remains in the cage structure and an unbonded H_2O is trapped in the center (64) (Fig. 29).

These interpretations are strengthened by the recent findings of Echt et al. (43) showing that the dissociation of metastable water-cluster ions in the vicinity of the 21-mer depletes the population of adjacent species and leaves the especially stable cluster as a distinct feature in the distribution. It is indeed likely that the abundance of the 21-mer is enhanced by dissociation of higher-order clusters at times earlier than the metastable observations allow, as is suggested by the slightly enhanced "normal" TOF spectrum reported by these authors.

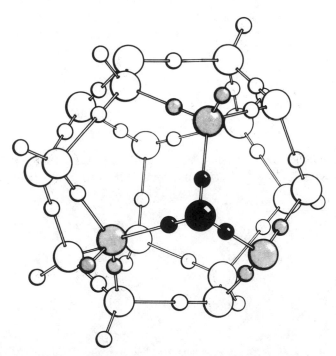

Figure 29.   Charged 20-molecule water clathrate in which oxygen atoms (large spheres) form a dodecahedron about central cavity. Shaded molecules represent a transient $H_3O^+(H_2O)_3$ structure contained within the clathrate cage. After Holland and Castleman (64).

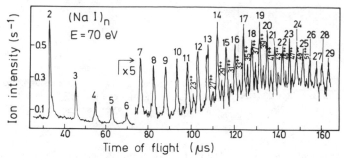

Figure 30. Mass spectrum of (NaI)$_n$ cluster ions produced by 70-eV electrons. After Sattler et al., reported by Martin (100).

3.3.3 Alkali–Halide-Cluster Distributions

Recently, Sattler and colleagues (e.g., see Ref. 100) reported mass spectra of NaI clusters grown by inert-gas condensation (see Fig. 30) and related the magic numbers observed to the multiring structure of the predominantly stable neutral particles. Similarly, Campana et al. (101) have observed CsI$^+$ clusters obtained by SIMS (see Fig. 31). Observed anomalies were interpreted by a bond breaking model, under the assumption that they arose in the ion production process. Thus in both cases the observed structures were thought to be present directly after production of the ions. However, Ens et al. (102) have recently shown that a smooth distribution (see Fig. 31) is obtained directly after production of the ions, and that the observed anomalies are due to subsequent metastable-ion decay (see also Ref. 101). This is in accordance with theoretical calculations of the binding energy per molecule as a function of cluster size n (Fig. 32) (see Ref. 100).

3.4 Reactivity

Ion association processes or ion clustering (i.e., the successive addition of the same neutral to an ion) represent a very active area of research in the field of gaseous

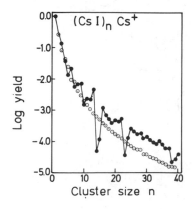

Figure 31. Relative yields of (CsI)$_n^+$ clusters obtained by SIMS. Filled circles, experimental results of Campana et al (101); open circles, experimental results of Ens et al. (102). For details see text and Ref. 102

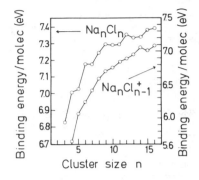

Figure 32. Calculated binding energy per molecule for the most stable forms of positively charged and neutral NaCl clusters. After Martin (100).

ion chemistry. Increasing attention is now being paid to problems related to the temperature and energy dependences of such reactions (103). Ligand-switching reactions (104) in general proceed rapidly (105) ($k > 10^{-10}$ cm^3/sec) when they are exothermic. Depending on the energetics of solvation, the exchanges may come to a halt at some intermediate stage of mixed solvation (e.g., see Ref. 106). In what follows we consider only reactive (cluster) ion–molecule reactions (107), including charge-transfer, proton-transfer, and ion–atom interchange processes.

The effect of clustering on the reactivity of ions is important for an understanding of plasmas, as well as of solvation chemistry in general (108). For example, Smith et al. (109) found that association of N_2 to N_2^+ does not seriously modify the reactivity of the N_2^+ except that it somewhat reduces the energy available for reaction. For ligand exchange, some (nonresonant) charge-transfer processes of $O_2^-(H_2O)_n$, and reactions of hydrated hydronium ions, increased hydration of the reactant ion does not significantly decrease the reaction constant [see, e.g., Fahey et al. (110)].

In contrast, for some (resonant) charge-transfer processes and several ion–atom interchange and/or nucleophilic displacement reactions, increased solvation of the reactant ion results in a significant decrease in reactivity (see, e.g., Ref. 111). Similarly, Bohme et al. (112) found that stepwise hydration leads to a decrease in the reaction rate constant for proton transfer from H_3O^+ to, for example, H_2S, with a concomitant change in ΔG^0. In the case of nucleophilic displacement reactions these results may be interpreted, according to Bohme and Mackay (113) and Henchman et al. (114), in terms of the qualitative model developed by Brauman and co-workers (115).

Another exciting example of the potential role of gaseous ion clusters is their influence on reactivity among neutrals. Apparently they can sometimes catalyze specific reactions. For example, Ferguson and co-workers (116) reported the observation of a new class of chemical reactions in which the reactivity of a neutral molecule clustered on an alkali ion with another gas-phase neutral molecule is greatly enhanced compared with that of an unclustered neutral. Kappes and Staley (117) reported the first case of catalysis in gas-phase positive-transition metal-ion chemistry. This is shown by the coupling of the last three reactions in Table 3. Such classes of reactions offer the prospect of elucidation of the fundamental mechanisms

TABLE 3   The Role of Ions in Reactions Between Neutrals[a]

Reaction	Rate Constant k	Reference
$Na^+ \cdot N_2O_5 + NO \rightarrow Na^+N_2O_4 + NO_2$	$\sim 3 \times 10^{-13}$ cm³/sec	Rowe et al. (116)
$Li^+ \cdot N_2O_5 + NO \rightarrow Li^+N_2O_4 + NO_2$	$\sim 1 \times 10^{-11}$ cm³/sec	
$N_2O_5 + NO \rightarrow N_2O_4 + NO_2$	$\sim 10^{-20}$ cm³/sec	
$Fe^+ + N_2O \rightarrow FeO^+ + N_2$	0.7×10^{-10} cm³/sec	Kappes and Staley (117)
$FeO^+ + CO \rightarrow Fe^+ + CO_2$	9×10^{-10} cm³/sec	
$N_2O + CO \rightarrow CO_2 + N_2$	not observed at room temperature	

[a]See text for discussion.

responsible for many catalytic reactions, and one can expect this area of research to be very productive in coming years.

3.5   Photodissociation

Studies of the interactions of photons with molecular ions are of both fundamental and practical interest, and although work in this field is relatively sparse, it is growing at an ever increasing rate. High-resolution studies can provide information about the location, shape, and symmetry of the ground and excited states of the ions. Furthermore, information on molecular bond energies and in some cases on the electron affinity of the parent neutral is obtained. Additionally, results on kinetic-energy-release measurements often provide information on energy transfer and on the dissociation dynamics of ion clusters.

Although there is a wealth of information on electron photodetachment spectroscopy of atoms and molecular ions [see Lineberger (118)] and the photodissociation of positive molecular ions (119), there are only limited data for cluster ions. For an interesting discussion of the infrared absorption of some molecular ions in matrices, the reader is referred to Andrews (120); of related interest is the work of Woodin et al. (121), who investigated the multiphoton dissociation of the dimer ion $H^+((C_2H_5)_2O)_2$ in the infrared region. Cluster ion–laser interaction is a developing field of research and a review of the various findings is beyond the scope of this chapter. The interested reader is referred to Märk and Castleman (1) and the references contained therein.

An example of the use of photodissociation data to gain an understanding of the condensed phase is provided by the results of Freiser and Beauchamp (122) on the photodissociation of clusters composed of C_6H_5X (where X = H, CN, NH_2, CHO, $COCH_3$, NO_2, OCH_3, O^-, and S^-) and with pyridine and ferrocene bound to H^+ and Li^+. These data have provided information on the changing bond energies of electronically excited ion complexes. Comparison of the excitation energies has provided useful information on the acidities and basicities of these complexes.

The first reported laser-induced fluorescence spectra of free jet cooled inert-gas clusters has been reported by Heaven et al. (123). The laser excitation spectrum of $C_6F_5H^+$ expanded in a mixture of 10% argon and 90% helium was obtained at stagnation pressures of 1.3, 2.7, and 4.0 atm. The extent of the red shift in the spectrum increases with increasing clustering to argon.

Further advances can certainly be expected based on data derived from both photodissociation studies of mass-selected cluster ions and investigations of fluorescence. One of the objectives of such studies is elucidating the changes occurring during the continuous course from the gaseous to the condensed phase. It is evident that techniques are now available for reaching this goal.

4 CONDENSATION PHENOMENA

In this last section we consider the applicability of gas-phase properties of cluster ions to the elucidation of bulk properties of matter. It is seen that data viewed as a function of the increasing degree of aggregation will provide a more detailed understanding of interphase physics, the development of surfaces, and, ultimately, solvation phenomena and formation of the condensed state. Four general topics are discussed in this section: (1) application of the data to understanding phase transitions; (2) the general problem of ion solvation; (3) the solvation of ion pairs; and (4) the use of cluster-ion data to clarify the properties of a neutral system, with metals chosen as the example.

4.1 Phase Transitions: Nucleation Phenomena

The general nucleation processes that are potentially responsible for gas-to-particle conversion may be classified into three general categories: homogeneous, heteromolecular, and heterogeneous. Heteromolecular nucleation is defined as that which is enhanced by the presence of a foreign molecule of composition different from that of the bulk nucleating phase or the presence of other attractive sites, such as electrically charged ones (ions). Customarily, both heteromolecular and heterogeneous phenomena are treated by extending the classical theory of homogeneous nucleation to interacting systems. As an outgrowth of recent work (124,125), a semiempirical correlation model was developed that enables the enthalpy contributions to the nucleation-energy barrier to be predicted for a number of classes of molecules condensing around electrically charged sites (ions). An important aspect of this work is its recognition of the importance of small-cluster stability and structure on the magnitude of the energy barrier (124,126).

Unlike the majority of nucleation experiments conducted with cloud chambers or supersonic jets, with the high-pressure mass spectrometric technique it is possible to obtain information on the initial clustering reactions leading to vapor nucleation. It has thus been possible to evaluate the limitations of the classical liquid-drop formulation of ion-induced nucleation (classical Thomson equation). Lee et al. (125) compared the available experimental data on hydration enthalpies with the appro-

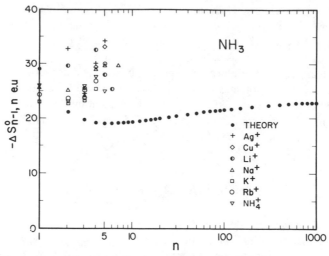

Figure 33.   Comparison of experimentally measured cluster entropies with predictions based on the classical Thomson equation (theory). After Holland and Castleman (127).

priate derivations of the free energy, and obtained excellent agreement for clusters containing as few as four to six water molecules. A similar result was obtained for the clustering of ammonia about a number of ionic species.

Large discrepancies, however, were found between the experimental and calculated entropies of cluster formation (125) (see Fig. 33). More negative values are found experimentally than are predicted by theories based on bulk liquid properties. This finding reflects the more highly ordered structure of small systems relative to that of liquids. The importance of these ordered structures is that they qualitatively explain the observation that although the Thomson equation agrees with experiment for some systems, large differences are found for others (see Ref. 127).

An important outcome of this recent work was the finding that a modified effective radius for the clustering neutral molecule could be used to correlate both the enthalpy and entropy contributions to the energy barrier of nucleation for a variety of systems. Figure 34 shows the good agreement obtained from this correlation procedure when an effective size for water is used for several solvating species. Satisfactory agreement for both ΔH and ΔS values of clustering are obtained by this empirical correlation procedure. It can be readily used for clustering of the ligands pyridine, methyl cyanide, methyl alcohol, and ammonia around virtually any positive charge.

4.2   Ion Solvation

Examples of gas-phase ion–molecule complexes that have analogies in the condensed phase are becoming legion. Several recent ones have been found in our measurements of the clustering of ammonia around alkali-metal and transition-metal ions. The clusters $Ag^+ \cdot (NH_3)_2$, $Cu^+ \cdot (NH_3)_2$, and $Ag^+ \cdot (pyridine)_2$ are examples of such com-

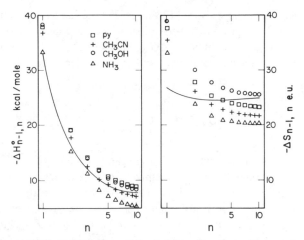

Figure 34. Charged-liquid-drop calculations of enthalpies and entropies of cluster formation compared with experimental gas-phase values. The calculations were made employing an "effective size" similar to that of water for all liquids. py, pyridine. After Holland and Castleman (127).

plexes in the gas phase (128). These cluster ions further demonstrate very sharp decreases in binding after the addition of the second ligand, which correlates well with their chemistry in the aqueous phase.

Studies of the successive clustering of ammonia about Li^+ and Na^+ (see Fig. 25) reveal a preferred coordination complex having four ligands attached to the central ion, while ammonia clusters with larger ions such as K^+ and Rb^+ do not display well-defined structures (91). These results are in very good agreement with Raman spectra published by Gans and Gill (129) based on liquid-phase measurements. Gas-phase complexes with solution-phase analogies were similarly found in transition-metal studies (128). In particular, in the case of Cu^+ and Ag^+ ions it is known that bonding strongly favors complexes with a linear structure. Gas-phase measurements show that ammonia binds much more strongly than water (11.5 kcal/mol for binding of NH_3 to Ag^+ and >20 kcal/mol for NH_3 binding to Cu^+). The difference in binding between ammonia and water cluster ions with Ag^+ and Cu^+ is a function of cluster size; after addition of the second ligand, water binds more strongly than ammonia, whereas ammonia binds more strongly to smaller sizes. This crossover, revealed by the gas-phase data, is consistent with the observed preferences of the Ag^+ and Cu^+ ions for forming diligand complexes in the aqueous phase.

It is useful to consider the extent to which data on the clustering of molecules to individual ions (e.g., binding energies) have a direct bearing on the energies of solvation of individual ions in the condensed phase. A number of authors have used thermochemical data to interpret trends and expectations concerning ions in liquids. The interested reader is referred a few relevant references (130–132).

One of the first attempts to predict solvation energies in the condensed state was based on the Born relationship, which gives the change in energy needed to create a cavity in a dielectric medium. Born (133) regarded the ion as consisting of a rigid sphere of radius r_i and charge q immersed in a structureless continuum with dielectric constant ϵ. The free energy of solvation ΔG_{solv}^0 is taken as the electrostatic potential difference between the ion in the solvent and in a vacuum. The Born relationship is

$$\Delta G_{solv}^0 = \frac{q^2}{2} (1 - \frac{1}{\epsilon}) \frac{1}{r_i}. \tag{15}$$

In principle, the corresponding enthalpy change can be obtained from an appropriate derivative of this equation with respect to temperature. Numerical evaluation requires data on the temperature coefficient of the dielectric constant of the solvent.

The Born relationship often leads to an overestimation of the heat of solvation when typical values of crystalline radii are used for the ion size. This failure has lead to proposed modifications based on structural considerations, as well as other attempts to employ detailed ion–dipole, ion–quadrupole, and higher-order interactions. In some treatments, adjustments of the crystal radii have been made.

Although it fails quantitatively, the Born equation renders a qualitatively correct picture of solvation and, with the indicated modifications, is often useful for thermodynamics purposes.

This approach has been moderately successful, but is obviously too simple for quantitative treatment of systems that display different trends of bonding to different ligands. Nevertheless, an extension of this model by Lee et al. (125) successfully uses microscopic surface-tension concepts to correct the general solvation energies obtained on the basis of the Born equation.

The most interesting finding is that all of the gas-phase negative- and positive-ion hydration data, as well as the positive-ion ammoniation data, can be related to the liquid-phase single-ion heats of solvation. The ΔH_{solv}^0 (standard heat of solvation) values were derived from the Randles separation method. It can be seen from Fig. 35 that the ratio of ΔH_{solv}^0 to the partial gas-phase enthalpies $\Delta H_{0,n}^0$ approaches a value that is independent of the nature of the ion. Hence, the single-ion solvation values can be deduced from knowledge of the first six cluster bond energies measured in the gas phase. These findings suggest that it may be possible to use gas-phase experiments to elucidate the molecular aspects of the condensed phase for other systems as well. Further evidence of the relationship of cluster data to ionic solvation comes from observations of discontinuities in otherwise smooth trends in ΔH with size (91,96,128,134), which indicate the possible formation of solvation shells.

The importance of measuring intrinsic basicities and acidities of ligands clustered around ions for an understanding of the origin of solution-phase ordering has been suggested by a number of investigators, including Beauchamp (135), Bartmess and McIver (136), Aue and Bowers (137), and Kebarle (130). Recent studies by Taft and co-workers (131,138) have addressed the necessity of understanding the solvation by a single molecule of a ligand in the gas phase if one wishes to understand the effects of molecular structure on solvation by a bulk solvent in the liquid phase. The importance of accounting for interaction among the ligands and solvent mol-

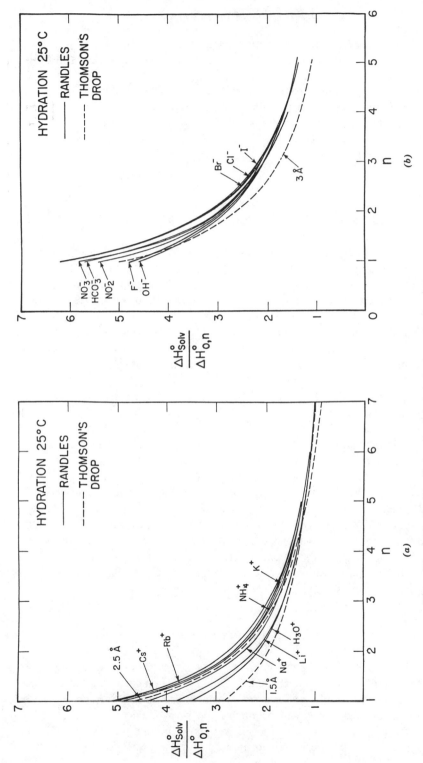

Figure 35. (*a*) Ratio of Randles' total enthalpy of solvation to the partial gas-phase enthalpy of hydration for positive-ionic cluster size *n*. (*b*) Same as *a*, but for negative-ionic cluster size *n*. From Castleman et al. (88).

ecules was discussed in a recent article by Arnett et al. (139). Castleman (11) has considered this effect for the ion OH^- solvated in water and clustered with CO_2 to form HCO_3^-.

To elucidate the nature of some S_n^2-type reactions in the solution phase, Bohme et al. (112) exploited a similar technique to investigate a few analogous reactions. In particular, they investigated reactions of the types

$$B^- \cdot S_n + AH \rightleftharpoons A^- \cdot S_m + (n - m)S + BH \qquad (16)$$

and

$$B^- \cdot S_n + CH_3Br \rightleftharpoons Br^- \cdot S_m + (n - m)S + CH_3B. \qquad (17)$$

In addition, they studied the influence of stepwise hydration on the kinetics of proton transfer from the hydronium ion to a variety of polar molecules.

Kebarle and co-workers (140,141) have reported that the strength of the hydrogen bond in the monohydrate $A^- \cdot H_2O$ complex increases with the gas-phase basicity of A^-. Similarly, they found that the strength of the hydrogen bond in $BH^+ \cdot H_2O$ increases with the gas-phase acidity of BH^+. These relationships accounting for the bonding of a water molecule in terms of the basicity of A^- or the acidity of BH^+ provided an explanation on a one-molecule solvation basis for the attenuation mechanisms observed in the solution phase. Thus, it was suggested that a substituent that increases the acidity of AH decreases the basicity of A^- and therefore decreases the strength of the hydrogen bonding in $A^- \cdot H_2O$. As with the cation clusters, there is much to be learned about solution of organic constituents.

For certain organic systems, Davidson et al. (140) have also made a comparison of proton-transfer free energies between the gaseous and solution phase. They concluded that the large attenuation due to the substituent effect in solution must result from an effect of the substituent on solvation that partially cancels the effect on the isolated molecule. For instance, an electron-withdrawing substituent, say CN, which increases the intrinsic acidity of, for example, phenol, is expected to affect unfavorably the solvation of cyanophenoxide ion and thus reduce the acidity increase of cyanophenol in aqueous solution. Cluster-ion research has a great deal to contribute to the field of solvation involving organic constituents; further discussion is beyond the scope of this review.

4.3 Solvation of Ion Pairs

The correlation between gas-phase bond energies of ion clusters and solvation energies in the liquid phase holds forth the promise that data on clustering for individual ions can be suitably combined for pairs of anions and cations in such a way that the solvation of electrolytes can be interpreted on a molecular basis (125). To establish the applicability of data for single ions to the interpretation of solvation of neutral electrolytes, researchers are now turning their attention to the study of

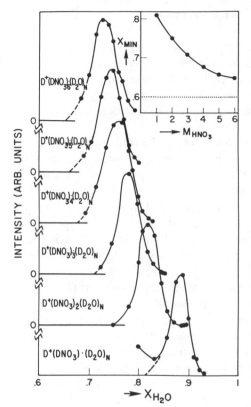

Figure 36. Linear plots of intensity vs. mole fraction of H_2O (X_{H_2O}) for clusters containing between one and six DNO_3 molecules. The maximum peak in each distribution is assigned a value of unity. The insert in the upper right-hand corner shows a plot of the extrapolated onset mole fraction (X_{min}) versus the number of DNO_3 molecules in the cluster. The horizontal dashed line corresponds to the bulk-phase azeotropic composition of nitric acid–water solutions. From Kay et al. (143).

cluster distributions and the changing dipole moments of systems that can form both charge-transfer complexes and solvated ion pairs. Most of the work has been theoretical and there are very few definitive experimental results in this area (see, e.g., Ref. 142).

One system that has received major experimental attention is HNO_3 solvated in water clusters. Results of recent studies of the cluster distribution strongly suggest that solvation of nitric acid begins at very small cluster sizes (143) (see Fig. 36). However, more recent experiments employing both a hexapole and a quadrupole electrostatic focusing technique failed to reveal large dipole-moment changes in clustering beyond the second hydration step. This finding is interpreted as suggesting that the nitric acid–water clusters have very large polarizabilities. Recent measurements have been made in our laboratory of the distribution of clusters of HCl with water, with results similar to those for the nitric acid–water system.

One of the most promising areas for research is systems composed of electroytes that have metal cations, since these should not display such rapid mobility and would be expected to display the expected change in dipole moment. Interestingly, the hydration of these systems has just recently been treated theoretically by Rode and co-workers (144).

4.4   Metal-Cluster Systems

Currently, studies of the properties of metal clusters are being pursued very actively in many laboratories, and the findings have wide-ranging implications for applied and fundamental work. The results bear on the field of condensed-matter physics, particularly on the problem of the onset of metallic conductivity, catalysis and the cluster-surface analogy, surface oxidation and corrosion, microelectronics, light scattering from aerosols, and atmospheric chemistry.

An interesting application pertaining to cluster ions is the study of the changing ionization potentials, and hence the work function, of metallic clusters as a function of their degree of aggregation. Of particular interest is the question of when small clusters begin to display bulk metallic properties. Systems composed of alkali metals have been the subject of considerable attention. Because of their relatively simple electronic configurations, comparatively straightforward experimental interpretation and comparison with theoretical calculations are possible. Furthermore, their relatively volatile nature facilitates the production of clusters in an adiabatic expansion process.

Pioneering work on sodium clusters was done by Robbins and co-workers (145,146), who first reported information on the ionization threshold for these clusters. Using more sophisticated techniques, Schumacher and co-workers (147,148) detected clusters of Na_x for x values up to 65 and reported ionization potentials for $2 \leq x \leq 14$. Castleman and co-workers (149) have remeasured the ionization-potential values for x up to 8, obtaining somewhat different values for x = 5, 6, and 8.

It is instructive to compare the experimental values of the ionization potential with those predicted by the classical equation that accounts for the influence of particle size on the ion-image potential contributions to the work function. The appropriate expression has been derived (150) from classical electrostatics equations, and involves terms which account for the image potential at a flat plane, a correction due to a surface of curvature, and the Coulomb effect in removing the electron from a nongrounded aggregate. This leads to

$$W(R) = W_\infty + \frac{3}{8} \frac{e^2}{R}, \tag{18}$$

where W represents the work function, R the radius of the equivalent sphere, and e the elementary charge (see Fig. 37). The results obtained with the classical expression are in fair accord with the experimental findings.

Two other investigators have reported findings relevant to the problem of the size dependence of the work function of metallic systems. Powers et al. (151) observed that the work function of Cu_9 is less than 5.58 eV, whereas that of Cu_{29} is greater than 4.98 eV. Using an appropriate covalent radius based on the lattice parameter for copper, the classical predictions are 6.68–6.87 eV and 6.02–6.15 eV, respectively. In the case of ion clusters, Rohlfing et al. (152) have reported the work functions for Fe_3 through Fe_5 to be greater than that for Fe_2, and those for

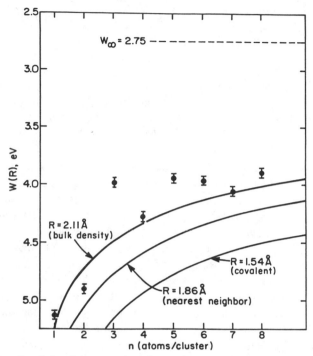

Figure 37. Plot of work function vs. number of atoms in a cluster. The conversion between number and equivalent spherical size was made using $r = 1.86$ Å for the nearest-neighbor distance in sodium and $r = 1.54$ Å based on the convalent radius. Experimental data are shown as filled circles. Note that the values for different crystal faces range from 2.65 to 3.10 eV. After Peterson et al. (149).

Fe$_{13}$ through Fe$_{17}$ to be greater than that for Fe$_{12}$. These findings are contradict to the predictions of the classical equation and other theoretical predictions (152).

Herrmann et al. (148) compared their results with predictions of several quantum mechanical calculations then available, finding rough agreement between experiment and theory. However, the predicted variation in ionization potential between odd- and even-n clusters was larger than was observed experimentally. Yet our own work shows that the potential variation between odd- and even-n systems, often said to be due to pairing of electrons in the even-n cluster systems, is not evident in clusters beyond the pentamer. More recently, Flad et al. (153) have calculated ionization energies for sodium clusters up to Na$_8$. Considering the range expected for the two structures considered in the calculations, the agreement between experiment and theory is reasonably good. Goddard and co-workers (154–156) have considered the ionization potentials and electron affinities of nickel clusters. Their results, based on an self-consistent field (SCF) method, suggest that the ionization potential of nickel should converge to the bulk value for clusters of more than 87 atoms, having reached the bulk value and then dipped below it at $n = 43$. However, the results suggested that the electron affinity does not converge to the bulk value even in clusters as large as Ni$_{87}$.

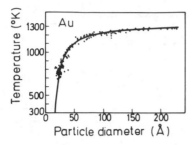

Figure 38. Experimental (dots) and theoretical (solid line) values of the melting-point temperature of gold particles as a function of particle diameter. After Buffat and Borel (160).

Recently we have turned our attention to an investigation of the photoionization of suboxides and subchlorides of alkali metals, with the aim of further comparison with theory (157,158). The ionization-potential values for the metal suboxides, M_xO ($1 < x < 4$), are found to fall dramatically with increasing x, leveling off to an essentially constant value for the trimer and tetramer species. The precipitous drop and leveling off of the ionization potentials were more abrupt than expected. The values do not differ greatly from these for the analogous metallic species, but in the case of sodium the value for the tetramer does fall below that for the corresponding pure metal cluster. It has been reported that the work function of sodium is lowered appreciably by the presence of impurities (159), and the discrepancy between early and recent measurements has been traced to surface contamination. Surprisingly, our results for potassium fail to reveal a similar behavior of the ionization potential.

Clearly, continued work on higher-order metallic suboxides of varying systems promises to provide interesting data from which to gain further insight into such problems as surface oxidation, the influence of oxygen on the metallic conductivity of thin films, and the chemistry and physics of surfaces in general.

In conclusion, it is worth mentioning that not all of the cluster properties approach the bulk properties as fast as the ones discussed in the above-mentioned examples. Although "macroscopic" properties begin to be exhibited at sizes ($n \simeq 100$) that seem surprisingly low for phenomena of long-range order (superconductivity, ferromagnetism, melting, etc.), bulk values are reached only at sizes above the microcluster size defined in the beginning of the chapter (>100 Å). This is illustrated by a plot of the melting-point temperature of gold particles as a function of particle diameter (Fig. 38) (160).

ACKNOWLEDGMENTS

This work was partially supported by the Fonds zur Förderung der Wissenschaftlichen Forschung (Austria) under project nos. S-18/05, S-18/08, 5148, and 5504. T. D. Märk gratefully acknowledges partial support of a visiting professorship (1983) in the Chemistry Department of the Pennsylvania State University through NSF grant ATM-8204010. A. W. Castleman, Jr., thanks the U. S. Department of Energy (grant DE-ACO2-82-ER60055), the U. S. Army Research Office (grant DAAG29-

82-K-0160), and the National Science Foundation (grant ATM-8204010) for providing funding. He is also particularly pleased to acknowledge the hospitality of the Institut für Experimentalphysik of the Leopold-Franzens-Universität during his appointment as Visiting Professor of Physics (June/July 1984), when much of this manuscript was completed.

REFERENCES

1. T. D. Märk and A. W. Castleman, Jr., *Adv. Atom. Mol. Phys.* **20,** 65 (1985).

2. G. D. Stein, *Phys. Teacher,* 503 (1979).

3. M. R. Hoare, *Adv. Chem. Phys.* **40,** 49 (1979).

4. A. C. Zettlemoyer, Ed., *Nucleation,* Marcel Dekker, New York (1969).

5. O. Sinanoglu, *J. Chem. Phys.* **75,** 463 (1981).

6. K. Sattler, *Adv. Solid State Phys.* **23,** 1 (1983), and in *Current Topics in Materials Science,* E. Kaldis, Ed., North Holland, Amsterdam (1985).

7. T. D. Märk, in *Proceedings of the 10th International Mass Spectrometry Conferences,* J. F. J. Todd, Ed., Wiley, New York (1986).

8. D. Smith and N. G. Adams, in *Gas Phase Ion Chemistry,* Vol. 1, M. T. Bowers, Ed., Academic Press, New York (1979), p. 1.

9. E. W. McDaniel, V. Cermak, A. Dalgarno, E. E. Ferguson, and L. Friedman, *Ion Molecule Reactions,* Wiley-Interscience, New York (1970).

10. M. T. Leu, M. A. Biondi, and R. Johnsen, *Phys. Rev.* **A7,** 292 (1973).

11. A. W. Castleman, Jr., in *NATO Advanced Study Institute: Kinetics of Ion Molecule Reactions,* P. Ausloos, Ed., Plenum Press, New York (1979), pp. 295–321; A. W. Castleman, Jr., *Adv. Colloid Interface Sci.* **10,** 73 (1979); A. W. Castleman, Jr., and R. G. Keesee, in *Swarms of Ions and Electrons in Gases,* W. Lindinger, T. D. Märk, and F. Howorka, Eds., Springer-Verlag, Vienna, New York (1984), pp. 165–193.

12. I. N. Tang and A. W. Castleman, Jr. *J. Chem. Phys.* **60,** 3981 (1974).

13. A. W. Castleman, Jr., P. M. Holland, D. M. Lindsay, and K. I. Peterson, *J. Am. Chem. Soc.* **100,** 6039 (1978).

14. R. G. Keesee and A. W. Castleman, Jr., *J. Phys. Chem. Ref. Data,* in press (1985).

15. P. Kebarle, *Ann. Rev. Phys. Chem.* **28,** 445 (1977).

16. E. E. Ferguson, in *Ion Molecule Reactions,* Vol. 2, J. L. Franklin, Ed., Butterworths, London, (1972), pp. 363–393.

17. E. E. Ferguson, F. C. Fehsenfeld, and A. L. Schmeltekopf, *Adv. Atom. Mol. Phys.* **5,** 1 (1969).

18. E. E. Ferguson, F. C. Fehsenfeld, and D. L. Albritton, in *Gas Phase Ion Chemistry,* Vol. 1, M. T. Bowers, Ed., Academic Press, New York, (1979), p. 45.

19. D. L. Albritton, *Atomic Nuclear Tables* **22,** 1 (1978).

20. M. Grössl, M. Langenwalter, H. Helm, and T. D. Märk, *J. Chem. Phys.* **74,** 1728 (1981).

21. O. Echt, P. O. Dao, S. Morgan, and A. W. Castleman, Jr., *J. Chem. Phys.* **82,** 4076 (1985).

22. E. W. Schlag and H. J. Neusser, *Acc. Chem. Res.* **16,** 355 (1983).

23. A. Ding, J. H. Futrell, and R. A. Cassidy, in *Proceedings of the 4th Symposium of Atomic and Surface Physics, Maria Alm, Innsbruck,* F. Howorka, Ed. (1984), pp. 237–242; in *Proceedings of the XIIIth International Conference on the Physics of Electronic and Atomic Collisions, Berlin,* A. Ding and J. Hesslich, Eds. (1983), p. 658.

24. J. B. Anderson, *in Molecular Beams and Low Density Gas Dynamics*, P. P. Wegener, Ed., Marcel Dekker, New York (1974), pp. 1–91.

25. J. B. Anderson, R. P. Andres, and J. B. Fenn, *Adv. Chem. Phys.* **10,** 275 (1966).

26. O. F. Hagena, in *Molecular Beams and Low Density Gas Dynamics*, P. P. Wegener, Ed., Marcel Dekker, New York, (1974), p. 93.

27. O. F. Hagena, *Surface Sci.* **106,** 101 (1981).

28. H. W. Liepman and A. Roshko, *Elements of Gas Dynamics*, John Wiley & Sons, New York (1957).

29. H. Ashkenhas and F. S. Sherman in *Proceedings of the 4th Symposium on Rarefied Gas Dynamics*, Vol. 2, (1966), p. 84.

30. J. B. Anderson and J. B. Fenn, *Phys. Fluids* **8,** 780 (1965).

31. O. F. Hagena and W. Obert, *J. Chem. Phys.* **56,** 1793 (1972).

32. T. A. Milne, A. E. Vandegrift, and F. T. Greene, *J. Chem. Phys.* **52,** 1552 (1976).

33. R. P. Andres, in *Nucleation*, A. C. Zettlemoyer, Ed., Marcel Dekker, New York (1968), pp. 69–108.

34. C. Y. Ng. *Adv. Chem. Phys.* **52,** 263 (1983).

35. W. Obert, *Rarefied Gas Dynamics*, Vol. II, R. Campargue, Ed., LEA, Paris (1979), pp. 1181–1190.

36. J. J. Breen, K. Kilgore, K. Stephan, R. Hoffmann-Sievert, B. D. Kay, R. G. Keesee, T. D. Märk, J. Van Doren, and A. W. Castleman, Jr., *Chem. Phys.* **91,** 305 (1984).

37. T. Ellenbroek, J. Toennies, M. Wilde, and J. Wanner, *J. Chem. Phys.* **75,** 3414 (1981).

38. P. M. Dehmer and S. T. Pratt, *J. Chem. Phys.* **76,** 843 (1982).

39. J. B. Hopkins, P. R. R. Langridge-Smith, M. D. Morse, and R. E. Smalley, *J. Chem. Phys.* **78,** 1627 (1983).

40. T. D. Märk, *Beitr. Plasmaphys.* **22,** 257 (1982); *Int. J. Mass Spectrom. Ion Phys.* **45,** 125 (1982).

41. H. Helm, K. Stephan, and T. D. Märk, *Phys. Rev.* **A19,** 2154 (1978).

42. H. Kuhlewind, H. J. Neusser, and E. W. Schlag, *Int. J. Mass Spectrom. Ion Phys.* **51,** 255 (1983).

43. O. Echt, D. Kreisle, M. Knapp, and E. Recknagel, *Chem. Phys Lett.* **108,** 401 (1984).

44. T. D. Märk, "Properties and Reactions of Cluster Ions," in *4th Symposium on Elementary Processes and Chemical Reactions in Low Temperature Plasma, Stara Lesna*, CSSR, V. Martisovits and P. Lukac, Eds. (1982), pp. 55–73.

45. T. D. Märk, Europhys. Conf. Abstr., *6D,* 29 (1982).

46. T. D. Märk, in *Electron–Molecular Interactions and Their Applications*, Vol. 1, L. G. Christophorou, Ed., Academic Press, New York, (1984), pp. 251–334.

47. T. D. Märk, in *Electron Impact Ionization*, T. D. Märk and G. H. Dunn, Eds., Springer-Verlag, Vienna, New York (1985), pp. 137–197.

48. For example, see *Proceedings of the 3rd International Symposium on Small Particles and Inorganic Clusters, Berlin, July 5–13, 1984*, to be published in *Surface Science* **156** (1985).

49. K. Sattler, J. Mühlbach, O. Echt, P. Pfau, and E. Recknagel, *Phys. Rev. Lett.* **47,** 160 (1981).

50. K. Stephan and T. D. Märk, *Chem. Phys. Lett.* **87,** 226 (1982); **90,** 51 (1982).

51. K. Stephan, T. D. Märk, J. H. Futrell, and A. W. Castleman, Jr., *Vacuum* **33,** 77 (1983).

52. K. Stephan, J. H. Futrell, K. I. Peterson, A. W. Castleman, Jr., H. E. Wagner, N. Djuric, and T. D. Märk, *Int. J. Mass Spectrom. Ion Phys.* **44,** 167 (1982).

53. S. L. Anderson, T. Hirooka, P. W. Tiedemann, B. H. Mahan, and Y. T. Lee, *J. Chem. Phys.* **73,** 4479 (1980).

54. N. J. Kirchner, J. R. Gilbert, and M. T. Bowers, *Chem. Phys. Lett.* **106,** 7 (1984).

55. A. J. Stace and A. K. Shukla, *Int. J. Mass Spectrom. Ion Phys.* **36,** 119 (1980).

56. A. J. Stace and A. K. Shukla, *Chem. Phys. Lett.* **85,** 157 (1982).

57. A. J. Stace and A. K. Shukla, *J. Am. Chem. Soc.* **104**, 5314 (1982).

58. A. J. Stace and C. Moore, *Chem. Phys. Lett.* **96**, 80 (1983).

59. P. M. Dehmer and J. L. Dehmer, *J. Chem. Phys.* **69**, 125 (1978).

60. E. D. Poliakoff, P. M. Dehmer, J. L. Dehmer, and R. Stockbauer, *J. Chem. Phys.* **76**, 5214 (1982).

61. U. Buck and H. Meyer, *Phys. Rev. Lett.* **52**, 109 (1984).

62. U. Buck and H. Meyer, *Ber. Bunsenges. Phys. Chem.* **88**, 254 (1984).

63. O. Echt, S. Morgan, P.D. Dao, E. J. Stanley, and A. W. Castleman, Jr., *Ber. Bunsenges. Phys. Chem.* **88**, 217 (1984).

64. P. M. Holland and A. W. Castleman, Jr., *J. Chem. Phys.* **72**, 5984 (1980).

65. V. Hermann, B. D. Kay, and A. W. Castleman, Jr., *Chem. Phys.* **72**, 185 (1982).

66. A. W. Castleman, Jr., and B. D. Kay, *Int. J. Mass Spectrom. Ion Proc.* **66**, 217 (1985).

67. J. B. Hopkins, D. E. Powers, and R. E. Smalley, *J. Phys. Chem.* **85**, 3739 (1981).

68. P. D. Dao, S. Morgan, and A. W. Castleman, Jr., *Chem. Phys. Lett.* **111**, 38 (1984).

69. P. D. Dao, S. Morgan, and A. W. Castleman, Jr., *Chem. Phys. Lett.* **113**, 219 (1985).

70. K. Stephan and T. D. Märk, *Int. J. Mass Spectrom. Ion Phys.* **47**, 195 (1983).

71. R. G. Cooks, J. H. Beynon, R. M. Caprioli, and G. R. Lester, *Metastable Ions,* Elsevier, Amsterdam (1973).

72. D. E. Hunton, C. A. Albertoni, T. D. Märk, and A. W. Castleman, Jr., *Chem. Phys. Lett.* **106**, 544 (1984).

73. P. H. Dawson, *Int. J. Mass Spectrom. Ion Phys.* **43**, 195 (1982).

74. J. H. Futrell, K. Stephan, and T. D. Märk, *J. Chem. Phys.* **76**, 5893 (1982).

75. J. Sunner and P. Kebarle, *J. Chem. Phys.* **85**, 327 (1981).

76. A. J. Illies, M. F. Jarrold, L. M. Bass, and M. T. Bowers, *J. Am. Chem. Soc.* **105**, 5775 (1983).

77. J. Gspann, *Surface Sci.* **106**, 219 (1981).

78. J. P. Flamme, T. D. Märk, and J. Los, *Chem. Phys. Lett.* **75**, 419 (1980).

79. K. Stephan, T. D. Märk, and A. W. Castleman, Jr., *J. Chem. Phys.* **78**, 2953 (1983); K. Stephan, T. D. Märk, E. Märk, A. Stamatovic, N. Djuric, and A. W. Castleman, Jr., *Beitr. Plasmaphys.* **23**, 369 (1983).

80. (a) K. Stephan, A. Stamatovic, and T. D. Märk, *Phys. Rev.* **A 28**, 3105 (1983); (b) K. Stephan and T. D. Märk, *Phys Rev.* **A32**, 1447 (1985).

81. H. Helm, K. Stephan, T. D. Märk, and D. L. Huesties, *J. Chem. Phys.* **74**, 3844 (1981).

82. K. Stephan, T. D. Märk, and H. Helm, *Phys. Rev.* **A26**, 2981 (1982).

83. T. Jentsch, W. Drachsel, and J. H. Block, *Chem. Phys. Lett.* **93**, 144 (1982).

84. O. Echt, K. Sattler, and E. Recknagel, *Phys. Lett.* **90A**, 185 (1982).

85. W. Henkes and G. Isenberg, *Int. J. Mass Spectrom. Ion Phys.* **5**, 249 (1970).

86. I. N. Tang, M. S. Lian, and A. W. Castleman, Jr., *J. Chem. Phys.* **65**, 4022 (1976).

87. R. G. Keesee, N. Lee, and A. W. Castleman, Jr., *J. Chem. Phys.* **73**, 2195 (1980).

88. A. W. Castleman, Jr., P. M. Holland, and R. G. Keesee, *Radiat. Phys. Chem.* **20**, 57 (1982).

89. A. W. Castleman, Jr., K. I. Peterson, B. L. Upschulte, and J. F. Schelling, *Int. J. Mass Spectrom. Ion Phys.* **47**, 203 (1983).

90. E. Clementi, *Determination of Liquid Water Structure Coordination Numbers for Ions and Solvation for Biological Molecules* (Lecture Notes in Chemistry), Springer-Verlag, Berlin, Heidelberg, New York (1976).

91. A. W. Castleman, Jr., P. M. Holland, D. M. Lindsay, and K. I. Peterson, *J. Am. Chem. Soc.* **100**, 6039 (1978).

92. F. F. Abraham, M. R. Mruzik, and G. M. Pound, *Faraday Disc. Chem. Soc.* **61**, 34 (1976).

93. M. D. Newton, *J. Chem. Phys.* **67**, 5535 (1977).

94. S. F. Smith, J. Chandrasekhar, and W. Torgensen, *J. Chem. Phys.* **87**, 1898 (1983).

95. M. Eigen and L. DeMaeyer, *Proc. Roy. Soc. A.* **247**, 505 (1958).

96. Y. K. Lau, S. Ikuta, and P. Kebarle, *J. Am. Chem. Soc.* **104**, 1462 (1982).

97. J. Q. Searcy and J. B. Fenn, *J. Chem. Phys.* **61**, 5282 (1974).

98. G. M. Lancaster, F. Honda, Y. Fukuda, and J. W. Rabalais, *J. Am. Chem. Soc.* **101**, 1951 (1979).

99. M. F. Vernon, D. J. Krajnovich, H. S. Kwok, J. M. Lisy, Y. R. Shen, and Y. T. Lee, *J. Chem. Phys.* **77**, 47 (1982).

100. T. P. Martin, *Phys. Rev.* **95**, 167 (1983).

101. J. E. Campana, T. M. Barlak, R. J. Colton, J. J. DeCorpo, J. R. Wyatt, and B. I. Dunlap, *Phys. Rev. Lett.* **47**, 1046 (1981); J. E. Campana and B. N. Green, *J. Am. Chem. Soc.* **106**, 531 (1984).

102. W. Ens, R. Beavis, and K. G. Standing, *Phys. Rev. Lett.* **50**, 27 (1982).

103. N. G. Adams and D. Smith, in *Reactions of Small Transient Species: Kinetics and Energetics*, A. Fontijn and M. A. A. Clyne, Eds., Academic Press, New York (1983), pp. 311–385.

104. I. Dotan, J. A. Davidson, F. C. Fehsenfeld, and D. L. Albritton, *J. Geophys. Res.* **83**, 4036 (1978).

105. D. K. Bohme, in *NATO Advanced Study Institute, Vimeiro, Portugal* (1982), pp. 111–134.

106. D. Smith, N. G. Adams, and E. Alge, *Planet. Space Sci.* **29**, 449 (1981).

107. E. E. Ferguson, in *30th Annual Conference on Mass Spectrometry and Allied Topics, Honolulu*, (1982), pp. 200–203.

108. R. T. McIver, *Sci. Am.* **243** (Nov.), 186 (1980).

109. D. Smith, N. G. Adams, and T. M. Miller, *J. Chem. Phys.* **69**, 308 (1978).

110. D. W. Fahey, H. Böhringer, F. C. Fehsenfeld, and E. E. Ferguson, *J. Chem. Phys.* **76**, 1799 (1982).

111. C. J. Howard, F. C. Fehsenfeld, and M. McFarland, *J. Chem. Phys.* **60**, 5086 (1974).

112. D. K. Bohme, G. I. Mackay, and S. D. Tanner, *J. Am. Chem. Soc.* **101**, 3724 (1979).

113. D. K. Bohme and G. I. Mackay, *J. Am. Chem. Soc.* **103**, 978 (1981).

114. M. Henchman, J. F. Paulson, and P. M. Hierl, *J. Am. Chem. Soc.* **105**, 5509 (1983).

115. W. N. Olmstead and J. I. Brauman, *J. Am. Chem. Soc.* **99**, 4219 (1977); M. J. Pellerite and J. I. Brauman, *J. Am. Chem. Soc.* **102**, 5593 (1980).

116. B. R. Rowe, A. A. Viggiano, F. C. Fehsenfeld, D. W. Fahey, and E. E. Ferguson, *J. Chem. Phys.* **76**, 742 (1982).

117. M. M. Kappes and R. H. Staley, *J. Am. Chem. Soc.* **103**, 1286 (1981).

118. W. C. Lineberger, in *Applied Atomic Collision Physics*, Vol. 5, H. S. W. Massey, E. W. McDaniel, and B. Bederson, Eds., Academic Press, New York (1982), Chapter 8, pp. 239–254.

119. J. T. Moseley, *J. Phys. Chem.* **86**, 3282 (1982); J. T. Moseley, in *Applied Atomic Collision Physics*, Vol. 5, H. S. W. Massey, E. W. McDaniel, and B. Bederson, Eds., Academic Press, New York (1982), Chapter 10.

120. L. Andrews, in *NATO Advanced Studies Institute Series B, Physics*, Vol. 90, J. Berkowitz and K. O. Groeneveld, Eds., Plenum Press, New York, (1983), pp. 153–182.

121. R. L. Woodin, D. S. Bomse, and J. L. Beauchamp, *J. Am. Chem. Soc.* **100**, 3248 (1978).

122. B. S. Freiser and J. L. Beauchamp, *J. Am. Chem. Soc.* **99**, 3214 (1977).

123. M. Heaven, T. A. Miller, and V. E. Bondybey, *J. Chem. Phys.* **76**, 3831 (1982).

124. A. W. Castleman, Jr., P. M. Holland, R. G. Keesee, *J. Chem. Phys.* **68**, 1760 (1978).

125. N. Lee, R. G. Keesee, and A. W. Castleman, Jr., *J. Colloid Interface Sci.* **75**, 555 (1980).

126. A. W. Castleman, Jr., and I. N. Tang, *J. Chem. Phys.* **57**, 3629 (1972).

127. P. M. Holland and A. W. Castleman, Jr., *J. Phys. Chem.* **86**, 4181 (1982).

128. P. M. Holland and A. W. Castleman, Jr., *J. Chem. Phys.* **76,** 4195 (1982).

129. P. Gans and J. B. Gill, *J. Chem. Soc. Dalton Trans.,* Part 1, 779 (1976).

130. P. Kebarle, *Modern Aspects Electrochem.* **9,** 1 (1974).

131. R. W. Taft, *Prog. Phys. Org. Chem.* **14,** 247 (1983).

132. P. Schuster, P. Wolschann, and K. Tortschanoff, in *Chemical Relaxation in Molecular Biology,* Vol. 24, I. Pecht and R. Rigler, Eds., Springer-Verlag, Berlin, Heidelberg, New York (1977), pp. 107–190.

133. M. Born, *Z. Phys.* **1,** 45 (1920).

134. R. Yamdagni and P. Kebarle, *J. Am. Chem. Soc.* **94,** 2940 (1972).

135. J. L. Beauchamp, *Ann. Rev. Chem. Phys.* **22,** 527 (1971).

136. J. E. Bartmess and R. T. McIver, Jr., in *Gas Phase Ion Chemistry,* Vol. 2, M. T. Bowers, Ed., Academic Press, New York (1979), pp. 88–119.

137. D. M. Aue and M. T. Bowers, in *Gas Phase Ion Chemistry,* Vol. 2, M. T. Bowers, Ed., Academic Press, New York (1979), pp. 1–51.

138. J. Bromilow, J. L. M. Abboud, C. B. Lebrilla, R. W. Taft, G. Scorrono, and V. Lucchini, *J. Am. Chem. Soc.* **103,** 5448 (1980).

139. E. M. Arnett, F. M. Jones III, M. Taagepera, W. G. Henderson, J. L. Beauchamp, D. Holtz, and R. W. Taft, *J. Am. Chem. Soc.* **94,** 4724 (1972).

140. W. R. Davidson, J. Sunner, and P. Kebarle, *J. Am. Chem. Soc.* **101,** 1675 (1979).

141. R. Yamdagni and P. Kebarle, *J. Am. Chem. Soc.* **93,** 7139 (1971).

142. A. Breiz, A. Karpfen, H. Lischka, and P. Schuster, *Chem. Phys.* **89,** 337 (1984).

143. B. D. Kay, V. Hermann, and A. W. Castleman, Jr., *Chem. Phys. Lett.* **80,** 469 (1981).

144. B. M. Rode and J. P. Limtrakul, *Z. Naturforsch.* **39a,** 376 (1984).

145. E. J. Robbins and R. E. Leckenby, *Nature* **206,** 1253 (1965).

146. E. J. Robbins, R. E. Leckenby, and P. Willis, *Adv. Phys.* **16,** 739 (1967).

147. A. Herrmann, E. Schumacher, and L. Wöste, *J. Chem. Phys.* **68,** 2327 (1978).

148. A. Herrmann, S. Leutwyler, E. Schumacher, and L. Wöste, *Helv. Chim. Acta.* **61,** 453 (1978).

149. K. I. Peterson, P. D. Dao. R. W. Farley, and A. W. Castleman, Jr., *J. Chem. Phys.* **80,** 1780 (1984).

150. D. M. Wood, *Phys. Rev. Lett.* **46,** 749 (1981).

151. D. E. Powers, S. P. Hansen, M. E. Guesic, D. L. Michalopoulos, and R. E. Smalley, *J. Chem. Phys.* **78,** 2866 (1983).

152. E. A. Rohlfing, D. M. Cox, A. Kaldor, and K. H. Johnson, *J. Chem. Phys.* **81,** 3846 (1984).

153. J. Flad, H. Stoll, and H. Preuss, *J. Chem. Phys.* **71,** 3042 (1979).

154. T. H. Upton and W. A. Goddard III, in *Chemistry and Physics of Solid Surfaces,* Vol. III, R. Vanselow and W. England, Eds., CRC Press, Boca Raton, Florida (1983), p. 127.

155. T. H. Upton, W. A. Goddard III, and C. F. Melius, *J. Vac. Sci. Technol.* **16,** 531 (1979).

156. C. F. Melius, T. M. Upton, and W. A. Goddard III, *Solid State Commun.* **28,** 501 (1978).

157. P. D. Dao, K. I. Peterson, A. W. Castleman, Jr., *J. Chem. Phys.* **80,** 563 (1984).

158. K. I. Peterson, P. D. Dao, and A. W. Castleman, Jr., *J. Chem. Phys.* **79,** 777 (1983).

159. R. J. Whitefield and J. J. Brady, *Phys. Rev. Lett.* **26,** 380 (1971).

160. P. Buffat and J. P. Borel, *Phys. Rev.* **A13,** 2287 (1976).

CHAPTER 13

Biomedical Applications
of Mass Spectrometry

DAVID L. SMITH

Department of Medicinal Chemistry
Purdue University
West Lafayette, Indiana

If the mass spectrometry community were divided according to discipline, the largest group comprise persons using mass spectrometry to investigate problems relating to biomedical research. Some users would be concerned primarily with qualitative applications; that is, they would ask, ''What is it?'' Others would use mass spectrometry to quantify substances whose structure and chemical properties have been previously determined. In either case their attention would be focused more often on the latest developments in sample preparation or data analysis, than on the details of instrument design or the mechanisms responsible for the mass spectrum. Although users of biomedical mass spectrometry are generally not involved with the development of new instrumentation, they do borrow heavily from those research groups that use mass spectrometry as a tool to investigate fundamental problems in chemical physics.

The tremendous amount of activity in biomedical mass spectrometry is due in part to the wide range of substances that can be accurately and precisely characterized by this technique; any substance having a vapor pressure greater than 10^{-10} torr may be analyzed by conventional ionization techniques such as electron impact (EI) and chemical ionization (CI). At present, these techniques are the most sensitive and give the most structural information. Electron impact mass spectrometry is also playing an expanding role in such peripheral areas as blood- and respiratory-gas analysis. The recently developed technique of fast-atom-bombardment (FAB) mass spectrometry has greatly extended the range of compounds amenable to analysis by mass spectrometry. This technique is particularly good for characterizing substances that are not sufficiently volatile to be analyzed by EI or CI, such as small, chemically labile substances and large, polymeric compounds. For example, MH^+ ions of peptides having molecular weights up to 10,000 daltons have been detected by FAB mass spectrometry. For those substances too large for FAB and related particle-desorption techniques, pyrolysis mass spectrometry has proved useful. Although the results are only qualitative, this latter technique is very powerful when accompanied by sophisticated, multidimensional data analysis.

In addition to its notable versatility, mass spectrometry probably offers greater combined sensitivity and specificity than any other analytical technique. For this reason, mass spectrometry is the preferred technique for elucidation of the structures of a wide variety of substances isolated from biological materials. These may be natural products; xenobiotic substances, including their metabolites; or therapeutic drugs and their metabolites. Once the structure of an isolate has been established, mass spectrometry may also be used as a quantitative tool for investigating its role or fate in biological systems.

This survey of biomedical applications of mass spectrometry describes only techniques that may be regarded as standard procedures in mass spectrometry laboratories working at the forefront of biomedical research. So that attention may be focused on the mainstream of biomedical applications of mass spectrometry, the highly specialized techniques used in only a few laboratories have been omitted, even though their potential importance may be very great. For a comprehensive review of biomedical applications of mass spectrometry, the reader should consult the review by Burlingame, Whitney, and Russell (1) or reference texts (2, 3).

1   STRUCTURE ELUCIDATION

In the typical structure-elucidation problem, the objective is to establish the primary structure of some substance that is known to have certain biochemical or chromatographic properties. While the sample does not need to be pure for quantitative analysis, the first step in structure elucidation is isolation and purification of the substance. This is often a major problem, since the substance of interest may be only a minor component of a complex mixture containing several other constituents of similar structure. It is common to have only a few micrograms of pure material at the end of a year of sample isolation and purification. Electron impact ionization continues to play an important role in structure elucidation because it yields important information about the molecular weight, the elemental composition, and the structural configuration of the atoms. Chemical ionization is generally useful only in those cases where EI ionization does not give an apparent molecular ion.

In EI ionization, the sample is first heated until the desired rate of vaporization is achieved, then simultaneously ionized and excited to some high electronic state. This electronically excited ion then undergoes radiationless decay to the ground electronic state to give a vibrationally excited molecular ion. Depending on the amount of excitation energy, this ion may undergo one or more fragmentation reactions to give a characteristic and reproducible mass spectrum. After adjustments for the adiabatic reaction conditions prevailing in the mass spectrometer ion source, preferred fragmentation paths may be rationalized on the basis of conventional organic chemistry reaction theory. The specificity of EI mass spectrometry is also due in part to the fact that the analysis is performed in the gas phase, where the solvent and sample matrix do not contribute to the mass spectrum.

An example of how conventional EI mass spectrometry may be used to determine the structure of a previously unidentified substance may be found in a recent publication by Swinton et al. (4). It was known that bacteriophage mu controls a DNA-modification process that results in modification of approximately 15% of the adenosine residues. In addition to information deduced from changes in biological activity, the fact that a chemical modification had occurred in the DNA was established by the appearance of an additional peak in the chromatogram of the hydrolyzed DNA. The structure of this modified nucleoside (Fig. 1) was established by mass spectrometry.

DNA isolated from culture cells was enzymatically digested to give the free nucleosides, which were isolated and purified by high-performance liquid chromatography (HPLC). Because nucleosides represent a class of compounds whose thermal instabilities and volatilities lie on the fringes of the acceptable ranges for conventional mass spectral analysis, the trimethylsilyl (TMS) derivative was prepared. This step replaces all exchangeable hydrogens with TMS groups and greatly increases the sample volatility. (The use of chemical derivatives to increase volatility continues to be important in spite of associated problems, which include possible modification of the primary structure during derivatization, incomplete reaction in the case of very small samples, matrix interference, and instability of the derivative. Although newer desorption techniques capable of analyzing the underivatized material are available, chemical derivatives remain attractive both because EI mass spectrometry gives the most structural information and because the biomedical analyst may prefer a chemical to an instrumental approach.)

The low-resolution EI mass spectrum of the TMS derivative of the isolated material is given in Fig. 2. The peak at $m/z = 596$ is accompanied by a peak at an m/z value 15 units lower, which likely results from the loss of a methyl radical. This is a common and characteristic fragmentation process for the molecular ion of a TMS derivative and indicates a molecular weight of 596 for the derivative. Peaks at $m/z = 170$, 155, and 103 indicate the presence of a sugar, which is consistent with the notion that the unknown is a modified nucleoside. The peak at $m/z = 336$ (M^+–sugar) shows that the modification occurs in the base moiety.

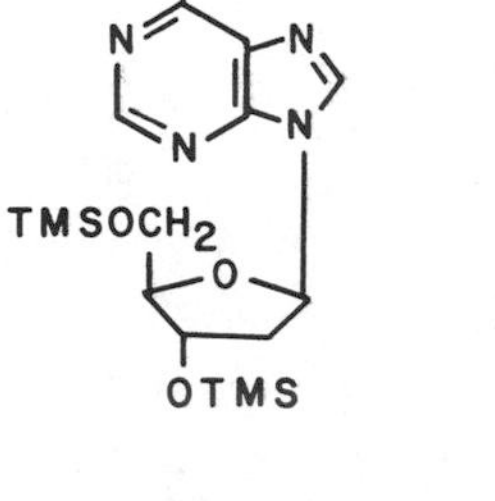

TMSNCH$_2$CONHTMS

TMSOCH$_2$

OTMS

Figure 1. Structure of α-N-(9-β-D-2′-deoxyribofuranosylpurin-6-yl) glycinamide isolated from modified DNA.

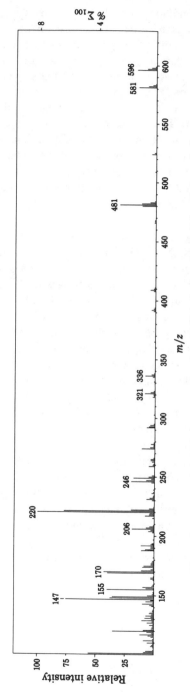

Figure 2. Low-resolution EI mass spectrum of the TMS derivative of α-*N*-(9-β-D-2′-deoxyribofura-nosylpurin-6-yl) glycinamide. From Ref. 4.

The number of TMS groups in the molecular and fragment ions may be determined from the mass shifts observed in a separate experiment in which the perdeuterated TMS derivative is used. Each TMS group has nine hydrogens and therefore increases the mass by nine mass units. Such a measurement shows that the numbers of TMS groups in the peaks due to the molecular ion and purine base are four and two, respectively. That is, the $m/z = 596$ peak increases to 632 and the $m/z = 336$ peak increases to 354. These results may be compared with those for unmodified adenosine, which has similar peaks at $m/z = 555$ and 208. Collectively, these results show that the modification adds 57 mass units to the purine base. In addition, the extra TMS group in the base indicates the presence of an additional exchangeable hydrogen.

The elemental compositions of the major ions were determined from high-resolution, exact mass data. Although this is a well-established technique and gives invaluable information, there are inherent limitations that are sometimes overlooked. Foremost among these pitfalls is that while the mass of an ion may be determined with an accuracy better than 5 ppm, the number of possible elemental compositions consistent with a given mass may be quite large.

A list of all possible elemental compositions as determined from the mass of the molecular ion, 596.2815, is given in Table 1. It is important to note that even though the number of silicon atoms was fixed at four, there are still nine compositions consistent with the 5-ppm mass window. Since this is the molecular ion, the nitrogen rule may be invoked to eliminate all compositions having an odd number of nitrogens. Of the remaining candidates (compositions 1, 4, 5, and 8), only number 1 is consistent with a modified adenosine. From this it can be shown that the modification must consist of addition of C_2H_3NO to adenosine. Note that if the number of possible silicons had been adjustable between zero and six, there would have been 57 possible compositions, and had there been a sulfur in the unknown, it would not have been considered by the computer. If the computer had been programmed to allow for the possible presence of one sulfur, the number of elemental compositions would have

TABLE 1   Possible Elemental Compositions for a Molecule of Mass 596.2815[a]

Composition	Error (ppm)	C	H	N	O	Si
1	0	24	48	6	4	4
2	2	22	46	9	3	4
3	2	26	50	3	5	4
4	2	25	44	10	0	4
5	2	23	52	2	8	4
6	4	27	46	7	1	4
7	4	28	52	0	6	4
8	3	20	44	12	2	4
9	5	21	50	5	7	4

[a] Only those compositions are listed for which the mass is within 5 ppm of the measured value, the number of nitrogens and oxygens is between zero and 12, and the number of rings plus that of double bonds is between zero and 20.

increased from nine to 20. It is evident that the mass must be determined with the highest possible accuracy if the elemental composition of an ion is to be established unambiguously.

Once it was demonstrated that the modification consists of addition of C_2H_3NO to the purine base of adenosine, the configuration of these elements, as well as the point of attachment to the purine base, remained to be established. Based on the elemental composition of the ion at $m/z = 481$ and the fact that the base moiety contains two TMS groups, it was concluded that the modification must have the form —CH_2CONH_2, as indicated in Fig. 1. Since it is not possible in this case to determine the point of attachment by mass spectrometry alone, a model compound having the proposed structure was synthesized. The mass spectrum of the TMS derivative of the model compound was identical to Fig. 2, confirming the structure of the isolated material. It is important to note that this structure determination was made using less than 15 μg of material.

2 GAS CHROMATOGRAPHY–MASS SPECTROMETRY AND QUANTITATION

Gas chromatography–mass spectrometry (GCMS) is a well-established technique for the quantification of a wide variety of substances. Although limited by the fact that the compound must be amenable to analysis by gas chromatography (GC), the technique has been used extensively because of its unsurpassed sensitivity and specificity. For compounds that are not sufficiently volatile for analysis by GC, reliable and sometimes elaborate methods of derivatization have been developed. The details of such methods depend on the nature of the substance, the available instrumentation, and the final objective; one successful approach is illustrated in a recent publication by Welch et al. (5).

A program for standardization of clinical methods has been undertaken by the National Bureau of Standards to establish definitive methods of analysis with which other, less rigorous methods may be compared. Because of its demonstrated high precision and high accuracy, isotope-dilution mass spectrometry (IDMS) has been selected as the definitive method for quantifying urea in blood serum.

Using an approach in which all parameters were optimized for maximum accuracy, the authors (5) demonstrated that the concentration of urea in frozen serum pools could be determined with a coefficient of variation of 0.19%. To accurately determine the level of urea in serum, weighed aliquots of dried serum and the internal standard, ^{18}O-labeled urea, were combined and thoroughly equilibrated. The internal standard is very important, because it accounts for sample losses during isolation and derivatization as well as for changes in the mass spectrometer sensitivity. The urea was isolated from the bulk of the serum by sublimation at reduced pressure and subsequently converted to 6-methyluracil. This step takes advantage of the fact that the TMS derivatives of uracils have excellent chromatographic properties. On conversion to the TMS derivative, the relative amounts of endogenous and ^{18}O-labeled urea present in the original serum sample were determined by

GCMS. The sequence of steps used for urea conversion and derivatization is illustrated in Fig. 3.

The derivatized 6-methyluracil was separated from extraneous material remaining in the sample by a 30-m support-coated open tubular (SCOT) capillary column. Although SCOT columns have chromatographic resolution somewhat inferior to that of the more frequently used wall-coated columns, they do have a large capacity and will elute large quantities of material without overloading. Since sample quantity was not a limiting factor, a splitter type of GC injector was used to load the column.

The analyses were performed on a Varian MAT CH7 single-focusing magnetic sector mass spectrometer that had been substantially modified specifically for the purpose of making high-precision measurements. A microcomputer was used to control the magnetic field, record the ion signals, and calculate the relative abundance of the endogenous urea. It is interesting that these authors achieved higher precision when the magnetic field was switched than when the more conventional approach of switching the accelerating voltage was used. The technique of selected-ion monitoring was used in lieu of scanning the entire mass spectrum. The former is usually the method of choice for quantitative measurements, because it gives the highest sensitivity and because it allows a maximum number of cycles between the analyte and internal-standard peaks.

Principal measurements were made using the EI mode, which gave a relatively intense M-15 peak. The ions used for quantitation in IDMS were chosen such that interferences from other fragments or isotopically different ions were avoided. For example, although the M^+ peak is intense, it could not be measured accurately because of an interference from the first isotope peak of the $M–H^+$ ion. For this reason, it is generally desirable to use an internal standard several mass units heavier than the analyte.

Because a definitive method must have very high accuracy, a somewhat unusual measurement protocol was used. A preliminary analysis of the sample was made to determine the approximate ratio of analyte to internal standard. Two calibration standards were then prepared with analyte/internal standard ratios closely bracketing that of the unknown. After injection of one standard, the sample and second standard were injected successively at 5.5-minute intervals. Since the sample elution time was approximately 30 minutes, the solvent front from the last injection eluted long before the derivatized urea of the first injection. This method allows the analyte and internal standard concentrations to be compared with standards having similar concentrations and does not require long-term stability of the instrument.

This procedure is concerned primarily with accuracy and is restricted neither by

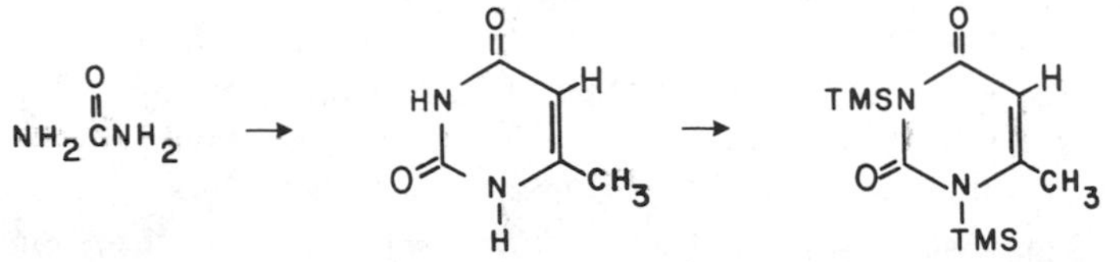

Figure 3.   Chemical conversion and derivatization of urea prior to analysis by GCMS.

sample size nor by the time required for analysis. In contrast, Smith et al. (6) have used IDMS to quantify thromboxane B_2 in human plasma. Since the normal level of thromboxane B_2 is about 7 pg/ml, all experimental parameters must be optimized to give the highest possible sensitivity. In addition, because clinical research investigations often require analysis of many samples, a short analysis time is desirable. Although the concentration of arachidonic acid metabolites, including thromboxane B_2, may be determined by different methods, the high specificity of IDMS makes it, in principle, the most accurate.

Although CI is used less frequently than EI, it is preferred for quantification of those compounds that give only weak ions in the high-mass range on EI ionization. While the total yield of ions in positive-ion CI may be 10 to 100 times less than in the EI mode, the absolute intensity of high-mass ions may be increased substantially with the appropriate choice of reagent gas. Since the likelihood of interference is less for high-mass ions, added specificity is the most important benefit of positive-ion CI for quantitative analyses. This method is also valuable for determining the molecular weights of compounds belonging to those classes of substances that give very weak molecular ions in the EI mode, such as acetates.

For a substance with a high electron affinity, negative-ion CI may yield more ions than either EI or positive-ion CI. The increased sensitivity of negative-ion CI is due primarily to the very large cross section for the electron attachment process. The range of compounds amenable to high-sensitivity, negative-ion CI has been expanded by making derivatives having functional groups that increase the probability of capturing and retaining an electron. In addition to having high senstivity, negative-ion CI has superior specificity, because most of the ions are of high mass and because only substances having a high electron affinity can effectively compete for the free electrons.

For the thromboxane B_2 assay, a Finnigan 4000 quadruple GCMS system fitted with a 30-m, fused-silica, wall-coated capillary column was used. It is interesting to reflect on the variety of schemes developed over the last 20 years for coupling GC and MS. For high-sensitivity measurements, the currently preferred method is simply to extend the capillary column into the vacuum chamber so that the column terminates inside the ion source. Although a portion of the end of the column is operating under subatmospheric pressure, the flow rate may be adjusted to retain full chromatographic performance. This feature lowers the elution temperature and is beneficial for the analysis of polar, thermally labile compounds.

Although the increased resolution of modern capillary columns benefits most analyses, such columns do require specialized sample-loading techniques. The simplest approach uses a sample splitter and is useful for analyses that are not restricted by the quantity of sample available. If the assay is sample limited, on-column sample injection may be preferred. This method uses a fine needle to place the sample inside the capillary column. Although this method makes effective use of the sample, extra attention must be paid to the volume of the injected sample and the injector temperature if full chromatographic resolution is to be achieved.

Whatever biomedical applications may lack in terms of sophisticated instrumentation is balanced by the use of elaborate procedures for sample preparation. In the

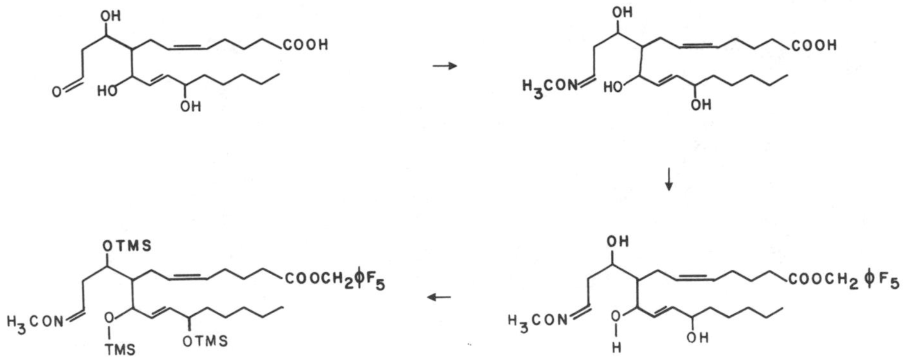

Figure 4. Three-step derivatization of thromboxane B_2.

case of thromboxane B_2, a series of three derivatization steps was used (Fig. 4). To achieve the highest possible sensitivity, it is essential that each step give a high yield, have few side reactions, and be effective for picogram quantities of material that may be present only as a minor constituent of a complex mixture. The importance of proper sample preparation for biomedical applications of mass spectrometry cannot be overemphasized. The sensitivity of the thromboxane B_2 assay has been improved by using the pentafluorobenzyl derivative, which substantially increases the probability of capturing an electron. This derivative has an interesting combination of properties. On capture of an electron, the chemical bond formed during derivatization is cleaved, with transfer of a hydride ion to the oxygen. Thus, the functionality that originally attracted the extra electron readily fragments from the original molecule, leaving behind the charge. This is clearly a desirable feature and contrasts with the behavior of other electrophilic derivatives, such as the trifluoroacetates, in which the derivative fragment retains the charge. The importance of this feature is illustrated in Fig. 5, which shows the negative-ion CI mass spectra of the d_0 and d_4 derivatized thromboxane B_2. Essentially all of the ion current is concentrated in the fragment peaks at $m/z = 614$ and 618, which result from loss of $-CH\phi F_5$ after hydride transfer to the thromboxane B_2 moiety.

Although most of the GCMS quantitative assays use low-resolution MS, specificity may be greatly increased if a high mass-resolving power is used. It is interesting that although use of high-resolution conditions will result in decreased instrument sensitivity, the detection limit may be lowered in those cases where it is determined by chemical noise. Figure 6 shows the signal obtained for 320 pg of the TMS derivative of the highly modified nucleic-acid base queuine. (See Fig. 7 for the structure of the TMS derivative of queuine.) The sample was loaded on a 12-m fused-silica capillary column with a falling needle injector. This type of injector is ideally suited to analyses of small quantities of low-volatility substances, since the entire sample may be readily loaded. The mass spectrometer was set to a resolution of 10,000 and repetitively swept over a 200-ppm mass interval centered about the exact mass of the ion of interest. A single-ion-pulse-counting amplifier and multichannel signal averager were used to record the arrival of ions in the specified mass

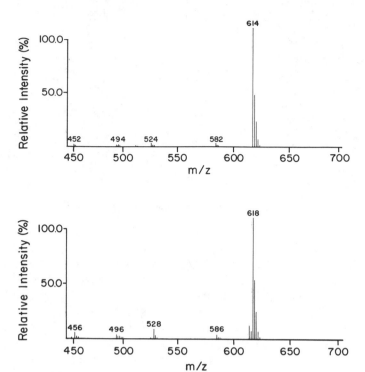

Figure 5. Negative-ion CI mass spectra of thromboxane B₂ (top) and d_4 internal standard (bottom). Reproduced from Ref. 6 with the kind permission of John Wiley & Sons.

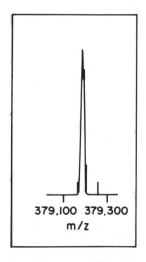

Figure 6. Time-averaged signal for $m/z = 379$ for 320 pg of the TMS derivative of queuine (resolution, $M/\Delta M = 10,000$). From Ref. 7.

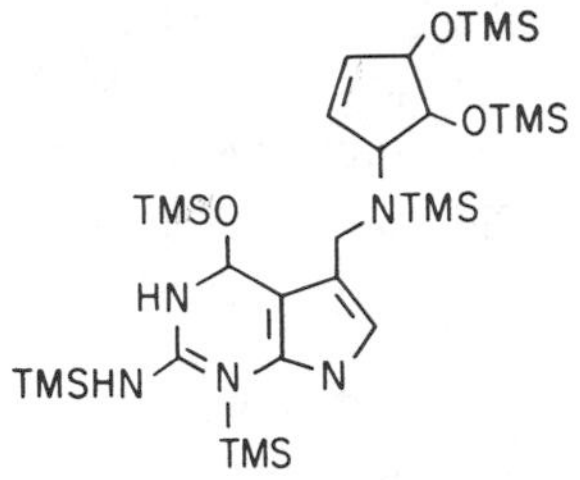

OTMS

OTMS

TMSO NTMS

HN

TMSHN   N   N

TMS

MOL. WT. 709

Figure 7. Structure of the TMS derivative of the highly modified nucleic-acid base queuine.

window during the time interval in which the derivatized queuine was known to elute from the column. For injection of 320 pg of the TMS derivative of queuine, a total of 149 ions having the expected mass were detected. Because of the high specificity of this technique, the signal-to-noise ratio is determined primarily by ion statistics and could be improved if a more sensitive mass spectrometer were used.

3 FAST ATOM BOMBARDMENT

Although EI mass spectrometry has and will continue to play a major role in biomedical research, its scope is limited by the requirement for vaporization of the sample prior to analysis. That is, to be analyzed by EI mass spectrometry, a substance must have an appreciable vapor pressure at a temperature below which it undergoes thermal degradation. Fast-atom-bombardment (FAB) mass spectrometry substantially reduces the requirement for sample volatility and has been used to obtain mass spectra of underivatized peptides with molecular weights of up to 10,000 mass units. Perhaps of greater impact is its ability to analyze substances that, although of relatively low molecular weight, are thermally labile. Examples of such applications include analyses of oligonucleotides, steroid sulfates, phospholipids, and antibiotics. The long-term effects of FAB on the field of mass spectrometry will likely be as great as those of coupling GC and MS.

An essential step in obtaining a FAB mass spectrum is to dissolve the sample in a suitable liquid matrix such as glycerol. The sample plus matrix is then inserted into the vacuum chamber and bombarded by a beam of high-energy atoms, which cause sputtering or desorption of ions from the surface of the matrix. The ions, which are now in the gaseous phase, may be analyzed by any of the usual mass spectrometric techniques, including high resolution and mass spectrometry/mass spectrometry (MS/MS).

Although details of the role played by the dispersing matrix are not fully understood, it is clear that dispersion of the sample in a liquid is an essential requirement of the FAB technique. A suitable dispersing matrix must have a relatively low vapor pressure, dissolve the sample completely, and give a simple mass spectrum with only a few peaks. Glycerol has been used most often, and new solvents giving superior results for specific classes of compounds have been reported recently. The most promising of these include thioglycerol, sulfolane, and mixtures such as di-thioerethretol–dithiothreitol or tetramethylurea–trimethylamine. Although a compound may be soluble in all of these solvents in a macroscopic sense, the success of a particular solvent is believed to be related to the charge state of the solute and to its tendency to form micelles on the surface.

In analyses of substances of biomedical interest, similar results are obtained whether charged or neutral particles are used to effect desorption of ions from the surface of the sample. Maximum sensitivity is achieved when high-intensity beams of particles having translational energies between 5 and 10 keV are used. Suitable beams of ions may be produced by thermal ionization, whereas a high-energy neutral beam is generated by charge-exchange reactions occurring in a high-voltage discharge source. Devices for producing beams of Cs^+ ions are somewhat easier to construct than those for production of neutral beams. In addition, the ion beam may be focused directly on the sample and may be conveniently varied over a wide range of intensities. To overcome the 4- to 8-kV potential barrier of the source in a sector mass spectrometer operating in the positive-ion mode, a high voltage is required to accelerate the Cs^+ ion beam initially to 10 or 15 keV. Use of a neutral beam circumvents this problem, but has the undesirable feature of passing a relatively high volume of gas into the ion source, which then must be evacuated with a high-speed pumping system.

Positive- and negative-ion FAB mass spectra tend to be complimentary with respect to structural information. Although the relative intensities of positive- and negative-ion FAB mass spectra are compound dependent, most neutral compounds give useful spectra in both modes. The positive-ion mass spectra usually contain MH^+ ions as well as molecular adduct ions due to cationation by the metal ions often present in biological samples. The negative-ion FAB mass spectra of the same compounds normally have corresponding $M\text{-}H^-$ ions. In the case of ionic substances, the unmodified cations and anions are observed in the positive- and negative-ion spectra, respectively. Collectively these ions may be used to establish the molecular weight of the substance. Inorganic salts such as Li^+, K^+, and Ag^+ may be used to induce changes in the mass spectra that differentiate MH^+ from MNa^+ ions. Depending on the nature of the compound, positive or negative fragment ions diagnostic of the structure may be present. The fragmentation pathways normally result in formation of ions with an even number of electrons, whose presence may be rationalized in terms of their thermodynamic stability. In the case of pure glycerol, a series of peaks due to $G_n + H^+$ or $G_n - H^-$ are observed up to $m/z = 1000$.

The ability of FAB mass spectrometry to determine the molecular weight of peptides of rather high mass is illustrated in Fig. 8, which shows the molecular ion region of the positive-ion FAB mass spectrum of the peptide glucagon. The significance of molecular-weight determination in this mass range has been discussed by Yergey et al. (9), who emphasize that the difference between the nominal mass and the exact mass of high-molecular-weight compounds may be surprisingly large. For example, the nominal mass ($H = 1, 0 = 16$, etc.) of the protonated molecular

Leu — Tyr — Lys — Ser — Tyr — Asp — Ser — Thr — Phe — Thr — Gly — Gln — Ser — His — H
|
Asp — Ser — Arg — Arg — Ala — Gln — Asp — Phe — Val — Gln — Trp — Leu — Met — Asn — Thr — OH

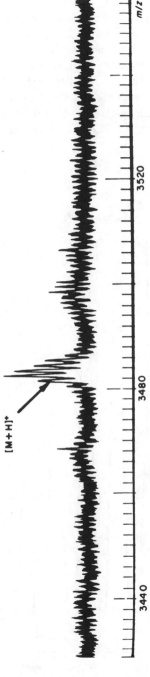

Figure 8. Positive-ion FAB mass spectrum of the peptide glucagon. Reproduced from Ref. 8 with the kind permission of John Wiley & Sons.

ion of glucagon, $(C_{153}H_{225}N_{42}O_{50}S)^+$, is 3481. In contrast, the exact mass, which accounts for the mass defects of the atoms, is 3482.6. This is the exact mass of the lowest-mass peak in the observed isotopic series comprising the MH^+ ion and is 1.6 mass units greater than the nominal mass.

Although we usually associate the most intense peak with the monoisotopic mass (i.e., the molecular weight as calculated from the exact masses of the lightest isotopes), this association may not hold for high-mass ions. The theoretical molecular ion distribution for glucagon given in Fig. 9 shows that the monoisotopic peak is smaller than both the first and second isotopic peaks, which are due to the presence

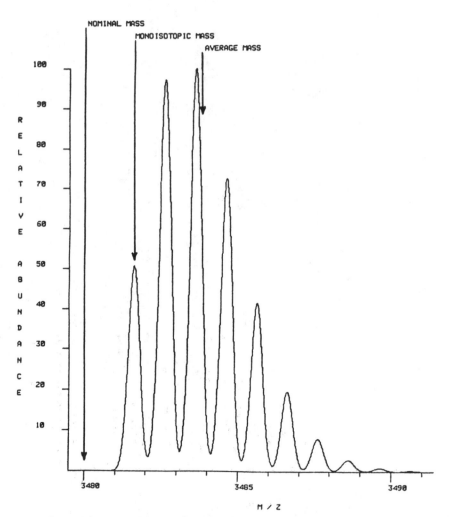

Figure 9.   Theoretical molecular ion distribution of glucagon. From Ref. 9.

of one or more of the naturally occurring minor isotopes such as ^{13}C, [17]O, ^{18}O, and [15]N. While the monoisotopic peak has a unique combination of isotopes, it is important to note that the isotopic peaks are composed of mixtures of isotopes that cannot be resolved by high resolution and therefore cannot be used to determine the elemental composition. For example, the monoisotopic peak of the trimer of glucagon at $m/z = 10,442.7$ is probably too small to be detected and the most abundant mass would be that represented by the sixth isotopic peak.

It was also shown (8) that 22 isotopically different ions of glucagon have both an exact mass within $\pm$ 0.02 mass units of the average and a relative intensity greater than 1% of that of the most intense peak. It would obviously be useless to attempt to deduce the elemental composition from the "exact mass" of the composite peak. It was suggested that the most useful information that may be derived from high-mass FAB spectra is the average mass as determined from the envelope containing all of the isotopic peaks. If this is the case, one may operate the mass spectrometer with less than unit mass resolution and benefit from the improved instrument sensitivity.

Gibson et al. (10) have used positive-ion FAB mass spectrometry to resolve uncertainties in the primary structure of human myelin, which is a major constituent of the myelin protein of the central nervous system. Human myelin contains 169 amino acids and has an approximate molecular weight of 18,500 daltons. Classical sequencing techniques have been used by several investigators to determine the amino acid sequence of this protein. The results of these investigations are in general agreement, but some areas of the sequence require further clarification.

Since the uncertainties in structure were located between amino acids 45 and 89, human myelin was first digested with bovine brain cathepsin D. The 45–89 segment was isolated and subsequently digested with trypsin, which cleaved the peptide wherever there was an arginine or lysine. The digest was separated into three multicomponent fractions by HPLC. This step not only reduced the number of peptides in the sample for mass spectrometric analysis, but also removed excess salt and enzyme. These fractions were dried, dissolved in 50% acetic acid and glycerol, and analyzed by positive-ion FAB mass spectrometry.

The FAB mass spectra contained several molecular ions MH[+], which could be correlated with peptides predicted to be present in the tryptic digest. These peptides covered the entire sequence from positions 45 to 89 and are summarized in Table 2. The presence of protonated molecules at $m/z = 1101$ and 1538 was interpreted as confirmation of there being a tyrosine at position 69. These results also permit evaluation of the previously postulated presence of phosphorylated serine at position 56. The peptide with MH[+] at $m/z = 1304$ includes the serine at position 56 and the m/z value is not consistent with phosphorylation of the serine. Since a peak corresponding to a similar peptide with an additional phosphate moiety was not observed at $m/z = 1384$, it was concluded that at least the major portion of the myelin protein does not have a phosphorylated serine at position 56.

In previous studies, it was not clear whether the sequence at positions 77 and 78 is Gly–His or the reverse, His–Gly. This region of the protein was included in the two tryptic peptides having ions MH[+] of $m/z = 1538$ and 1622. Note that in

TABLE 2   Molecular Weights of Tryptic Peptides
of Human Myelin Basic Protein as Determined
from FABMS[a]

m/z Value of MH^+	Peptides	HPLC Fraction
551	$Phe_{45}-Arg_{49}$	1
581	Unidentified	1
904	$Phe_{45}-Lys_{53}$	2
934	Unidentified	2
1101	$Thr_{66}-Lys_{75}$	2
1185	$Thr_{80}-Phe_{89}$	3
1304	$Arg_{54}-Arg_{65}$	1
1538	$Thr_{66}-Arg_{79}$	2
1662	$Ser_{76}-Phe_{89}$	3

[a] From Ref. 10.

addition to peaks for MH^+ ions, FAB mass spectra of peptides may also contain peaks that are diagnostic of the amino acid sequence. A scheme illustrating this very important feature is given in Fig. 10. In favorable cases, the entire peptide may be sequenced from these fragments. Investigation of the fragment ions from the two peptides with protonated molecular ions at $m/z = 1538$ and 1622 showed that the sequence is His(77)–Gly(78).

Phospholipids are another class of compounds for which FAB mass spectrometry offers the potential for greatly improved analysis. The analysis of phospholipids is difficult, because a large number of homologues differing only in fatty-acid chain length and saturation usually occur together. In addition, the presence in phospholipids of the phosphate ester and an amine group results in low volatility and thermal instability. Although a variety of mass spectrometric methods have been used to analyze phospholipids, they unfortunately rely on sample pretreatment to remove the ionic portion of the molecule or they analyze the volatile products of pyrolysis.

The use of FAB mass spectrometry to quantify dipalmitoylphosphatidylcholine (DPC) in human amniotic fluid as a measure of fetal-lung maturity has been demonstrated by Ho et al. (11). A known quantity of internal standard into which nine deuteriums had been incorporated was added to the amniotic-fluid sample. After equilibration of the sample and internal standard, the DPC was extracted with chloroform. An aliquot containing a few micrograms of DPC was dried on the FAB probe tip and covered with 1 μl of thioglycerol. Intensities of protonated-molecule peaks for both DPC and DPC-d_9, $m/z = 735$ and 744, respectively, were recorded and averaged for five consecutive scans. From these results, the quantity of the DPC present in the original sample of amniotic fluid was determined.

Since this was one of the first reported attempts to use FAB mass spectrometry as a quantitative tool, care was taken to establish its reliability. The ratio between the molecular ion peaks at $m/z = 735$ and 744 was measured for samples in which

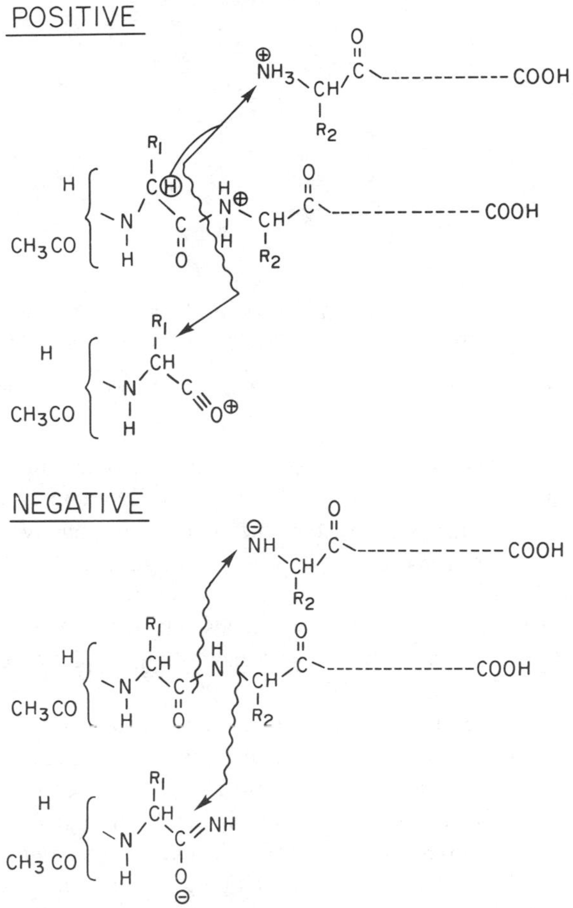

POSITIVE

NEGATIVE

Figure 10. Fragment ions that are often found in FAB mass spectra of peptides and are diagnostic of the amino acid sequence.

the ratio of d_0 to d_9 DPC was varied from 0.01 to 100. A plot of the intensity ratios vs. the corresponding concentration ratios was linear (correlation coefficient of 0.9994), indicating that isotope-dilution FAB mass spectrometry could be used as a quantitative tool.

Although this method requires the availability of sophisticated instrumentation and highly specialized personnel compared with the conventional methods relying on chromatographic separation and UV detection, it does provide considerably improved accuracy as well as a description of the DPC fatty acid constituents. Because of this, FAB mass spectrometry will likely be used as a reference method with which to compare other, less sophisticated techniques, and will be particularly useful for investigations that require its analytical superiority.

The sensitivity of FAB mass spectrometry, expressed in terms of the number of ions produced from a given amount of sample (in coulombs per gram) varies sub-

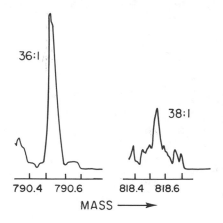

Figure 11. Time-averaged signals for MH$^+$ ions produced from 10 ng of phosphatidylserine by high-resolution positive-ion FAB mass spectrometry (resolution, M/ΔM = 10,000). From Ref. 7.

stantially and unpredictably from one compound to another; nonetheless, rather intense beams of MH$^+$ ions are often observed. In favorable cases the absolute sensitivity of FAB approaches that of EI ionization. Unfortunately, the background signal in FAB mass spectrometry tends to be very high and reduces the signal-to-noise ratio. For this reason, the FAB mode has not proved as suitable for trace analysis, as have EI and CI. The limitations imposed by the ever-present background in the FAB mode may be partially offset by using high resolution or MS/MS.

The ability to analyze phosphatidylserine (PS) by high-resolution FAB mass spectrometry is illustrated in Fig. 11, which shows the MH$^+$ ion peaks at m/z = 790.5 and 818.5. The generalized structure of PS is given in Fig. 12. The spectrum illustrated in Fig. 11 was obtained from 10 ng of PS that comprised fatty acids having a total of 36 or 38 carbons and one double bond. The experimental conditions were configured to optimize both the sensitivity and the specificity of the method and were similar to those used for Fig. 6. Use of high resolution ($M/\Delta M$ = 10,000) greatly reduces the interferences from background ions, which are usually present in the FAB mode. The presence of chemical noise is illustrated by the nonzero base line in Fig. 11. Chemical interferences due to the glycerol may be manipulated to advantage by use of deuterated (d_3 or d_5) glycerol.

Although the sensitivity decreased 20-fold when the resolution was increased from 1000 to 10,000, the detection limit was improved by a factor of 10. Use of high resolution also enables accurate mass measurements and subsequent determination of the elemental composition of the ion. Similar results have been obtained for other phospholipids, including phosphatidylcholine, phosphatidylinositol, and sphingomyelin.

High-specificity techniques have been used by Millington, Roe, and Maltby (12)

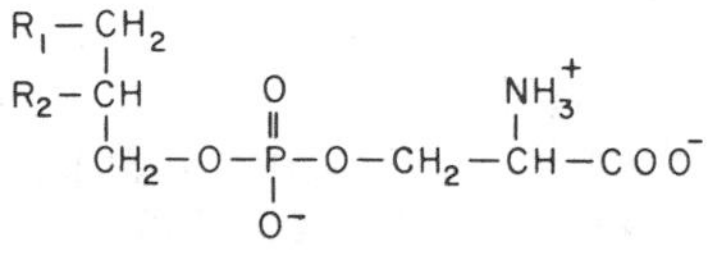

```
R₁ – CH₂
  |
R₂ – CH        O          NH₃⁺
  |            ||          |
  CH₂ – O – P – O – CH₂ – CH – C O O⁻
               |
               O⁻
```

Figure 12. Generalized structure of phosphatidylserine.

to investigate the possible clinical use of L-carnitine as a transporter of fatty acids. Positive-ion FAB mass spectrometry was used to characterize synthetic acylcarnitines, to identify major acylcarnitines present in urine, and to develop methods for their quantitation. Reference FAB mass spectra of authentic acetylcarnitine and propionylcarnitine were investigated using B/E (B = magnetic field, E = electric sector field) linked scans to substantiate postulated fragmentation pathways; high-resolution, accurate mass measurements were used to determine the elemental composition of the fragments.

Since the acylcarnitines were expected to occur as mixtures, the FAB mass spectrum of a sample composed of equimolar amounts of a homologous series of fatty acids (C-2 through C-6) was analyzed to determine possible differences in the sensitivity of the technique for these different substances. A severe tendency to discriminate against the lower-molecular-weight components was observed. This was attributed to changes in the surface activity as the length of the alkyl chain in the acyl group increases. For this reason, as well as because extraneous material (e.g., inorganic salts) present in biological samples often has profound effects on FAB mass spectra, it was suggested (12) that quantitation in the FAB mode will be possible only via the technique of isotope dilution.

The search for acetylcarnitine and propionylcarnitine in the urine of patients receiving L-carnitine was facilitated by chromatographic fractionating of the urine. An aliquot of the fraction that had the highest concentration of acylcarnitines was mixed with glycerol and analyzed by positive-ion FAB mass spectrometry. Figure 13 shows the normal ion scan of this fraction. Daughter ions of the $m/z = 218$ ion

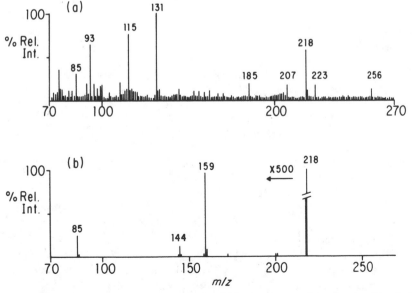

Figure 13. The normal and B/E linked positive-ion FAB mass spectra of propionylcarnitine in fractionated urine. Reproduced from Ref. 12 with the kind permission of John Wiley & Sons.

were selectively recorded using a B/E linked scan and are included in Fig. 13. This MS/MS scan has an excellent signal-to-noise ratio and shows that propionylcarnitine is present in the urine. This point was further substantiated by high-resolution, accurate mass measurements of both the MH$^+$ ion and its fragments.

The final step in this investigation was developing a method to quantify acetyl- and propionylcarnitine in the urine of persons receiving therapeutic doses of L-carnitine. Acetylcarnitine (d_3) and propionylcarnitine (d_5) were added to the urine samples and used as internal standards. The ratios of MH$^+$ ions due to deuterated and nondeuterated acetylcarnitines were determined by selected-ion monitoring at a mass resolution of 7000. High resolution was necessary to separate matrix peaks due to sodium- and potassium-cationated glycerol dimers from the peaks for acetylcarnitine MH$^+$ ions having the same nominal mass. Replicate analyses of different aliquots of the same urine sample indicated a coefficient of variation of 8%.

The profound dependence of FAB mass spectra on the sample matrix is in contrast to the EI mode, where instrument sensitivity varies in a rather predictable manner from one substance to another and is generally independent of any extraneous substances in the sample. Stated in another way, EI sensitivity is a fundamental property of the substance in the gas phase, whereas FAB sensitivity is dependent on the solute–matrix interaction. Although considerable effort has been expended to reduce or control this interaction to obtain mass spectra characteristic of the solute, an untapped potential of FAB mass spectrometry may lie in its ability to investigate these troublesome solute–matrix interactions.

REFERENCES

1. A. L. Burlingame, J. O. Whitney, and D. H. Russell, *Anal. Chem.* **56**, 417R (1984).

2. F. W. McLafferty, *Interpretation of Mass Spectra*, 3rd ed., University Science Books, Mill Valley, California (1980).

3. I. House, D. H. Williams, and R. D. Bowen, *Mass Spectrometry: Principles and Applications,* 2nd ed., McGraw-Hill, New York (1981).

4. D. Swinton, S. Hattman, P. F. Crain, C. S. Cheng, D. L. Smith, and J. A. McCloskey, *Proc. Nat. Acad. Sci. USA* **80**, 7400 (1983).

5. W. J. Welch, A. Cohen, H. S. Hertz, F. C. Reugg, R. Schaffer, L. T. Sniegoski, and E. White V, *Anal. Chem.* **56**, 713 (1984).

6. B. J. Smith, D. A. Herold, R. M. Ross, F. Marquis, J. Savory, M. R. Wills, and C. R. Ayers, *Biomed. Mass Spectrom.* (submitted, 1986).

7. D. L. Smith and J. A. McCloskey, unpublished results.

8. M. Barber, R. S. Bordoli, R. D. Sedgwick, A. N. Tyler, G. V. Garner, D. B. Gordon, L. W. Tetler, and R. C. Hider, *Biomed. Mass Spectrom.* **9**, 269 (1982).

9. J. Yergey, D. Heller, G. Hansen, R. J. Cotter, and C. Fenselau, *Anal. Chem.* **55**, 353 (1983).

10. B. W. Gibson, R. D. Gilliom, J. N. Whitaker, and K. Biemann, *J. Biol. Chem.* **259**, 5028 (1984).

11. B. C. Ho, C. Fenselau, G. Hansen, J. Larsen, and A. Daniel, *Clin. Chem.* **29**, 1349 (1983).

12. D. S. Millington, C. R. Roe, and D. A. Maltby, *Biomed. Mass Spectrom.* **11**, 236 (1984).

INDEX